LEGO® MINDSTORMS™ INTERFACING

Check out these other exciting robotics titles from TAB Electronics:

Gordon McComb *Robot Builder's Bonanza*
Gordon McComb *Robot Builder's Sourcebook*
Myke Predko and Ben Wirz *TAB Electronics Build Your Own Robot Kit*
John Iovine *Robots, Androids and Animatrons*
Chris Hannold *Combat Robots Complete*
David Shircliff *Build a Remote Controlled Robot*

2003

LEGO® Mindstorms™ Interfacing

Don Wilcher

Illustrations by Don Wilcher
Photos by Matta

McGraw-Hill
New York Chicago San Francisco Lisbon
London Madrid Mexico City Milan New Delhi
San Juan Seoul Singapore Sydney Toronto

The McGraw-Hill Companies

Cataloging-in-Publication Data is on file with the Library of Congress.

1 2 3 4 5 6 7 8 9 0 DOC/DOC 0 9 8 7 6 5 4 3 2

P/N 140206-3
PART OF
ISBN 0-07-140205-5

The sponsoring editor for this book was Scott Grillo and the production supervisor was Sherri Souffrance. It was set in Century Schoolbook by MacAllister Publishing Services, LLC.

Printed and bound by RR Donnelley.

This book is printed on recycled, acid-free paper containing a minimum of 50 percent recycled de-inked fiber.

McGraw-Hill books are available at special quantity discounts to use as premiums and sales promotions, or for use in corporate training programs. For more information, please write to the Director of Special Sales, Professional Publishing, McGraw-Hill, Two Penn Plaza, New York, NY 10121-2298. Or contact your local bookstore.

This book is for my lovely kids, Tiana, D'Vonn, and D'Mar, who will be tomorrow's Big Thinkers, as well as my beautiful wife Mattalene, who made the suggestion that I should pursue my dream of becoming a Technical Book Author.

ABOUT THE CD

The CD included with this book has on it the software applets (small program applications) discussed in the chapters. The freeware explained for these applets in the appropriate chapters are included as well. Each folder has the appropriate freeware or applet in it. The freeware programs are self-extracting, therefore, requiring a few clicks for installation on the setup dialog boxes. The book, along with the appropriate software applets, provides a rich learning environment for the hobbyist and experimentalist to explore the many circuit interfaces discussed in the chapters. It's encouraged by the author for the reader to experiment with the software to devise new ways to control a Scout or RCX-based Mindstorms robot.

CONTENTS

PREFACE

The LEGO Mindstorms is a truly wonderful constructor kit for people of all ages. By taking a constructionist's point of view, learning and developing technical skills through play enables the robotics developer to create magnificent and intelligent robots. The design and function of these machines are only limited by the developer's imagination and his or her knowledge of circuits, mechanics, and computer programming. After reviewing various LEGO Mindstorms web sites and reading several books on *programmable brick* (P-Brick) robots, one question continues to intrigue me: What design and development process was used to create their machines? Based on reading postings on the *LEGO Users Group Network* (LUGNUT) robotics web site, these developers have myriad backgrounds and personalities. The skill level ranges from first-time Mindstorms builders to doctorates conducting research on intelligent agents and behavior to engineers developing mechatronics prototypes. The books written on the Mindstorms products focus on robot construction techniques, software languages, or sensor interfacing.

To better assist in the "design and build" of robotic hardware and software interface controls, a systems approach may be used. This book presents a system-engineering methodology for the design, construction, and testing of robotic hardware, software, and mechanical interfacing. In order to carry out this objective of the engineering method when designing and building robotic systems, this book tries to answer six questions:

1. How can an engineering development process for building a Mindstorms robot consisting of hardware and software components using modeling, simulation, and prototyping tools be created?
2. How can a diagnostic/control tool be built using freeware, Microsoft Office Suite, and an open-source language for Mindstorms robots?
3. How can alternative data acquisition tools using a Texas Instruments (TI) graphics calculator, calculator-based laboratory (CBL) and supporting sensors, Vernier Electricity Software, a serial interface box, and a traditional test-instrumentation be used to capture data from a Mindstorms robot?
4. How can inspiration in the science and engineering disciplines assist in the development of electronic controls and software for Mindstorms robots?

5. How can reverse engineering help in the understanding of wireless/radio control circuits for robotic monitoring and control applications?
6. How can a Mindstorms robot's mechanical system interfacing be assisted using electronic controls and software?

By using these six fundamental questions, the book is written with the following outline:

- **Chapter 1, "Wireless Basics"** This section discusses transmitter-receiver circuits and how they work together as a wireless interface for the LEGO Scout P-Brick. The Scout *integrated development environment* (IDE) tool is explained with several testing procedures for an experimental wireless Scout controller. Reverse-engineering techniques are illustrated using a salvaged *resistor capacitor* (RC) car transmitter-receiver pair as an experimental wireless test module.
- **Chapter 2, "Developing GUIs: Software Control Basics"** This chapter introduces how to design and build a *graphical user interface* (GUI) using Microsoft Excel *Visual Basic for Applications* (VBA). A robotic control panel built using Excel VBA and various testing procedures are presented. Using low-cost software and instrumentation, the basics of data acquisition are also outlined.
- **Chapter 3, "Electromechanical Controls Interfacing"** This discussion focuses on controlling the LEGO Mindstorms RCX and Scout P-Brick using electric switches and relays. Testing procedures and a simple VBA switch monitor-detector are presented.
- **Chapter 4, "Electronic Switching Circuits"** Transistors, *Interactive C* (IC) switches, and timer controllers are presented in this section. TI's CBL, the *digital control unit* (DCU), and a Basic Stamp interface for Mindstorms control are discussed as well.
- **Chapter 5, "Sensor-Interfacing Basics"** A diode interface circuit used to connect analog signals to the Mindstorms input port is explained. Passive and active experimental sensors along with measurement techniques using alternative data acquisition and traditional test instrumentation are also explored.
- **Chapter 6, "Using Procedural Languages for Mindstorms Robot Control"** Software methods in writing code using nontraditional programming languages are investigated. *Not Quite C* (NQC) code is created using the discussed software methods along with IC and *LEGO Assembler* (LASM) programming languages fundamentals.

- **Chapter 7, "Client-Server Controller for Mindstorms Robots"** Python, an open-source *object-oriented programming* (OOP) language, is explained by creating a server for passing data to an Excel spreadsheet (client). Object-oriented methods and ActiveX controls are used to process Python data for controlling a Mindstorms robot.
- **Chapter 8, "Simulator Controls for Mindstorms Robots"** The Timer Control Circuit simulated in the Electronics Workbench is used as a virtual switch for controlling a Mindstorms robot. Advanced VBA GUI creation techniques are discussed by building a simulated Timer Control Circuit.
- **Chapter 9, "Remote-Control Techniques"** The Basic Stamp to LEGO Scout or RCX P-Brick is discussed. A master-slave controller using the Basic Stamp to LEGO Mindstorms interface is presented along with testing procedures for investigating an input diagnostic tool using the two embedded devices.
- **Chapter 10, "Virtual Prototyping and Control Using ActiveX Controls"** This section discusses software design methods for creating advanced GUIs using instrumentation ActiveX controls. Several instrumentation test panels for controlling a Mindstorms robot are illustrated.
- **Chapter 11, "Virtual Test Box (VTB) Development: Math-Based Controllers for Mindstorms Robots"** By using electronic circuit equations for calculating decibels, the voltage across the resistor, and so on, a math-based VTB can be created for robot control. This section is a compilation of Chapters 7 through 11 with hardware circuits for Mindstorms robot *input/output* (I/O) monitoring and control.
- **Chapter 12, "VBA Prototypes: Developing Mindstorms Tools with Advanced Programming Techniques"** This chapter presents the advanced techniques of GUI design. It includes the animation of electromechanical relay switching contacts and their connection for controlling a Mindstorms robot. This section explains how VBA prototypes can be used in a *Proof Of Concept* (POC) for a robotics invention.

Additional resources in circuit analysis, *computer-aided design* (CAD) freeware for LEGO mechanical design, mini tutorials on VBA, Python, and LASM programming languages are provided.

Each section provides a hands-on system-engineering method for robotic hardware and software controls design and development through experimental lab projects. The purpose of the lab projects is to illustrate a development process that the LEGO Mindstorms builder can use to capture design requirements and manage his or her robotics creations using advanced system-engineering methods. The book's main purpose is to illustrate POC developmental methods in hardware and software products for LEGO Mindstorms robots using experimental prototyping techniques. This book enables the reader to breadboard circuits and attach them to the target P-Brick using alligator test leads. This approach enables voltage and current values to be measured easily and the electronic circuits to be modified without the hassle of using a soldering iron. This process, I believe, provides valuable tools for creating wonderful LEGO Mindstorms robots.

ACKNOWLEDGMENTS

I would like to thank Scott Grillo, Vice-President and Publisher of McGraw-Hill, for giving me the chance to write an innovative and creative book on interfacing techniques for controlling LEGO Mindstorms Robots. Without his gentle encouragement, this book would have never materialized. I also would like to thank Beth Brown of MacAllister Publishing Services, LLC, who helped tremendously in answering my questions during the editing phase of the book. Beth's understanding of first time Technical Authors lack of knowing the editing process made reviewing and correcting the manuscript, photo images, and illustrations enjoyable as well as educational. Finally, I would like to thank Forrest Mims for his spiritual encouragement and constant guidance while writing this book. Forrest knew that one day I would write a book that reflects the inner talents that God's grace has given me. Finally, I thank the earlier pioneers at MIT's Media Lab whose research papers on the early predecessor of the RCX programmable brick intrigued me to the point of buying a LEGO Mindstorms kit and using it with my Wonders of Automotive Engineering class held at DaimlerChrysler's Jeep-Truck Engineering Facility bright and early on Saturday Mornings.

Don Wilcher

ABOUT THE AUTHOR

Don Wilcher is an Electronics Systems Engineer for Visteon Corporation responsible for automotive electronic design concepts supporting new electrical/electronic systems development. Mr. Wilcher has contributed articles to *Nuts and Volts* magazine and various engineering trade publications.

CHAPTER 1

Wireless Basics

Antenna, Transmitter, and Receiver Fundamentals

In a wireless system, electrical energy can move in two ways. It can move along a conductor (a group of electrons moving through a metal wire) or it can travel in the air as invisible waves. In a common wireless system, the electrical energy starts out as current or electrons flowing through a conductor. The electrons are then changed into electromagnetic waves using an oscillator and power amplifier circuits inside of the transmitter. The electromagnetic waves are then received by the receiver's antenna and are converted back into electrons. The receiver's electronic circuits will then process the electrons for audio, video, or digital operations. Figure 1-1 shows the energy-conversion process of a transmitter and receiver.

Although the diagram is very simplistic, it serves as a fundamental model for illustrating the energy-conversion process for a transmitter and receiver. This conversion model explains the basic atomic elements used in today's wireless devices. Throughout the book, this model will be referenced when we discuss advanced topics such as infrared communications and *Radio Frequency* (RF) electronics. Therefore, a basic communication model would consist of the key elements shown in Figure 1-1. Of course, additional building blocks will be added to this diagram such as a local oscillator or transistor relay switching circuit for automatic frequency transmission and reception applications. This communication model can be seen as a macroscopic view inside a transmitter and receiver. Another way to model a wireless system from a block diagram perspective is shown in Figure 1-2.

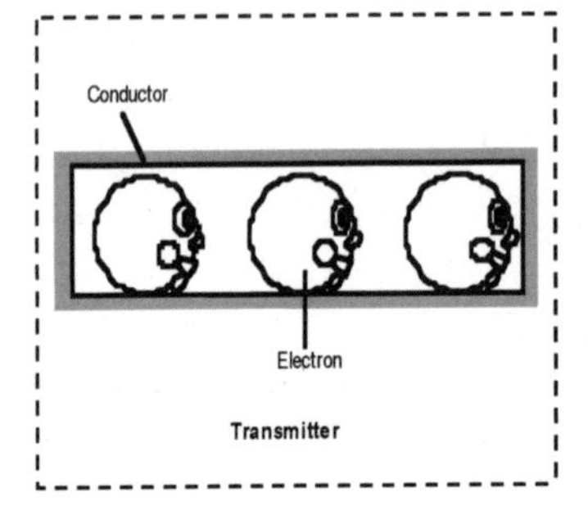

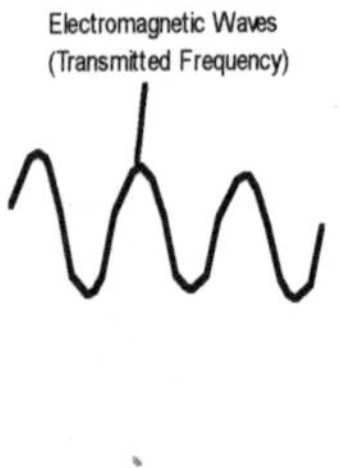

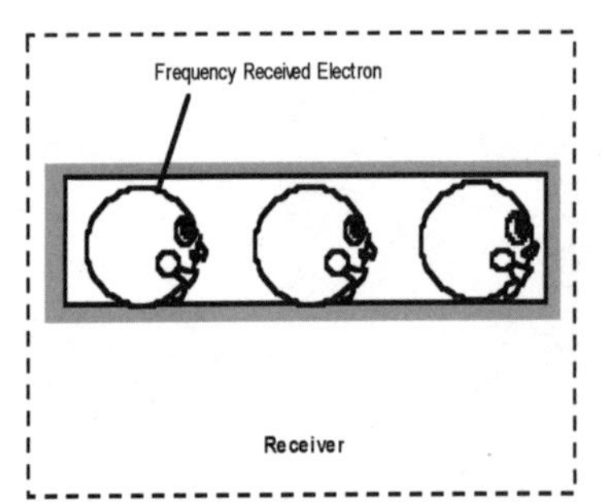

Figure 1-1 Diagram showing the electron-to-electromagnetic-to-electron conversion process

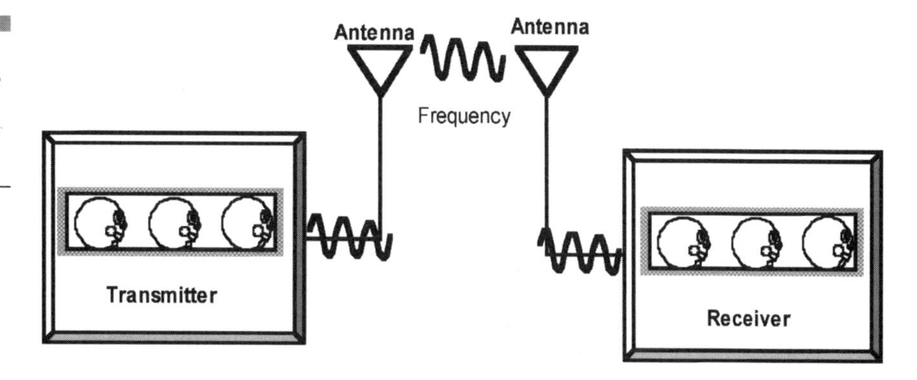

Figure 1-2
Block diagram of the transmitter-receiver communications system

The Antenna

Every wireless device has an antenna. The antenna's job is to change the electrons flowing up a conductor into airborne electromagnetic energy (the transmitter) or change the airborne energy into electrons flowing down a conductor into an audio, video, analog, or digital signal (the receiver). Without the antenna, the wireless device would not operate properly. The symbol used in a block diagram is a funnel. Visually, the top of the funnel represents the energy being transmitted and the bottom displays the electrons being received.

When electromagnetic energy leaves an antenna, the electrons radiate out from the antenna. But how does a transmitter and receiver work and what are the basic building blocks for each electronic circuit?

The Transmitter

A transmitter is an electronic circuit that produces a signal for communications, remote control, and monitoring applications. The signal consists of an electromagnetic wave made up of light and sound. A basic transmitter consists of an oscillator, a modulator, and a power amplifier. The oscillator is an electronic circuit that produces a fixed RF signal. This signal is transformed into radiated energy (that is right for wireless applications) by adding an information signal. The process of combining the RF energy with the information signal is known as *modulation*. The circuit used to create the modulated signal is known as a *modulator*. The modulated signal will be amplified so the imposed information can be transmitted to the receiver. A power amplifier is used for the purpose of increasing the modulated signal's strength, thereby enabling the airborne energy to be detected and processed by the receiver. The power amplifier is connected to a wire known as the *antenna*. Its function is to capture the amplified modulated signal to send it out into the air. Figure 1-3 shows a block diagram of a basic transmitter.

The Receiver

A receiver is an electronic circuit that captures a transmitter's modulated energy, processes it, and converts it into useful information such as audio, video, analog, or digital. A basic receiver consists of an RF amplifier, a detector, and a power amplifier. Once the modulated energy is trapped via the antenna, the RF amplifier will boost the signal to an appropriate level for processing. The detector is used to take the amplified RF signal and convert it to a DC voltage. The power amplifier, as discussed in the section "The Transmitter," amplifies the detector's signal to an appropriate voltage level that is right for interfacing to other semiconductor devices like transistors, *field effect transistors* (FETs), and microcontrollers. A FET is a monolithic or single formed water semiconductor amplifying device that has a high impedanze (Z) gate electrode used to control the flow of electrical current. Figure 1-4 shows a block diagram of a basic receiver.

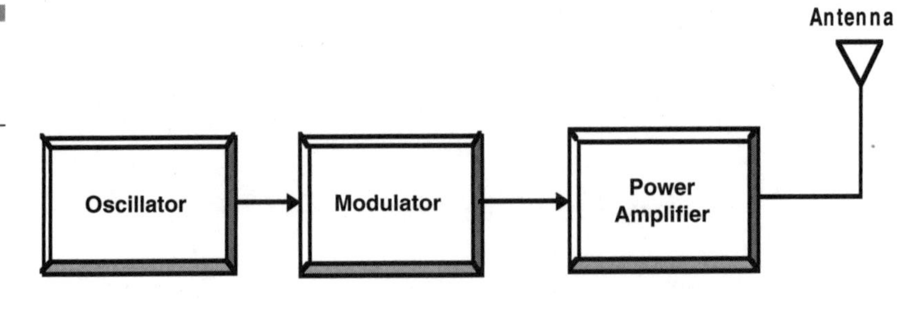

Figure 1-3
Building blocks for a basic RF transmitter

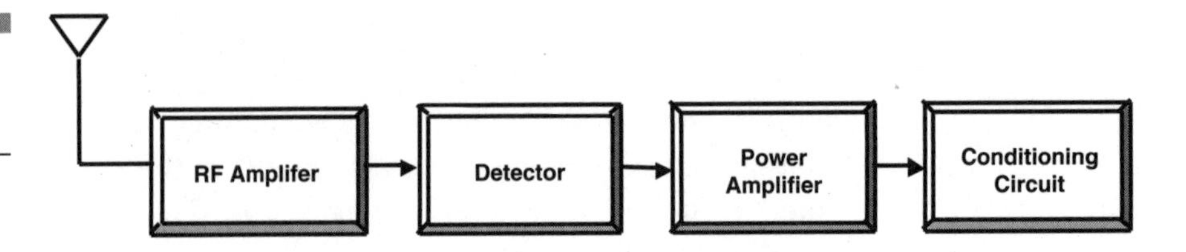

Figure 1-4
Building blocks for a basic receiver

An Experimental Wireless Controller

Now that we have a basic understanding of the building blocks used in wireless applications, the following experiment will demonstrate the fundamental operation of an RF device using a salvaged *Radio Control* (RC) car transmitter-receiver pair, a LEGO® Mindstorms Scout programmable brick, and a few electronic components. The objectives of the project are as follows:

- To build an experimental analog transmitter-receiver controller with a digital front end
- To demonstrate software diagnostics testing using *LEGO Assembler* (LASM) to toggle an output and read an input touch sensor
- To build an experimental test bed for validating the basic circuits used in RF circuits
- To explore alternative testing and software tools to assist in developing a solid foundation of wireless and data communications technology
- To build a walking beetle bot to explore mechanical measurement techniques to obtain speed data

So let's begin our experimental investigation of the programmable wireless controller by reviewing the block diagram shown in Figure 1-5.

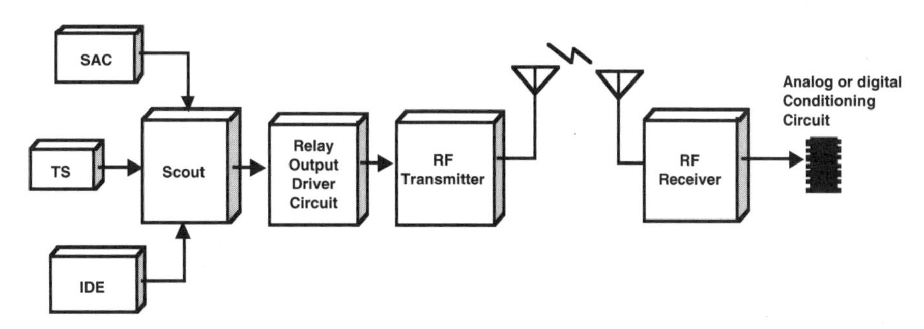

Figure 1-5 Block diagram for the experimental wireless controller

Review of the Experimental Wireless Controller Block Diagram

The digital front end for the analog RF transmitter consists of the following building blocks:

- *Stand-Alone Code* (SAC)
- Touch sensor
- *Integrated Development Environment* (IDE)
- Scout

We will start our discussion with the most obvious building block—the Scout. The Scout provides the physical interface to the outside world for the experimental wireless controller. The touch sensor provides the input signal for the Scout to process. SAC is the robotic behavior program created by the user. The user-created code processes the touch sensor status for the particular output function. To toggle an output or read the status of an input, the IDE is used. By sending LASM commands using an interactive command window, the Scout's *input/output* (I/O) can be monitored and controlled. Also, code can be written and downloaded to the Scout using this IDE platform. With all of the building blocks discussed, it becomes quite obvious that a digital front end or programmable driver can been developed for the analog transmitter. The transmitter's signal can be programmed to transmit when a particular event or natural stimulus such as an obstacle or light has been detected.

The relay output driver circuit switches the analog RF transmitter on and off using an electromechanical relay. The contacts of the electromechanical relay are wired to ground. When the Scout's output voltage switches from ground to +8.64 volts, the relay driver circuit is active. The contacts on the electromechanical relay close, thereby providing a 0-volt signal to the analog transmitter. The circuit sends RF energy into the airwaves only to be intercepted by the receiver. The receiver processes the received signal and depending on the final product, further conditioning circuits for either analog or digital applications will be used.

Experimental Test Procedure Introduction

In this section, the experimental wireless controller will be discussed for the purpose of understanding its basic operation. Also, this experiment will demonstrate how to add a programmable platform to an analog RF transmitter. The following are key items to note when wiring the RC transmitter to the Scout Control Relay and the receiver to an electrical device:

- Determine whether the RF transmitter being used switches ground or +VBattery for signal transmission. A manual inspection of the spring lever switch will help you make this determination. Once determined, the contacts should be wired to the appropriate voltage source.
- The output voltage wires of the receiver can be found by tracing them to the motor. Since the motor can change direction based on reversing the voltage connected across it, make note of the positive and negative wires using a voltmeter.

Reverse Engineering a Salvaged Radio Transmitter-Receiver Pair
Before proceeding with the lab procedures for creating an experimental wireless controller, the subject of reverse engineering must be discussed. Reverse engineering is an educational process that involves taking a product apart to see how it works. In the real world of product design and development, engineers create components and systems by learning from existing products. By dissecting a component or system to its finite element, an engineer enhances his or her knowledge of product design and how parts are placed within the environment they're designed for. Tools such as microscopes, magnifying glasses, oscilloscopes, data sheets, books, and other related products came from knowledgeable people in specific fields whose own experiences were used to reverse the design of a product.

The switching input signal required to transmit the RF signal from the transmitter salvaged out of an RC car is shown in Figure 1-6. The transmitter was reverse-engineered using knowledge in basic electronics circuit interfacing. A careful examination of the placement of components on the *Printed Circuit Board* (PCB) and the spring contacts for switching the RF signal provides a clue into the RF signal's transmission physical switching mechanism. A mental picture of the block diagram for the circuit was drawn on a sheet of blank paper. An electronic circuit interface that could switch the transmitter automatically was then created. After the circuit was designed, it was wired across the existing contacts using alligator test leads. When power was applied and the SAC was run on the Scout programmable brick, the signal was transmitted from the transmitter by pressing the touch sensor.

Figure 1-6 Photo of the modified RC transmitter

Bill of Materials (BOM)
Here is the list of materials required to build the Experimental Wireless Controller.

- LEGO Scout
- Scout IDE tool
- Touch sensor
- R1-100 ohm resistor
- Q1-2N3904 NPN transistor
- D1-1N914 silicon diode
- One salvaged RC transmitter-receiver pair
- Solderless breadboard
- *Volt-ohm-milliammeter* (VOM) or *Digital Multimeter* (DMM)
- *Direct Current* (DC) power supply

Test Procedure 1

1. Wire the wireless controller, as shown in Figure 1-7.
2. To connect the transistor relay driver circuit to the Scout's output block, make a wire harness assembly (shown in Figure 1-8).

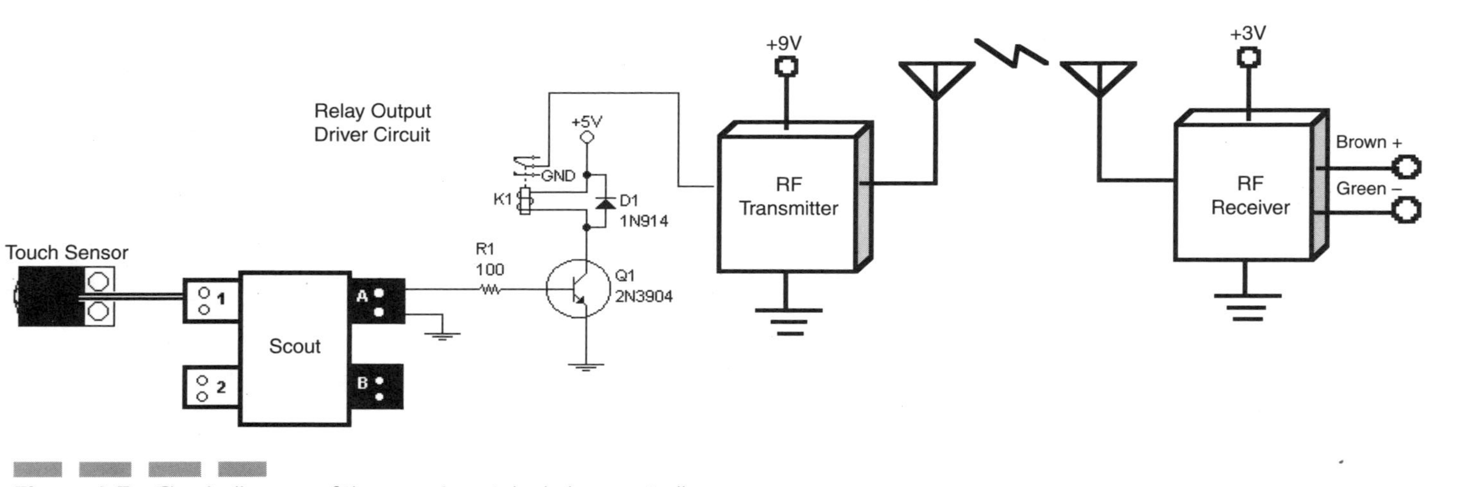

Figure 1-7 *Circuit diagram of the experimental wireless controller*

3. Connect the Scout to the transistor relay driver circuit using the wire harness assembly made in Step 2.
4. Add a touch sensor to input 1.
5. Program the LEGO Scout using the SAC, as shown in Figure 1-9.
6. Attach the RF transmitter to the *normally opened* (NO) contacts of relay K1, as shown in Figure 1-7. Note: Depending on the type of transmitter used, the input switching signal may be +VBattery or ground.
7. Using an external DC power supply, adjust the output voltage for +5 volts.
8. Apply the +5-volt DC source to the external relay circuit, as shown in Figure 1-7.
9. Run the SAC created in Step 5 for the LEGO Scout by pressing the green Run button.
10. Press the touch sensor attached to input 1 of the LEGO Scout.

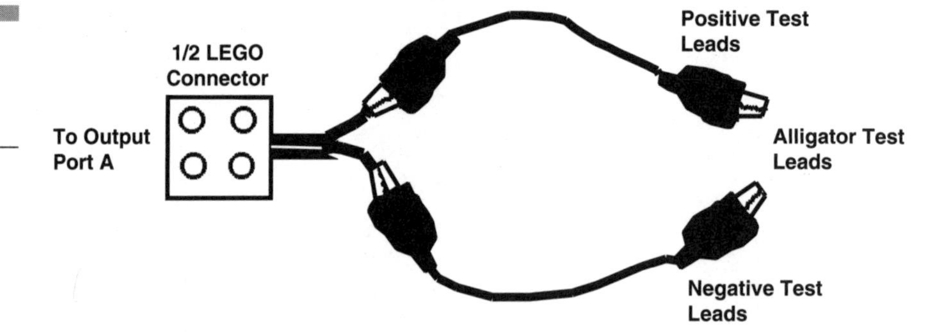

Figure 1-8
Wire harness for attaching to Scout's output A port

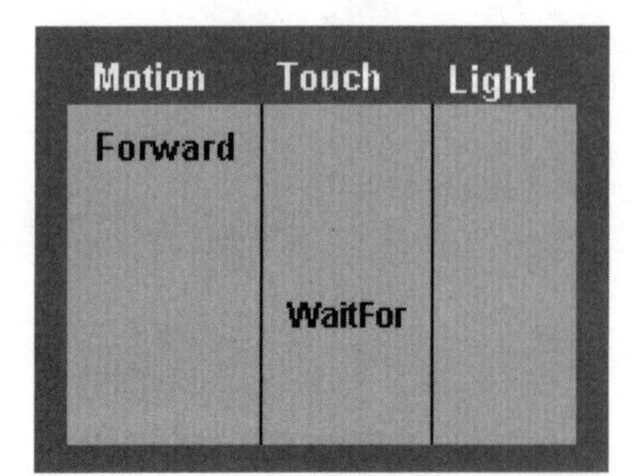

Figure 1-9
SAC for Scout wireless controller

11. The contacts of relay K1 should click once indicating a closed switching event.
12. Press the touch sensor again. The contacts of relay K1 should click again indicating an open switching event.
13. Attach a voltmeter to the output wires of the RF receiver. Note the output voltage and polarity of the signal. Record this value in the data table.
14. Repeat Step 11 and notice the output voltage and polarity of the RF receiver. Record this value in data table 1 shown below.
15. Repeat Step 12 and notice the output voltage and polarity of the RF receiver. Record this value in data table 1 shown below.
16. Keep this test setup for Test Procedure 2.

Test Procedure 2 Objectives

The objective of this test procedure is to validate the operation of the experimental wireless controller using the LEGO Scout IDE as a testing front-end tool for exploring the I/O of the programmable brick. The Scout IDE tool should be installed on your notebook or desktop prior to doing the following test procedure. The Scout IDE tool can be obtained from the AshbySoft web site at www.cix.co.uk/~ashbysoft/.

Test Procedure 2

1. Double-click on the LEGO Scout icon to open the IDE tool on your desktop or notebook. The *Graphical User Interface* (GUI) should be displayed on your screen. See Figure 1-10.

Data Table 1

Data table for recording RF receiver voltage values based on touch sensor status

Touch Sensor	Voltage Receiver
Press	
Release	

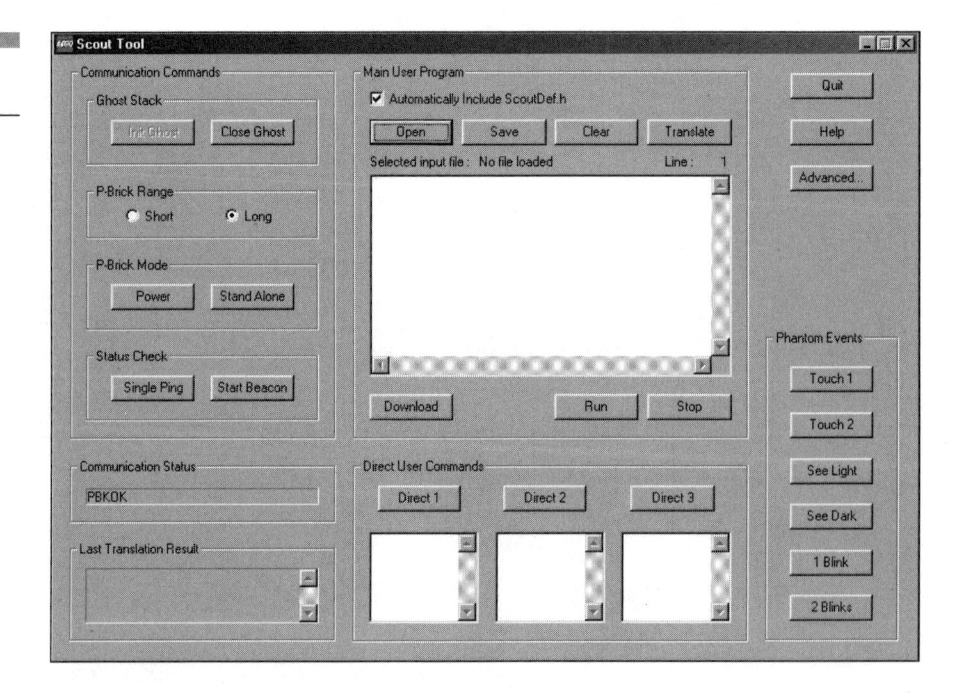

Figure 1-10
The Scout IDE tool

2. Check the Communication Status box on the GUI of the IDE for PBOK.
3. Next, run the SAC of Step 9 of Test Procedure 1.
4. Click the Touch 1 button on the Phantom Events.
5. Record the receiver's output voltage reading from the voltmeter onto the data table.
6. Repeat Steps 4 and 5. Click the Advanced button on the IDE tool. Another GUI panel (see Figure 1-11) will appear.
7. Click the Enable Monitoring button. Notice the Battery text box and the *Light-Emitting Diode* (LED) bar graph. The Battery text box will display the RCX battery voltage and the bar graph shows the ambient light surrounding the Scout brick.
8. Click the X in the right-hand corner of the Advanced Monitoring box to return to the IDE tool.
9. Keep the wireless controller circuit intake for the next experimental investigation.

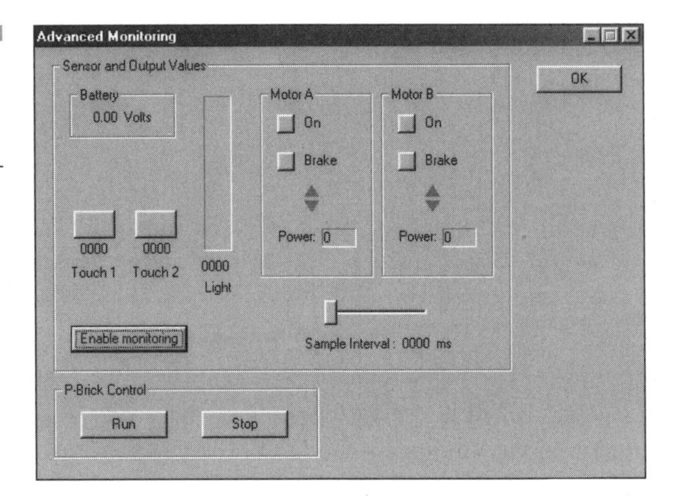

Figure 1-11
The Advanced Monitoring GUI for the Scout IDE tool

Test Procedure 3 Objectives

The objective of this test procedure is to validate the operation of the experimental wireless controller using the LEGO Scout IDE as a testing front-end tool for exploring direct user commands of the programmable brick.

Direct User Commands Investigation The IDE tool has provisions for sending assembly commands directly to the Scout. This feature enables the developer to investigate the operation of direct user commands prior to using them in a software control application. Direct user commands can be used to test the Scout's output port switching procedure of a given robotics or intelligent machine application. Therefore, IDE becomes a simple diagnostics tool for verifying the correct output device functionality. Also, these direct user commands can provide an immediate front-end software interface for wireless controller applications. The following testing procedure will illustrate the ease with which Scout output devices (such as playt, plays, VLL, and motors) can be immediately activated using simple mnemonic commands.

Test Procedure 3

1. With the IDE tool open and using Figure 1-12, type in the following direct user commands: out 2,1 (for Direct 1) and playt 100,100 (for Direct 2).
2. With the Scout in SAC mode, click on the Direct 1 button on the IDE panel. The transmitter's input relay contacts should click on. The receiver's output voltage should be equivalent to what was recorded in the data table from the previous test procedure.
3. Click on the Direct 2 button on the IDE panel. The Scout should produce a 100 Hz tone for 1 second.
4. Type out 1,1 under the previous direct user command Direct 1 button. Click on the Direct 1 button and notice the transmitter's input relay contacts. The receiver should be off via the DC voltmeter reading after sending this command from the IDE panel.
5. Keep the wireless controller interface circuit intake for the following robot project.

Project: Remote Tracking Beetle Bot

Proof of Concept (POC) Study

The objective of this robot is to demonstrate the *Proof Of Concept* (POC) of product development. The design task of the *Remote Tracking* (RT) beetle bot is to build a robot, whereby movement can be detected using a wireless

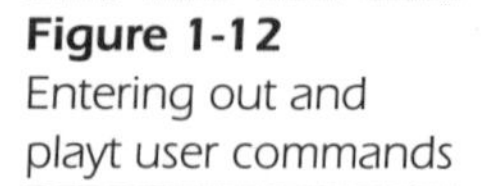

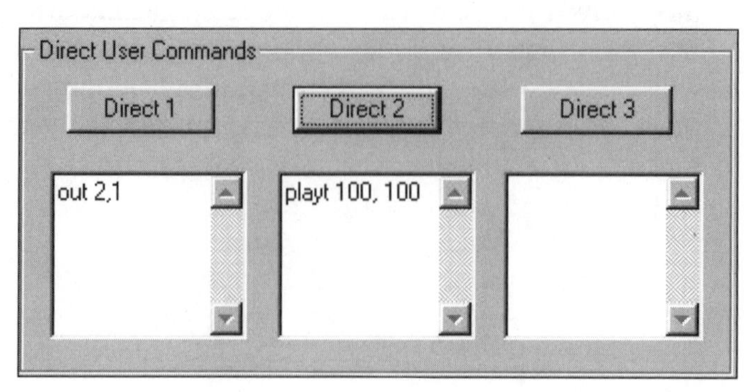

Figure 1-12 Entering out and playt user commands

device and touch sensor. The RT beetle bot's locomotive motion will be provided using two motors. The movement of the bot will be like that of a walking beetle.

Operation of RT Beetle Bot

The bot's sensing strategy is based on a simple event-driven state machine. When the robot's touch sensor is pressed momentarily, it moves forward and an RF signal is transmitted from the transmitter. This signal will be intercepted by the receiver and registered via a positive voltage displayed on the DMM. The presence of the positive voltage indicates that the beetle bot is on the move. When the touch sensor is pressed again, the robot stops. The DMM will then display a negative output voltage.

The RT Beetle Bot Electrical Architecture

Figure 1-13 shows the wiring diagram for the RT beetle bot transmitter. Output ports A and B provide the locomotive mechanical motion using two small LEGO electric motors. The RF transmitter is attached to output port A using a transistor relay driver circuit (see Figure 1-7 for a detailed driver circuit schematic). As discussed previously, when the touch sensor is pressed momentarily, the SAC provides the embedded state machine for the Scout programmable brick processing the digital input signal and convert-

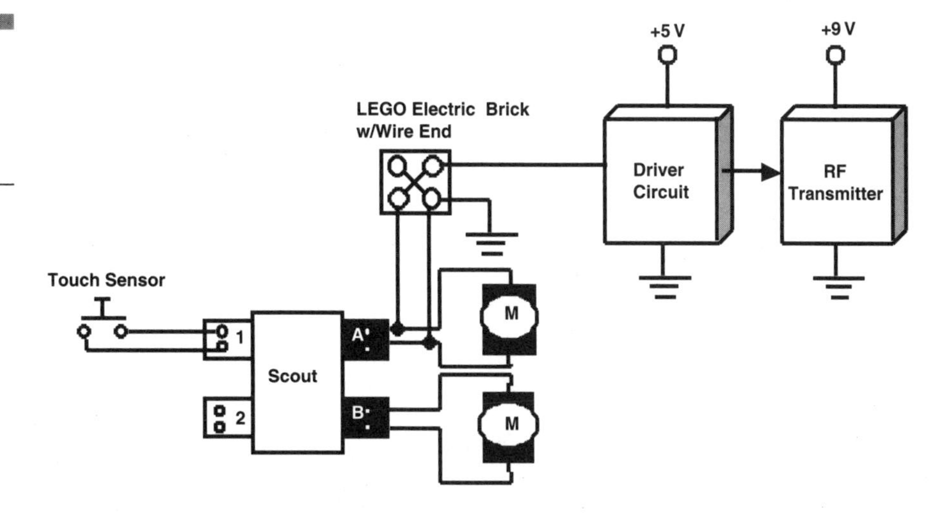

Figure 1-13 Electrical wiring diagram of the RT beetle bot's transmitter

ing it to a DC output voltage. This voltage is used to drive the transistor into saturation, thereby energizing the electromechanical relay's coil. The contacts are switched on, enabling a ground signal voltage to activate the RF transmitter for signal transmission.

The process is reversed when the robot's touch sensor is pressed a second time, thereby enabling the RF transmitter's signal to be shut off. When creating the beetle bot's electrical architecture, a simple structural analysis technique called *Decomposition* was used to partition its input touch sensor from the two electromechanical motor outputs. Decomposition enables the developer to manage the I/O of a robotic system by breaking the machine into small manageable subsystems.

By drawing the system block diagram, the robot's entire electrical/electronics interface can be viewed completely. Creating software becomes easy because each input and output component connected to the Scout can be converted into a state machine. The state machine provides a first-layer mechanism into code development because conditional statements and Boolean equations can be mapped into program constructs directly. In Chapter 6, "Using Procedureal Languages for Mindstorms Robot Control," this state-machine-to-programmable-brick-I/O-interface conversion will be demonstrated using the LEGO RCX and *Not Quite C* (NQC) programming language. For the RT beetle bot, a simple SAC (as shown in Figure 1-14) will provide a state machine that accomplishes the Wait For function in the LEGO Scout.

Figure 1-14 The Wait For state machine for the RT beetle bot

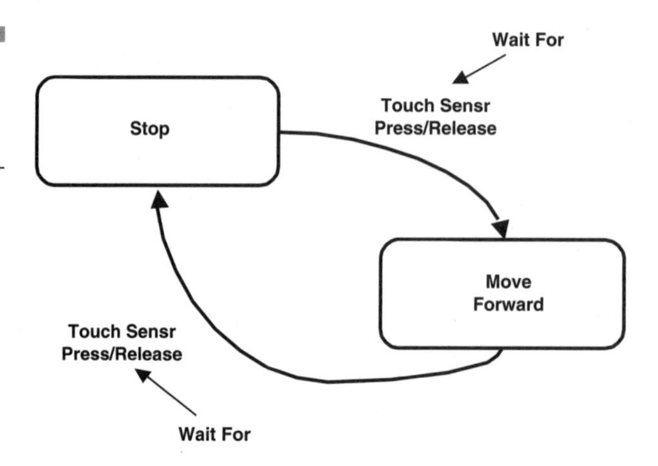

RT Beetle Bot Mechanics

In order to provide motion for the beetle bot, a mechanical drive system is required. The drive system used in the beetle bot is a simple set of gears meshed together to reduce the speed of the LEGO small DC motor by one-fifth. To accomplish the desired speed reduction of 5:1, a simple mechanics equation is used. The gear ratio is a divisional relationship between the driver gear and the driven gear. The gear ratio is represented mathematically as

$$Gear\ Ratio = \frac{DriverGear}{DrivenGear}$$

Figure 1-15 shows a LEGO *Computer-Aided Design* (CAD) model of the beetle bot's walking mechanism. By substituting the motor gear 8t (tooth) for the driver gear and the 40t (tooth) gear for the driven gear, the gear ratio can be calculated.

$$WalkingMechanismGearRatio = \frac{8\text{t}}{40\text{t}}$$

$$WalkingMechanismGearRatio = \frac{1}{5}$$

Using the LEGO Speed Computer and the gear ratio formula, it's easy to determine the RT beetle bot's walking mechanism speed in *revolutions per minute* (rpm). Table 1-1 is a data table showing calculated versus measured values obtained from the bot.

The walking mechanism speed was calculated by taking one-fifth of the motor speed—in this case, 380 rpm/5, which equals 76 rpm.

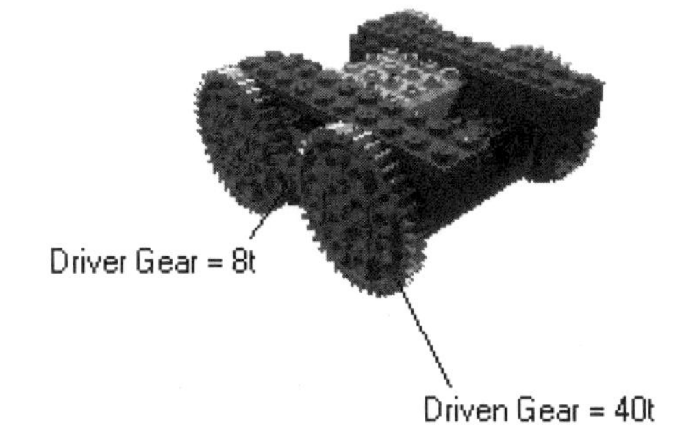

Figure 1-15 LEGO CAD model of the walking mechanism for the RT beetle bot

Table 1-1

Data table used to record calculated and measured beetle bot mechanics

Component	Measured	Calculated
Motor speed	380 rpm	Not applicable
Walking mechanism speed	80 rpm	76 rpm
Gear ratio	Not applicable	One-fifth

How Do You Measure the Speed of the Walking Mechanism? The speed values recorded in the data table were obtained using the LEGO Speed Computer (Cat. No. 5206). By using the LEGO Speed Computer, the RT beetle bot's walking mechanism's velocity can be easily measured. This speed data value will give an empirical indication of the robot's gear ratio and how it plays an important role in managing the walking mechanism's forward movement of locomotion. Figure 1-16 shows a schematic for rpm measurement setup using the LEGO Speed Computer. The rotation pickup sensor is attached to one of the 40t gear using a 6-stud axle. With the LEGO Speed Computer's mode selected for rpm, a velocity measurement can be obtained from the walking mechanism. This same speed measurement technique can be used to record the LEGO 9-volt motor by removing the 8t gear and replacing it with the rotation pickup sensor. The pictures in Figures 1-17 and 1-18 show how to attach the rotation pickup sensor of the speed computer to the beetle bot.

RT Beetle Bot Testing Procedure

The objective of the testing procedure is to validate the RT beetle bot's walking mechanism using the Scout IDE tool and the LEGO Speed Com-

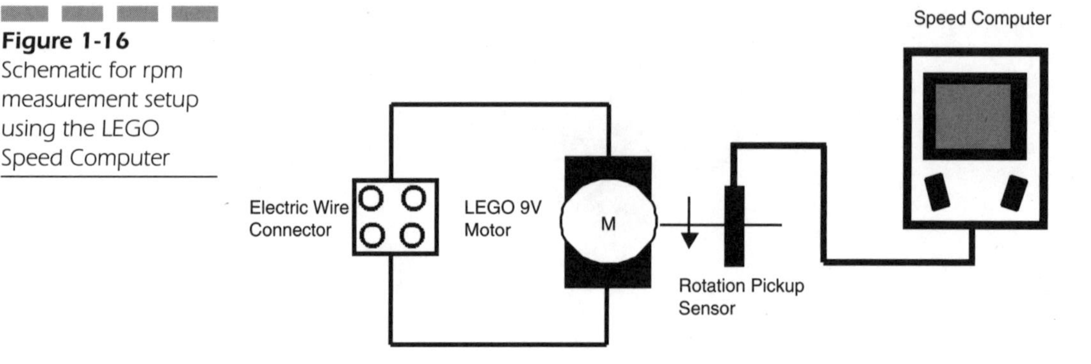

Figure 1-16 Schematic for rpm measurement setup using the LEGO Speed Computer

Figure 1-17
Photo A: measuring the walking mechanism with the LEGO Speed Computer

Figure 1-18
Photo B: measuring LEGO 9-volt motor rpms with the LEGO Speed Computer

puter. The results of the test should coincide with values shown in the speed data table.

BOM

Here is the list of materials required to build a test fixture for measuring the velocity (rpms) of the RT Beetle Bot's walking mechanism using a LEGO Speed Computer

- Scout IDE tool
- Mindstorms Infrared Tower

- The Robotics Discovery Bug Constructopedia
- The LEGO Speed Computer (Cat. No. 5204)
- A 6-stud axle
- The experimental wireless controller from Test Procedure 3
- Assorted LEGO bricks from the Robotics Discovery Set

Test Procedure

1. Build the beetle bot's walking mechanism using the pictorial instructions on pages 13 to 16 and page 53 of the Robotics Discovery Bug Constructopedia. (See Figure 1-19 for the complete bot.)
2. Attach the LEGO Speed Computer to the beetle bot using photo A in Figure 1-17.
3. Press the Mode button on the LEGO Speed Computer to display rpm.
4. Build a simple test stand, as shown in Figure 1-20.
5. Mount the beetle bot on top of the test stand.
6. Connect the two LEGO 9-volt motors to output ports A and B of the Scout.

Figure 1-19 Finished RT beetle bot with LEGO Speed Computer

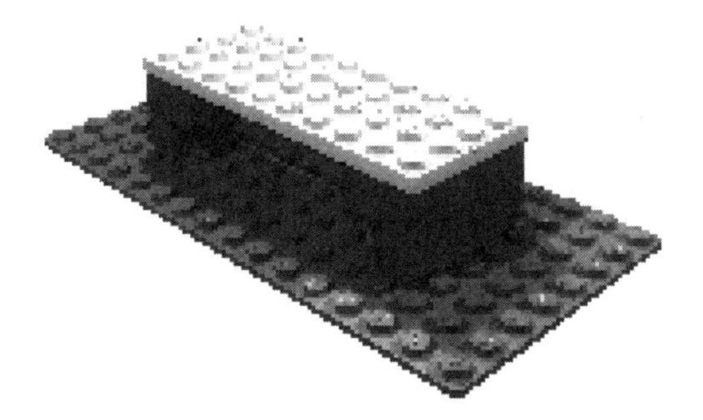

Figure 1-20 A simple mounting test stand for beetle bot speed measurements

7. Connect a touch sensor to input port 1 of the Scout.
8. Connect the experimental wireless controller from Test Procedure 3 (see Figure 1-7) to output port A.
9. Double-click on the LEGO Scout icon to open the IDE tool on your desktop or notebook.
10. Run the SAC of Figure 1-9.
11. Click the Touch 1 button on the Phantom Events.
12. The RT unit's walking mechanism should be moving. The LEGO Speed Computer's *Liquid Crystal Display* (LCD) should be recording the speed of the walking mechanism in rpms.
13. Let the walking mechanism run for a few seconds. Click the Touch 1 button on the Phantom Events section of the IDE GUI. The robot's walking mechanism should stop.
14. The LCD reading of the LEGO Speed Computer should flash 3 times and then stop.
15. Record the value onto the blank data table shown in Table 1-2.
16. Remove the rotation pickup sensor, the 8t gear, the 40t gear, and the 6-stud axle from the bot.

Table 1-2 Data table for recording measured and calculated speed values

Component	Measured	Calculated
Motor speed		
Walking mechanism speed		
Gear ratio		

17. Attach the rotation pickup sensor to the LEGO 9-volt motor shaft using photo B shown in Figure 1-18.
18. Repeat Steps 11 through 14 again. Record the motor's speed in the appropriate section of the data table.

Check Output Ports of the RT Beetle Bot This mini test procedure using the advanced monitoring feature of the Scout IDE tool will validate the output ports of the RT beetle bot.

Mini Test Procedure

1. Click the Touch 1 button on the Phantom Events.
2. Click the Advanced button on the Scout IDE tool. The Advanced Monitoring GUI should be displayed on the desktop or laptop monitor.
3. Click the Enable Monitoring button. The green LEDs for both motor A and B should be on. The LEDs on the Scout should be on as well.
4. Press the touch sensor. The blue LED should be on with a RAW value of 12 displayed on the GUI.
5. Release the touch sensor. The blue LED should be off with a RAW value of 255 displayed on the GUI.
6. Click the Disable Monitoring button to end the test.

Wrap-Up of the RT Beetle Bot Project

This chapter provides an introduction to wireless technology using a salvaged RC car transmitter-receiver pair. This RF device was connected to the LEGO Mindstorms Scout programmable brick using a transistor relay driver interface circuit, thus creating a wireless controller. Although simple in design, the wireless controller system built in this exercise demonstrated the principles of RF electronics and communications. A transmitter unit was analyzed to determine how the RF signal was turned on or off. By performing a reverse-engineering task, the radio control transmitter was modified to enable the transistor relay driver interface, controlled by the Scout programmable brick, to start/stop the unit's RF transmission signal automatically. By using both SAC and the Scout IDE tool, the embedded wireless controller was tested. The key to this embedded wireless controller was based on creating a software control front-end interface using the Scout programmable brick. Also, a simple introduction to LASM programming language was presented by entering the direct user commands (out) to turn on and off the appropriate output port.

A POC method of prototyping an RT robot was demonstrated by building the beetle bug using assembly instructions in the Robotics Discovery Bug Constructopedia. The key element behind this project is in developing a systems approach to building a robot for a given behavioral task using structural analysis. Decomposing the robots Wait For behavior enabled its electrical/electronics interface to be created. Then a series of test procedures were conducted on the robot not only to validate correct sensing operation, but also to obtain speed data on its walking mechanism. A mechanical measurement of the walking mechanism's motor and gear ratio was measured using the LEGO Speed Computer and simple divisional arithmetic, respectively. The Scout IDE tool assisted in this mechanical measurement activity by running the SAC via the Touch 1 Phantom Event button. The RT beetle bot's motor outputs were tested as well as the touch sensor and monitored via the Advanced GUI within the Scout IDE tool. The appropriate LEDs verified the correct I/O operation of the bot by visually lighting up on the GUI.

So what does all this bot wireless interfacing and testing stuff mean to the hobbyists? Basically, this chapter provides an introduction to product design by developing an RT robot using a salvaged transmitter-receiver pair, several electronic components, LEGO bricks, and a Scout programmable brick. The principle behind a POC study lies in prototyping a cost-effective model for experimental investigation. To develop an idea requires creating a physical prototype with the intended attributes of the production component or system. The testing procedures outlined in this chapter along with the tools used for quantitative measurements illustrate one key area of product design that is quite crucial in transforming an idea into a tangible object. Robots are good tools for learning engineering design because all the technical disciplines of software, electronics, and mechanics are intertwined and therefore you establish a multidisciplinary approach to the POC development process.

CHAPTER 2

Developing GUIs: Software Control Basics

Developing controls for robotic applications not only requires knowledge in electronics, but it also requires an understanding of software design and layout development. *Graphical User Interface* (GUI) creation (pronounced "goo wee") is more of an art than a science because the placement of controls is critical in order for the user to effectively have control over the intended object. Consumer electronics products such as remote controls, CD players, stereo systems, microwave ovens, and dishwashers are examples of push-button control layouts. The placement of the controls and how effectively the customer can use them without any discomfort fall under the technical discipline of Human Factors, or *Ergonomics*. This chapter does not intend to delve heavily into ergonomics, thereby making the reader an expert on the subject. Rather, it intends to explain how GUI development requires the proper placement of controls on the target robot display panel so the control device can be used and function effectively. The goal of this chapter is to make the reader aware that software controls are an important interface in the development of robotic applications. Therefore, the following topics will be discussed in this chapter:

- The hardware/software interface block diagram
- What is *Visual Basic for Applications* (VBA)?
- Building the robot control panel in VBA
- Designing a robot control panel using VBA
- What is a *Robot Digital Assistant* (RDA)?
- Building a Mindstorms RDA?
- VBA and *Object-Oriented Programming* (OOP) basics
- An alternative solution for *Data Acquisition* (DA) and measurement

The Hardware/Software Interface Block Diagram

Robotics consists of an interdisciplinary relationship between three key elements: electronics hardware, mechanics, and software. Electronics hardware and software provide the intelligence for processing input data and redirecting the processed information to the appropriate output mechanisms. Therefore, hardware and software must work together harmoniously in order for the robot to perform the tasks it was designed to do. Creating a virtual robot control panel that is used to assist the robot in carrying out its designed task requires a systems development approach.

Figure 2-1 illustrates a hardware/software interface block diagram that is used to develop a virtual robot control panel. As shown in the diagram, the top layer consists of software used to create the control panel and embedded tasks for the robot. The bottom layer relates to the physical controls of sensory detection and the electromechanical propulsion of the robot. Notice the crossover between the software and hardware interface layers in the diagram. Figure 2-2 shows the data flow descriptions, represented by arrows, in the modified diagram.

In order for the robot to perform the embedded behavior or task written in RCX code, the robot control panel must send out commands or a *Software Communication Protocol* (SCP) to the LEGO RCX using the Infrared Tower. With the Infrared Tower connected to the serial port or *Universal Serial Bus* (USB) (version 2.0 Robotics Invention System), the SCP is transmitted to the infrared transceiver mounted in front of the RCX (using a control panel created either by *Visual Basic* [VB] or VBA). The tasks written in *Robotics Command Explorer* (RCX) code will run on the programmable brick waiting for signal processing from either an analog or digital sensor. Once the data is processed, the robot's mechanism will be activated, putting

Figure 2-1 Hardware/software interface block diagram

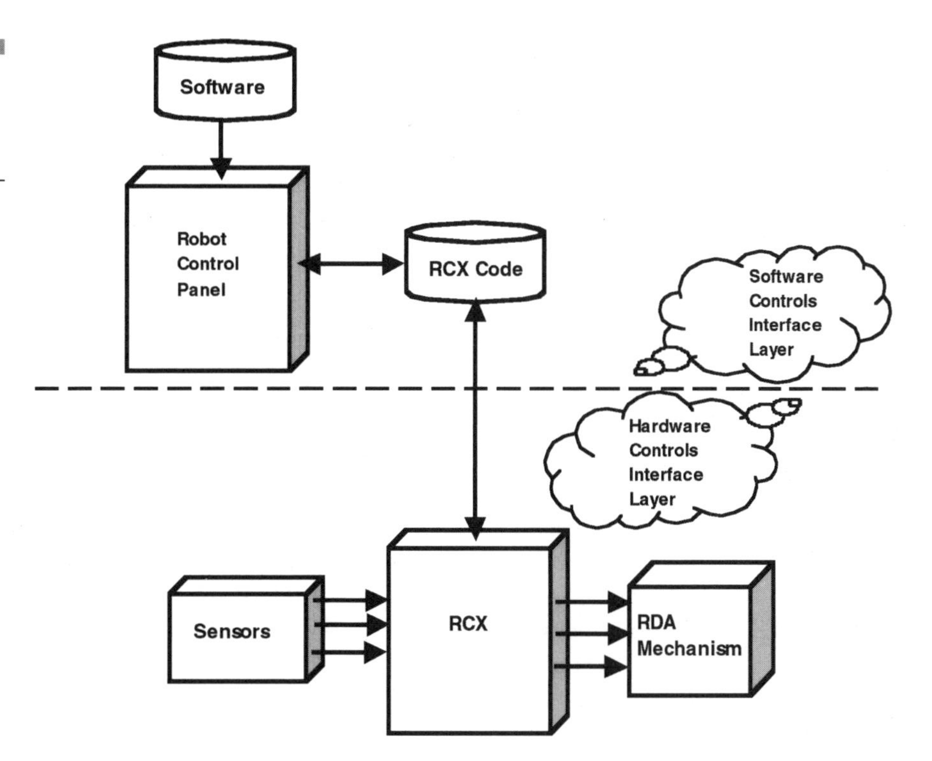

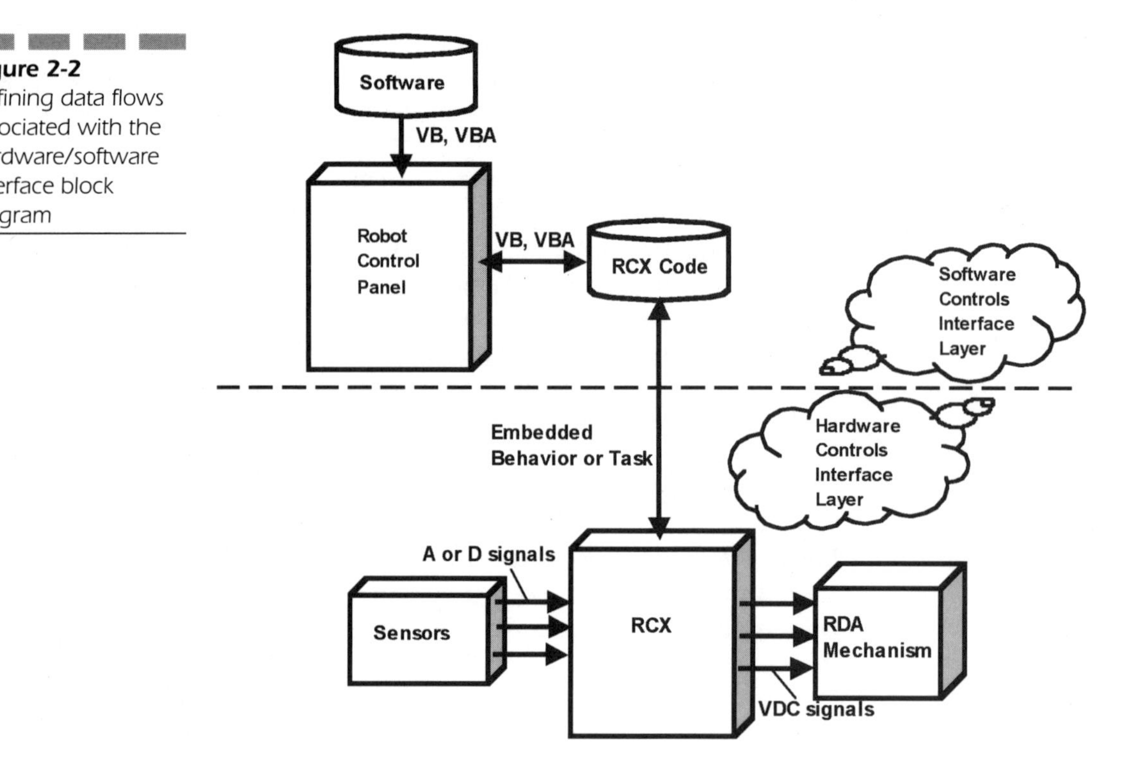

Figure 2-2 Defining data flows associated with the hardware/software interface block diagram

the machine into motion. Therefore, hardware and software cannot exist without the other when designing and building robots. The two layers work together seamlessly to achieve the task the designer has programmed the robot to perform. This basic understanding of interfacing robot hardware and software controls will enable you to develop a control panel GUI for robotics control.

What Is VBA?

When Microsoft introduced VB for Windows in 1991, the product made it easy for people to write stand-alone Windows applications. VBA is a scripting language that comes bundled with Microsoft Office Suite (versions 5.0–2000). The concept of VBA is to enable the developer to automate commonly used tasks performed in Excel, Word, Access, or Power Point environments. The principle behind *Visual Basic for Applications* (VBA) is to use object models for developing software applications for Windows. Unlike VB, VBA is not executable by itself; therefore, it needs to run on a main office application host such as Excel, Word, Access, or Power Point. If you are familiar with VB, then the following discussions on ActiveX controls, properties, methods, and *Object Oriented Programming* (OOP) will be a review. If you are not familiar with VB, detailed instructions for writing software code and panel layout will be provided.

VBA is a wonderful prototyping tool for developing software controls for LEGO Mindstorms robots because the language is graphical based. By placing the controls on a UserForm and defining the task of each visual object, a simple controller for manipulating a Mindstorms robot can be built and debugged in less than an hour. Advance control panels can be created using instrumentation ActiveX controls, thereby enabling the robot designer to use visual indicators for monitoring sensors and motors. Using VBA with Excel creates an analytical interface for performing data analysis and mathematical calculations on sensors. Data tables can be created using logged sensor data and plotted graphically with engineering graphs. The hands-on lab discussion that follows in the next section will explain how to build a robot control panel using Excel VBA.

Building the Robot Control Panel in Excel VBA

Building a robot control panel in VBA is relatively easy. The main requirement in programming using VBA is to always think of buttons as objects that have the capability to change the robot's physical appearance using *properties* and its functions using *methods*. The underlining principle behind programming in VBA is OOP. In the following lab exercise, OOP will be presented during the course of the project by building a simple control panel for the target robot.

The construction of the panel relies on placing VBA objects or buttons known as ActiveX controls on a UserForm. An ActiveX control is a registered object used to enhance and enable the interactivity of a Windows application. There are several standard controls used in Excel that enable the developer to create an interactive application for the user. For example, instead of entering data into cells within an Excel spreadsheet, the UserForm and ActiveX controls enable the user to automatically load data into the appropriate cells using one button. Calculations can be carried out in the same manner by using buttons to read data from cells and displaying results back onto the worksheet. The following procedure walks through the process of accessing the *Visual Basic Editor* (VBE) and Control Toolbox toolbar. This is followed by a list of steps that describe the construction of the robot control panel.

Lab 1: Accessing the VB Toolbar

1. Open the Excel Software Application.
2. Click View from the menu bar located underneath the Excel title bar. A pull-down menu will be displayed on the spreadsheet.
3. Click Toolbars and notice the sub-pull-down menu items.
4. Click Visual Basic from the selection items. (See Figure 2-3.)
5. After selecting Visual Basic from pull-down menu selections, the toolbar will be visible on the spreadsheet. (See Figure 2-4.)

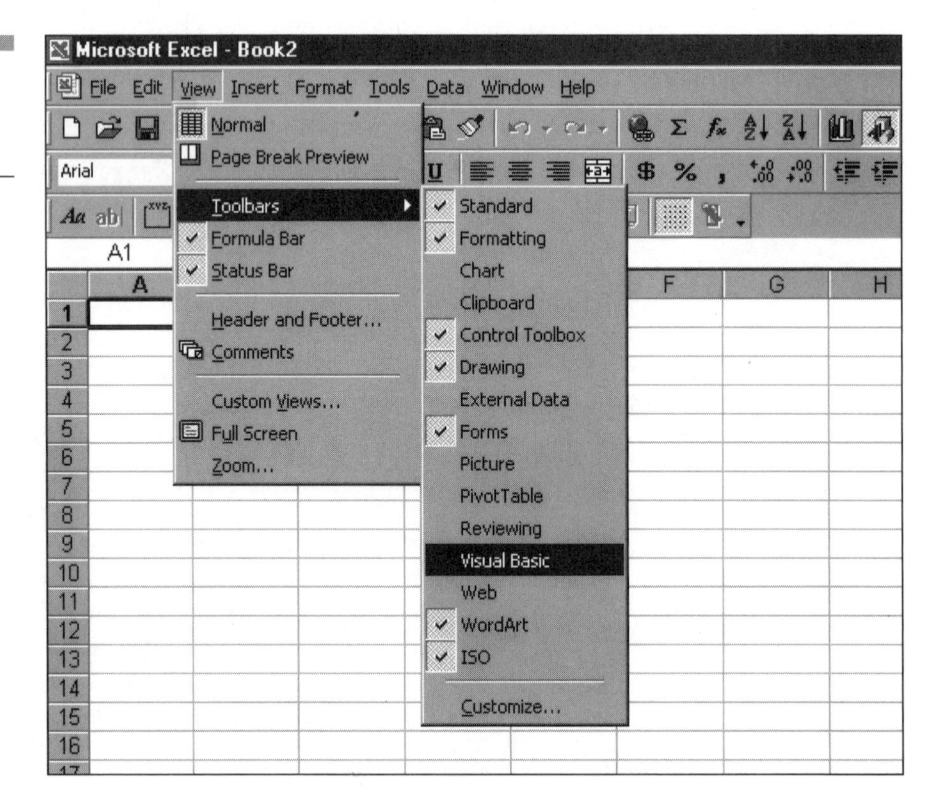

Figure 2-3 Adding the VB toolbar to the Excel application

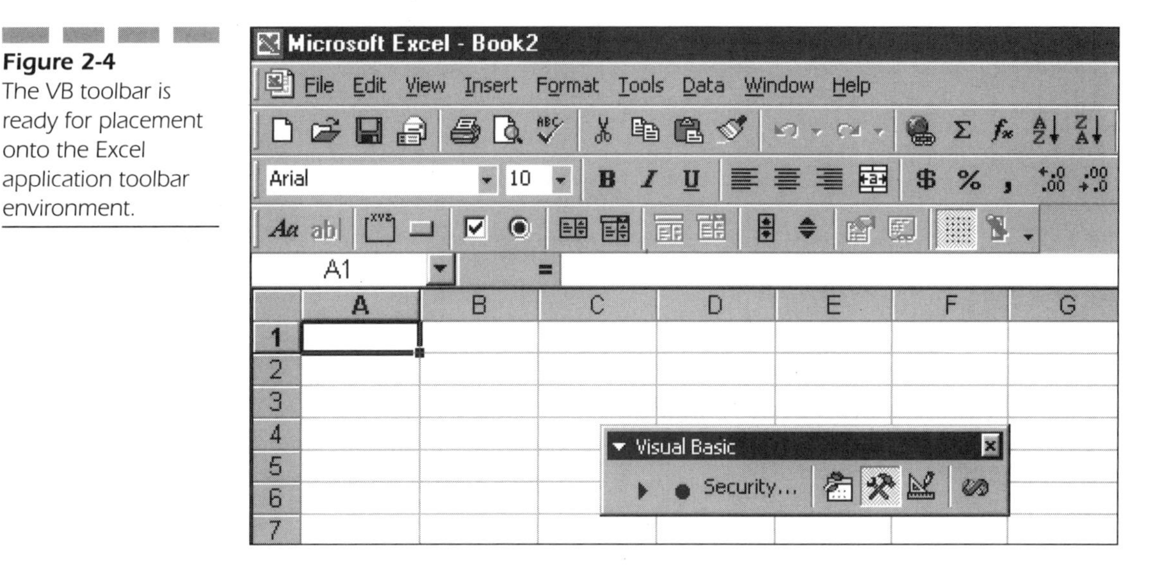

Figure 2-4
The VB toolbar is ready for placement onto the Excel application toolbar environment.

6. For permanent placement, locate the VB toolbar onto the Excel application, as shown in Figure 2-5.
7. Keep Excel open for the addition of the Control Toolbox toolbar.

Lab 2: Adding the Control Toolbox Toolbar for Excel

1. Click View from the menu bar located underneath the Excel title bar. A pull-down menu will be displayed on the spreadsheet.
2. Click Toolbars and notice the sub-pull-down menu items.
3. Click Control Toolbox from the selection items. (See Figure 2-6.)
4. After selecting Control Toolbox from pull-down menu selections, the toolbar will be visible on the spreadsheet. (See Figure 2-7.)
5. For permanent placement, locate the Control Toolbox toolbar onto the Excel application, as shown in Figure 2-8.

With the main toolbars in place, the robot control panel is ready to be built using the following design/prototype technique.

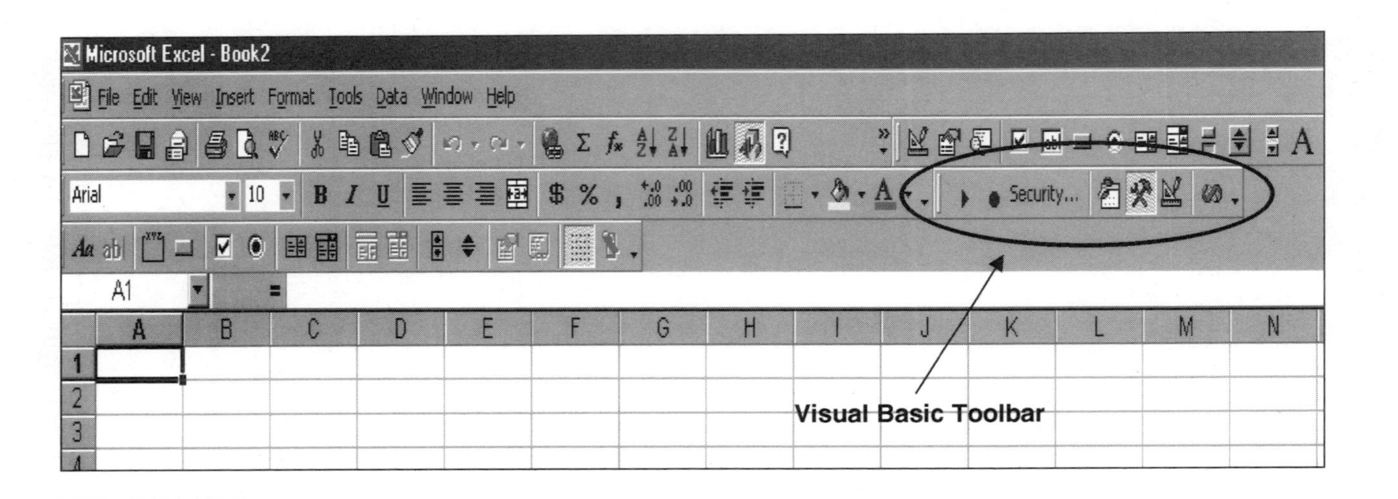

Figure 2-5 The VB toolbar is placed at the desired location within the Excel application toolbar environment.

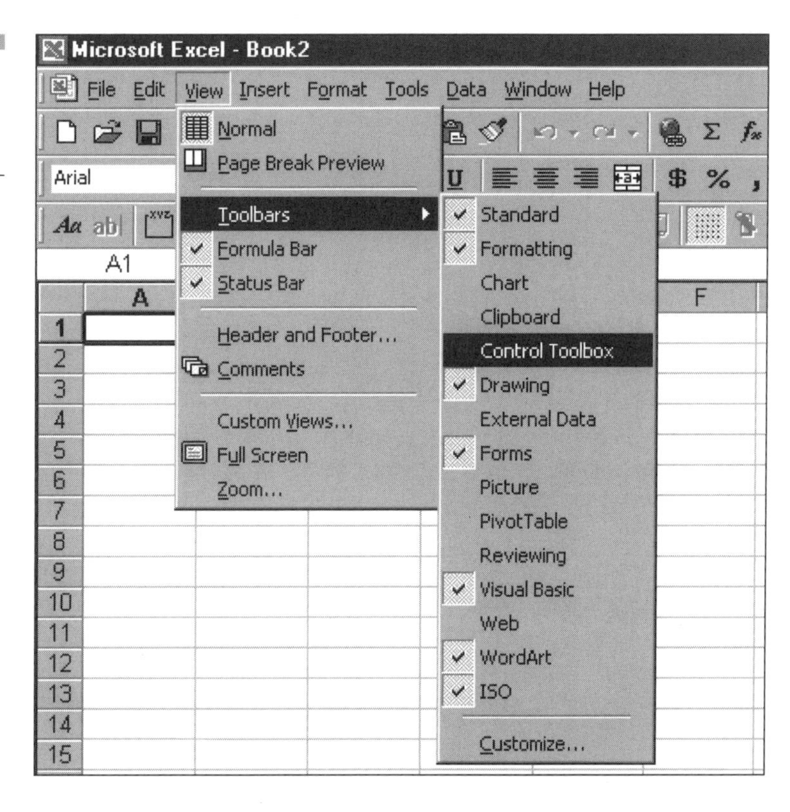

Figure 2-6
Adding the Control Toolbox toolbar to the Excel application

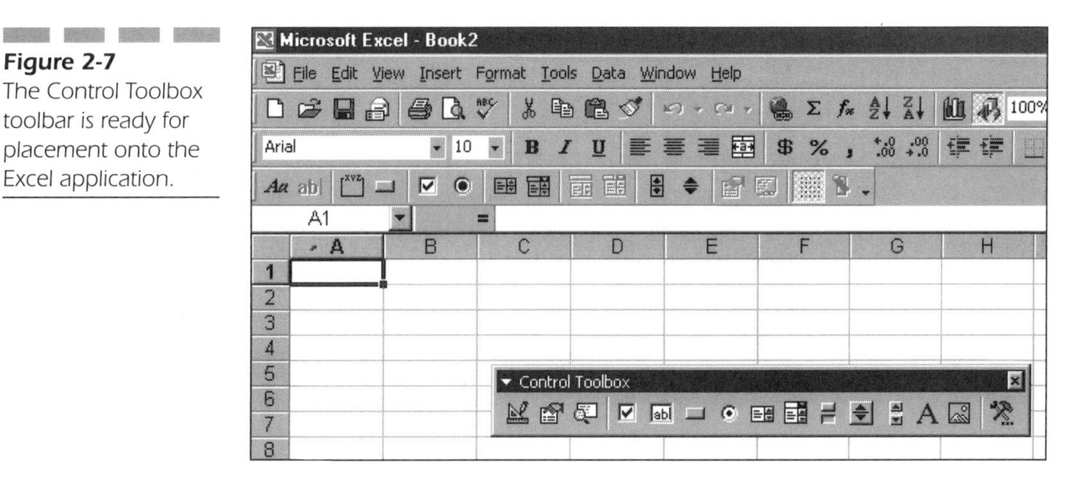

Figure 2-7
The Control Toolbox toolbar is ready for placement onto the Excel application.

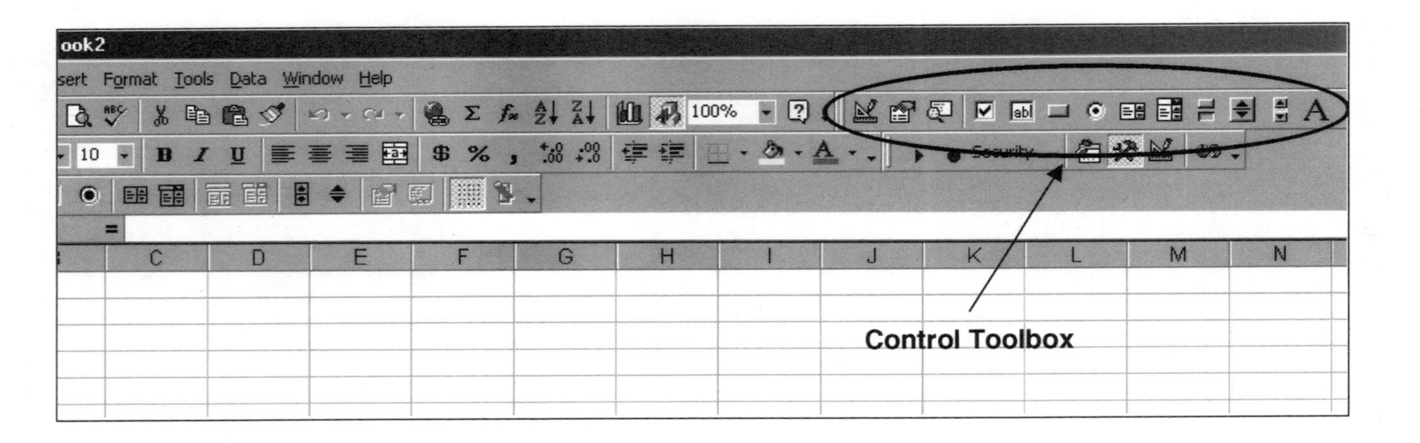

Figure 2-8 The Control Toolbar is placed at the desired location within the Excel application.

Designing a Robot Control Panel Using VBA

The main goal in building a software robot control panel is to design a virtual prototype that can be built using software components or graphical objects. The layout of buttons, sliders, and indicators can be executed as easily as taking a sheet of lined paper and sketching out a control panel design. Figure 2-9 shows a simple sketch of a proposed control panel for a *mechatronic* voltage-regulator-controlled robot named *V-Bot*. The concept behind sketching a design before building it is to provide a design template for quickly laying out the controls on the UserForm. This creates a prototype using software that enables you to easily manipulate virtual objects and define functions for each of the placed buttons using VBA code. Once the virtual panel is built, it is ready to be evaluated and tested through a testing procedure or document.

The following lab project illustrates this prototyping concept by using Figure 2-9 as a design template for building the control panel for V-Bot. As stated earlier, the VB toolbar is an essential Excel component of prototyping software controls for LEGO Mindstorms robots. If you are an experienced Excel user and have the VBA toolbar available, then proceed to the lab project procedures. If you have not added VBA to the Excel toolbar environment, then follow the steps in Labs 1 and 2.

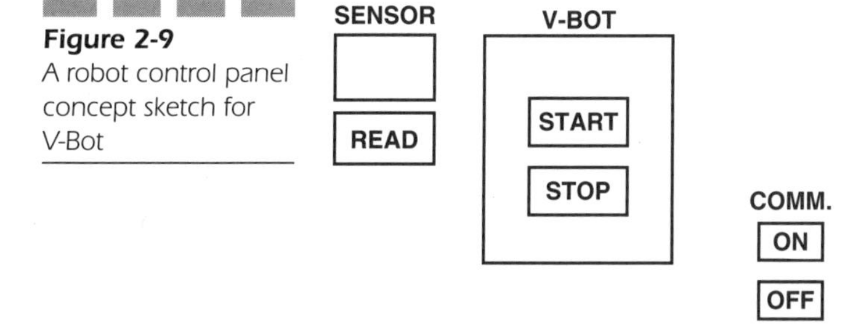

Figure 2-9
A robot control panel concept sketch for V-Bot

Constructing the V-Bot Control Panel Lab Project Objectives

The purpose of this lab project is to build a virtual prototype control panel to control and monitor a voltage regulator robot named V-Bot. The concepts discussed in the section "Designing a Robot Control Panel Using VBA" will be illustrated using the following hands-on lab project procedure. Figure 2-10 shows the system block diagram for V-Bot. An explanation of how the control panel works will be given in the section "What Is an RDA?" later in this chapter.

V-Bot Control Panel Build Procedure

1. Open a new Excel worksheet.
2. Click the VBE icon on the VBA toolbar. (See Figure 2-11.)
3. The VB *Integrated Development Environment* (IDE) window will be displayed, as shown in Figure 2-12.

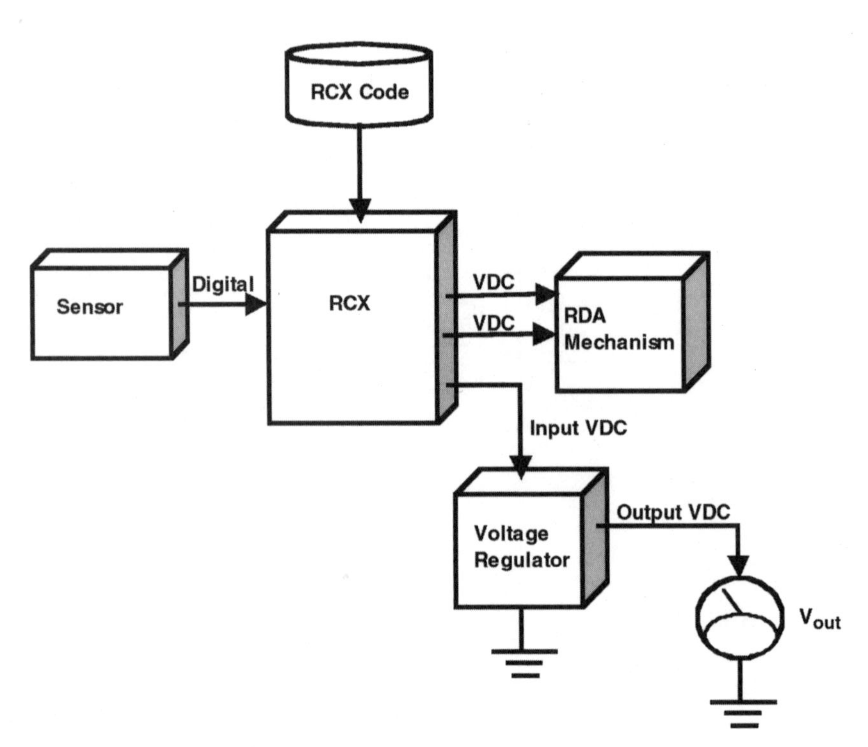

Figure 2-10
System block diagram for V-Bot

Figure 2-11
VBE icon

4. Click Insert from the menu bar.
5. Click UserForm from the pull-down menu selection items. (See Figure 2-13.)
6. A UserForm will be displayed on the VB IDE window, as shown in Figure 2-14.
7. Save the spreadsheet as V-Bot Control Panel on your hard drive.

NOTE: *Notice the LEGO logo on the toolbox in Figure 2-14. This icon is an ActiveX control for providing software objects for the LEGO RCX brick functions and features. The following subprocedure outlines the steps to take to add this ActiveX control to the toolbox.*

Subprocedure for Adding the LEGO ActiveX Control to the Toolbox

1. Click Tools from the VBE menu bar, as shown in Figure 2-15.
2. Click References from the pull-down menu.
3. A References-VBAProject dialog box will appear on the screen. (See Figure 2-16.)
4. Click Browse.
5. The Add Reference dialog box will appear on the screen, as shown in Figure 2-17.
6. Click the down arrow within the Files of type drop-list box.
7. Scroll down to ActiveX Controls (*.ocx) and click.
8. Look in Program Files. Find and double-click the LEGO Mindstorms folder. (See Figure 2-18.)
9. Double-click the system folder.
10. Double-click the Spirit.ocx icon, as shown in Figure 2-19.
11. The References-VBAProject dialog box will appear. Scroll down until the LEGO PBrickControl, OLE Module reference is displayed. (See Figure 2-20.) Click on the check box and click OK.

Figure 2-12 The VB IDE

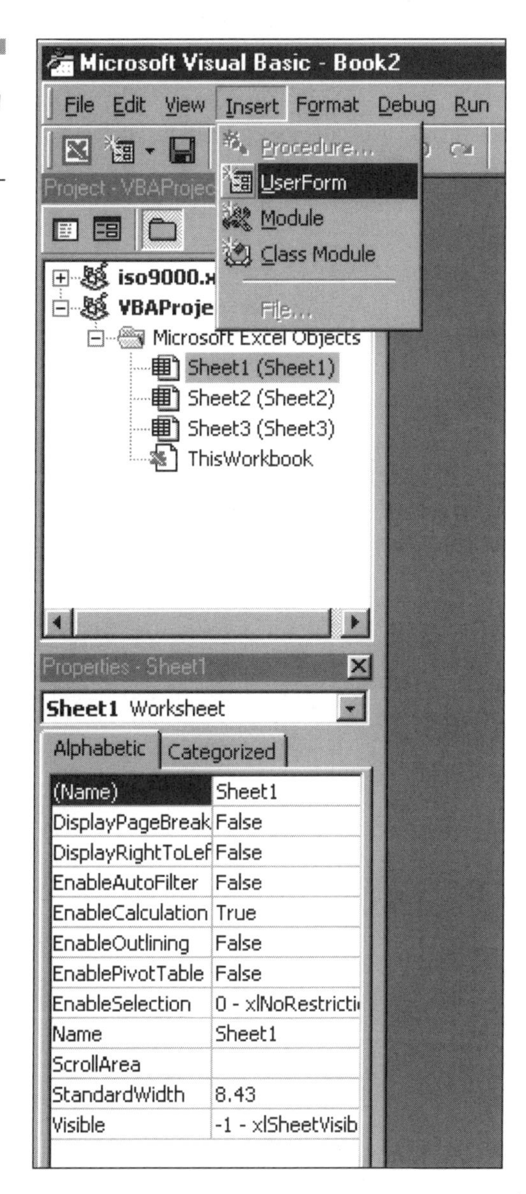

Figure 2-13 A UserForm obtained from the Insert pull-down menu

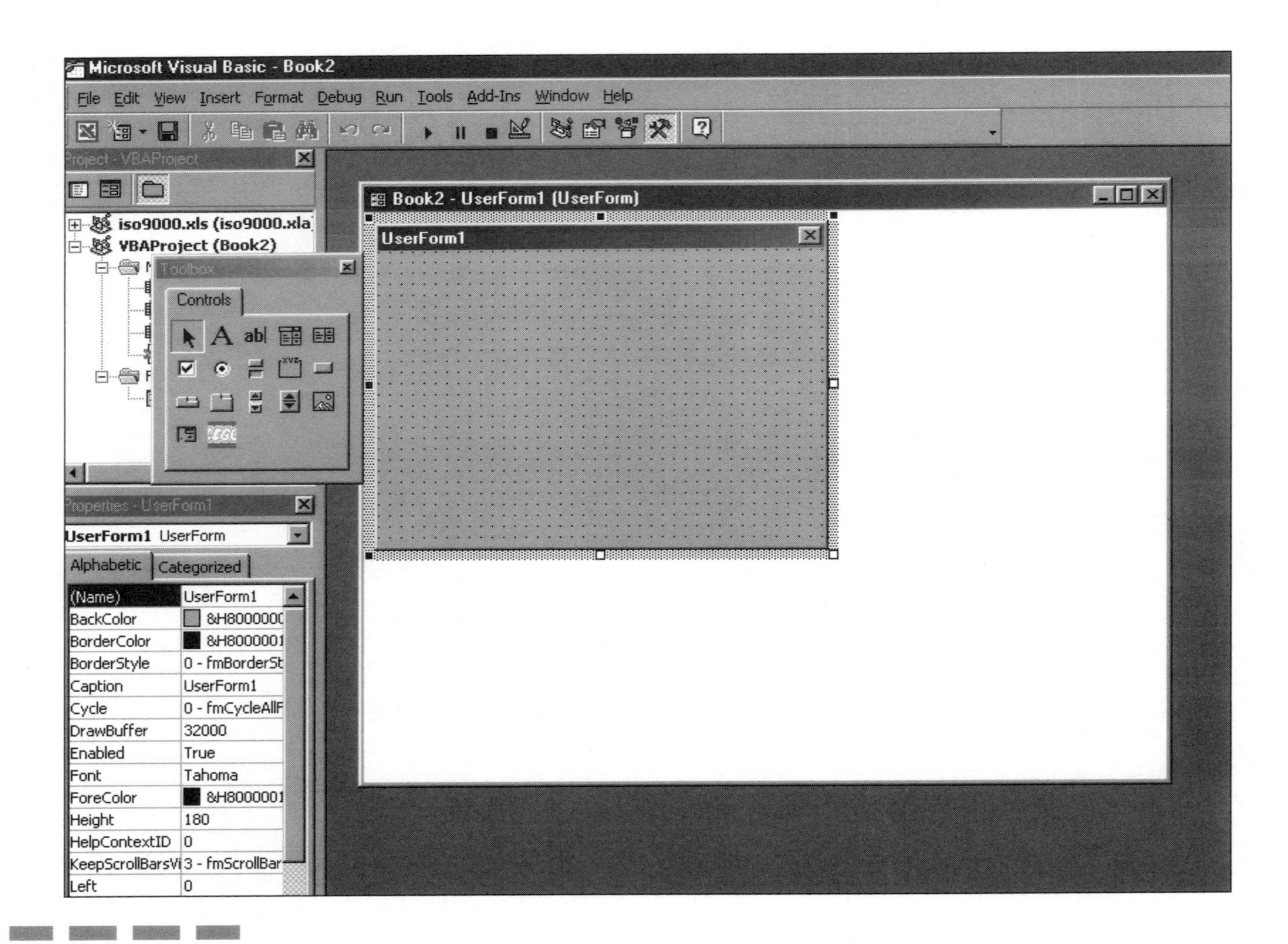

Figure 2-14 A UserForm placed on the VB IDE window

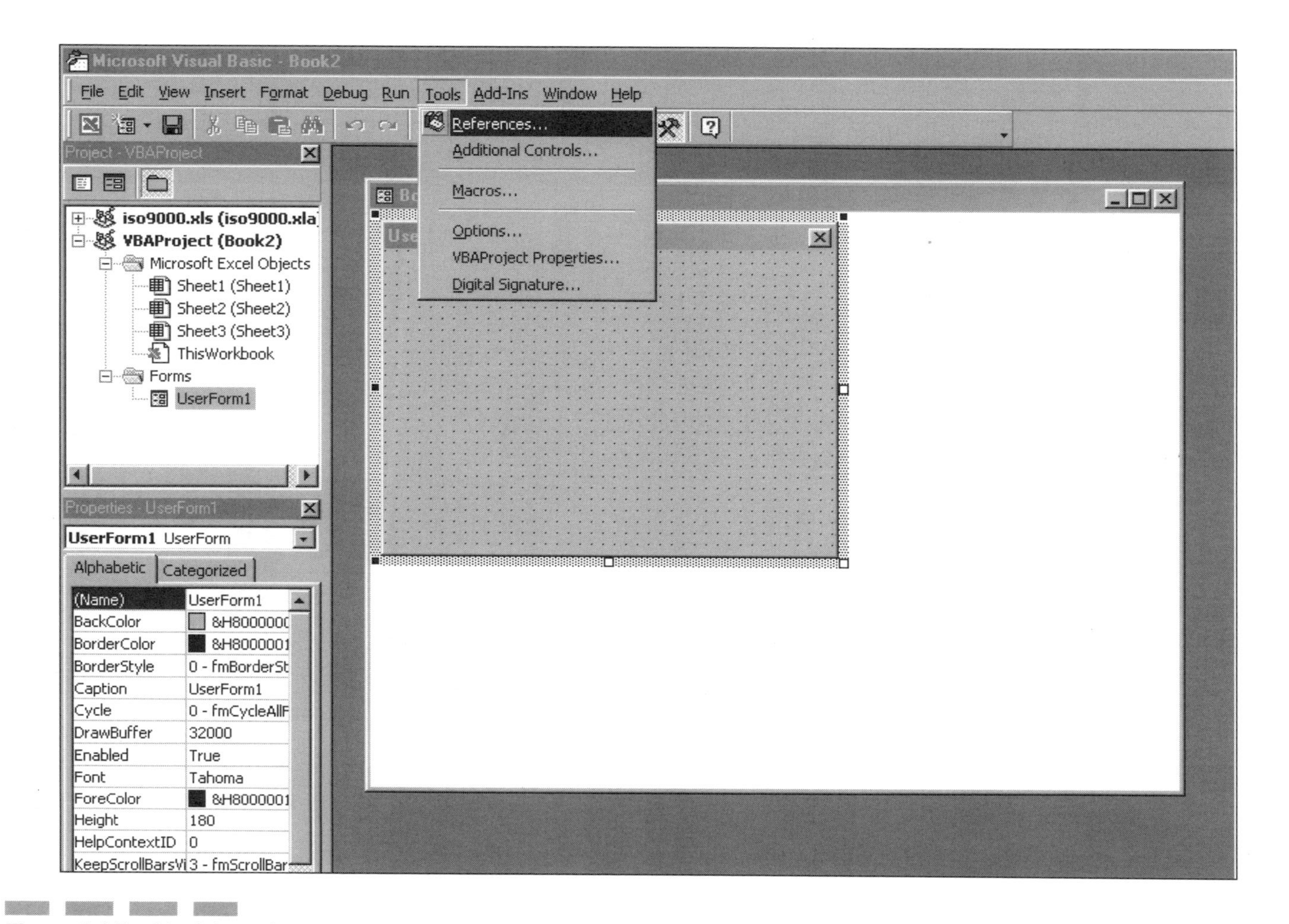

Figure 2-15 Obtaining references for the LEGO RCX ActiveX control

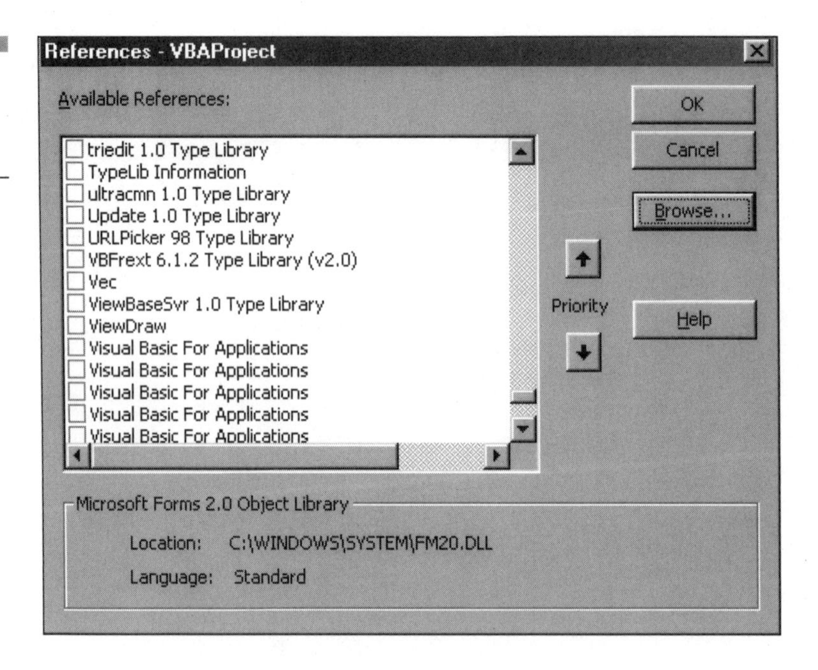

Figure 2-16 The References-VBAProject dialog box

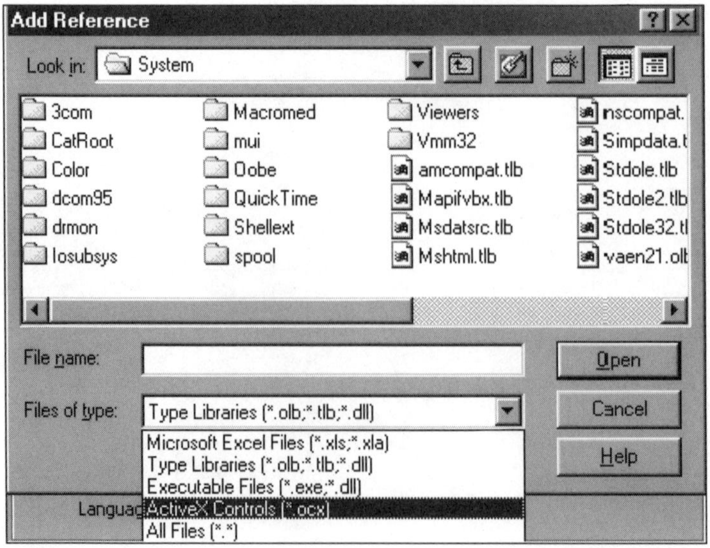

Add Reference
Look in: System
3com
CatRoot
Color
dcom95
drmon
Iosubsys
Macromed
mui
Oobe
QuickTime
Shellext
spool
Viewers
Vmm32
amcompat.tlb
Mapifvbx.tlb
Msdatsrc.tlb
Mshtml.tlb
nscompat.
Simpdata.t
Stdole.tlb
Stdole2.tlb
Stdole32.t
vaen21.olb
File name:
Open
Files of type: Type Libraries (*.olb;*.tlb;*.dll)
Cancel
Microsoft Excel Files (*.xls;*.xla)
Type Libraries (*.olb;*.tlb;*.dll)
Executable Files (*.exe;*.dll)
ActiveX Controls (*.ocx)
All Files (*.*)
Help
Languag

Figure 2-17 The Add Reference dialog box

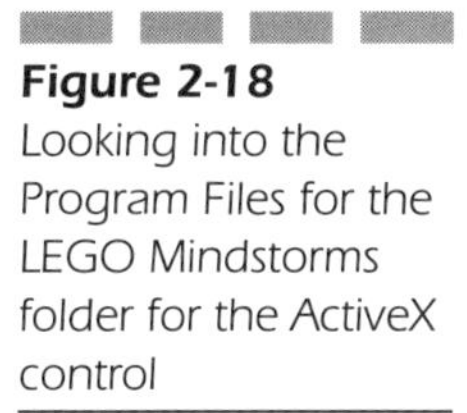

Figure 2-18 Looking into the Program Files for the LEGO Mindstorms folder for the ActiveX control

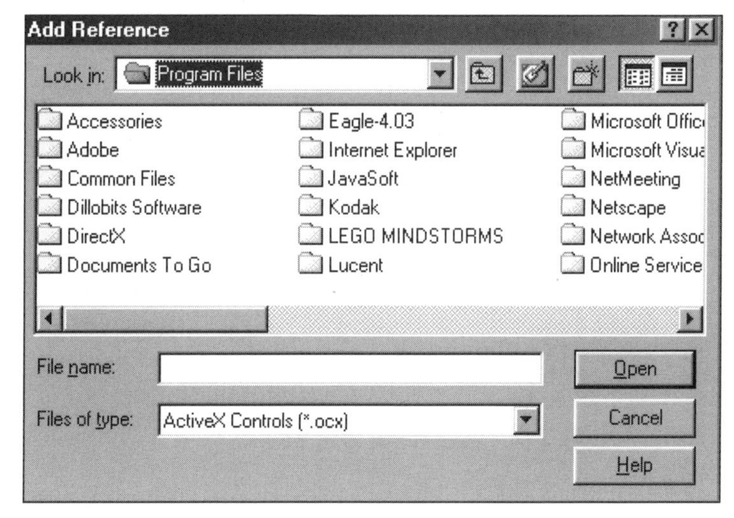

Figure 2-19 Opening the Spirit.ocx LEGO Mindstorms RCX references

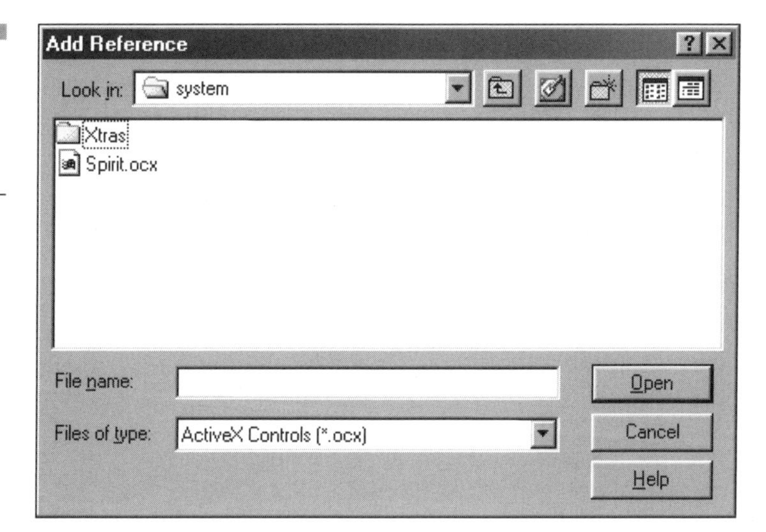

12. Click Tools from the VBE menu bar.
13. Click Additional Controls.
14. The Additional Controls dialog box will appear. Scroll down until the Spirit Control is found. Click on the box and click OK. (See Figure 2-21.)

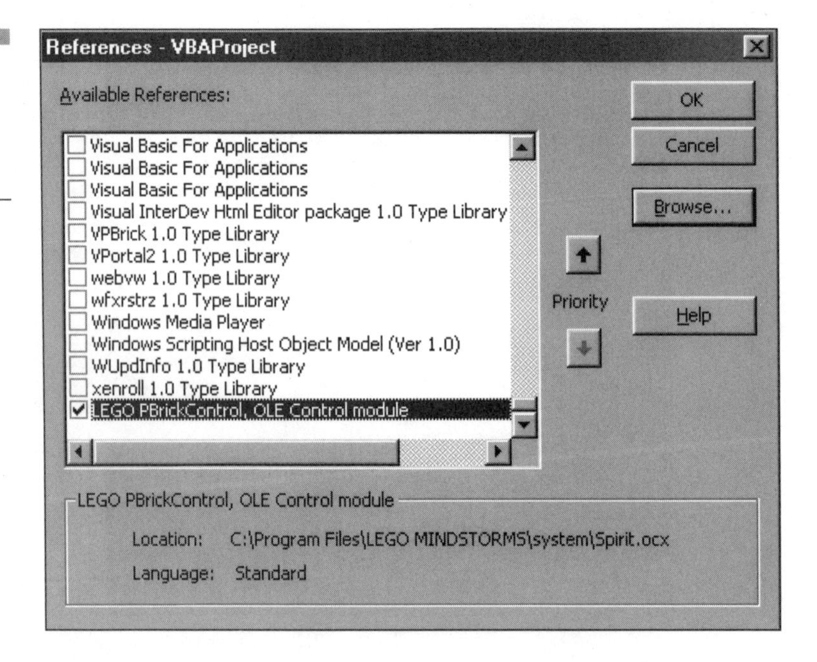

Figure 2-20 Adding the LEGO PBrickControl, OLE Module to the Excel VBA

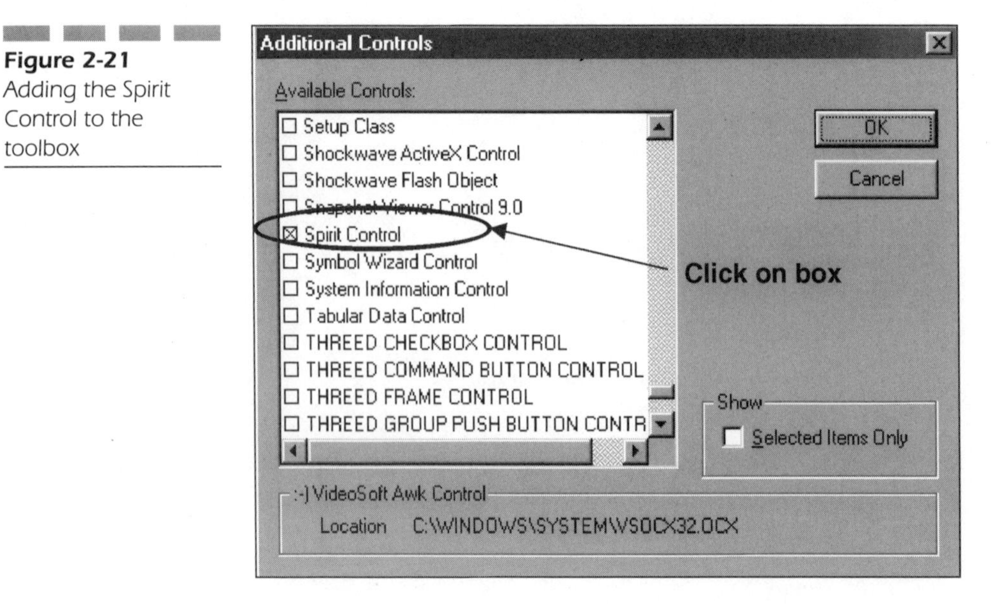

Figure 2-21 Adding the Spirit Control to the toolbox

Figure 2-22
The LEGO ActiveX control added to the VBA toolbox

If the procedure was followed carefully, the LEGO ActiveX control should be displayed on the VBA toolbox, as shown in Figure 2-22. If not, repeat the procedure until the icon is successfully displayed on the toolbox.

Now with the LEGO ActiveX control available, the V-Bot control panel can be built. The next lab session will provide further instructions on building the robot control panel.

V-Bot Control Panel Build Procedure 2

1. Open the Excel spreadsheet file V-Bot Control Panel.xls saved on your hard drive.
2. Click the VBE icon on the VBA toolbar. (See Figure 2-11.)
3. A blank UserForm1 along with the toolbox should be displayed on the screen.
4. Using the concept sketch of Figure 2-9 as a design guide, begin adding the other ActiveX controls from the toolbox to UserForm1. See Figure 2-23 for how the final panel should look.

NOTE: *The captions of each ActiveX control are renamed by changing the caption property of the VB object placed on the UserForm. Figure 2-24 shows how to rename the Label caption shown on the robot control panel of Figure 2-23.*

5. After the robot control panel is built in VBA, a simple test for displaying the GUI on the Excel spreadsheet can be performed. Click Run from the menu bar.

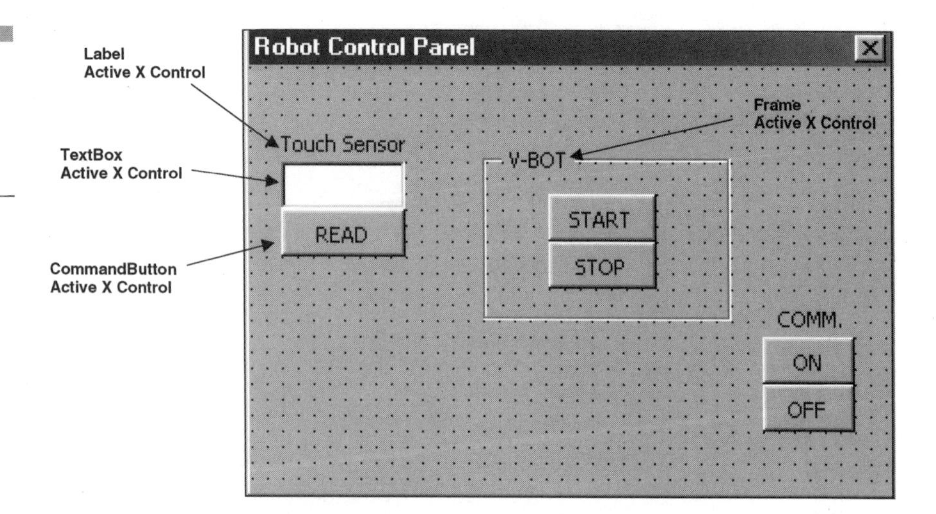

Figure 2-23 A finished robot control panel for V-BOT with typical ActiveX controls

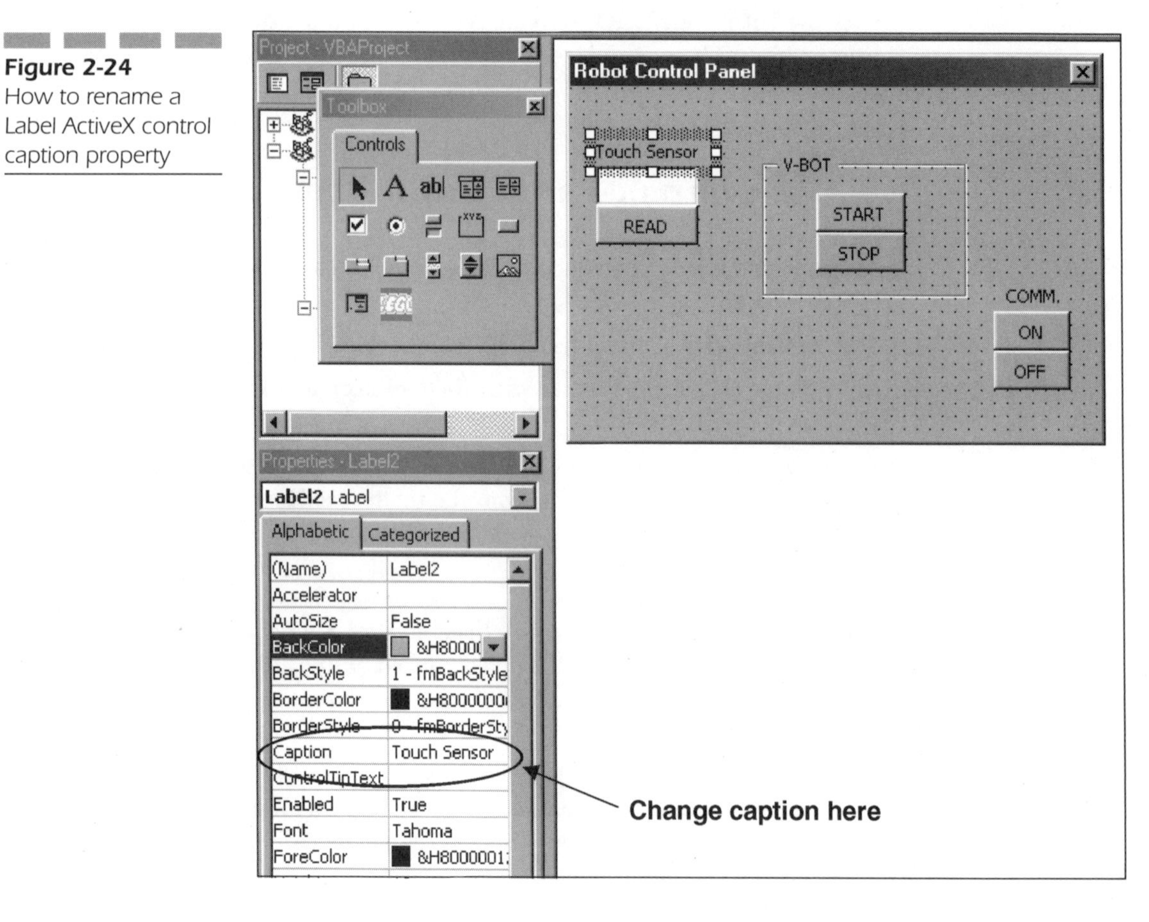

Figure 2-24 How to rename a Label ActiveX control caption property

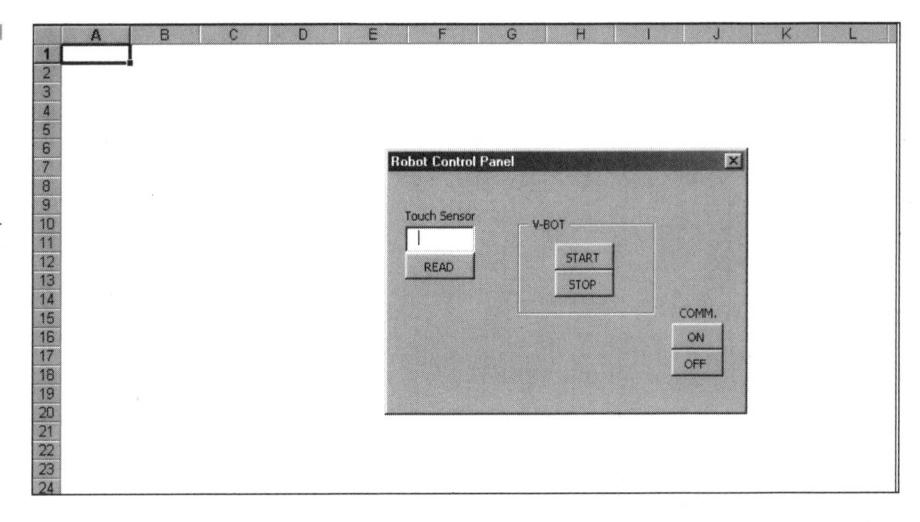

Figure 2-25
Displaying the robot control panel on an Excel spreadsheet using the Run feature in the VBE

6. Click Run Sub/UserForm from the pull-down menu. The GUI will be displayed on the Excel spreadsheet, as shown in Figure 2-25. In this figure, the cells are turned off. This function can be performed by clicking on the Toggle Grid icon located on the Forms toolbar within Excel. If this toolbar is not displayed, follow Steps 1 through 5 of Lab 2 and click Forms from the pull-down menu.

After the robot control panel GUI has been built and tested to be visually displayed on the Excel spreadsheet, the final task for the project is to add a button to the worksheet. This button enables the control panel to be displayed without having to go into the VBE and execute Run from the menu bar. The next section describes the procedure for adding an ActiveX CommndButton onto an Excel spreadsheet to call up the robot control panel GUI for V-Bot.

Calling Up a VBA-Built Robot Control Panel GUI Using an Excel Spreadsheet Button

This procedure explains how to call up the robot control panel shown in Figure 2-23 using an ActiveX control CommndButton located on an Excel spreadsheet.

1. With the completed GUI displayed in the VBE, click the Microsoft Excel icon located on the menu bar. A blank worksheet should be displayed on the screen.

2. Click on the CommndButton icon located on the Control Toolbox toolbar. A small right-angle triangle sitting on top of a ruler with a pencil should be displayed on the worksheet. This icon represents an active design-mode session.
3. Take the mouse and drag it to display the CommndButton on the spreadsheet. Figure 2-26 shows the CommndButton on the worksheet.
4. Place the cursor on the CommndButton and right-click to display the pull-down menu.
5. Click Properties to display its Property window.
6. Look for the Captions Property entry box and delete CommandButton1. Type V-BOT in the empty entry box. Notice the text displayed on the button. Figure 2-27 shows the new caption for the button.
7. Click on the X in the upper-right corner of the Properties window to remove the dialog box.

Now that the button is placed on the Excel spreadsheet and the robot control panel is completed, it is time to write the embedded code used to display the GUI on the worksheet based on a single click event of the mouse. To accomplish this, follow the steps in the next lab procedure.

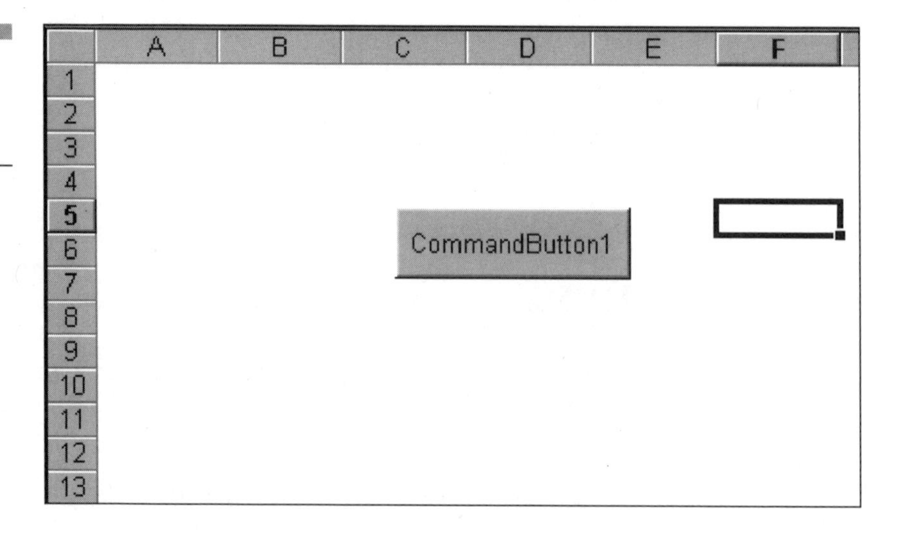

Figure 2-26 CommndButton placed on an Excel spreadsheet

Figure 2-27 Changing the caption for the CommandButton1 to V-BOT Panel

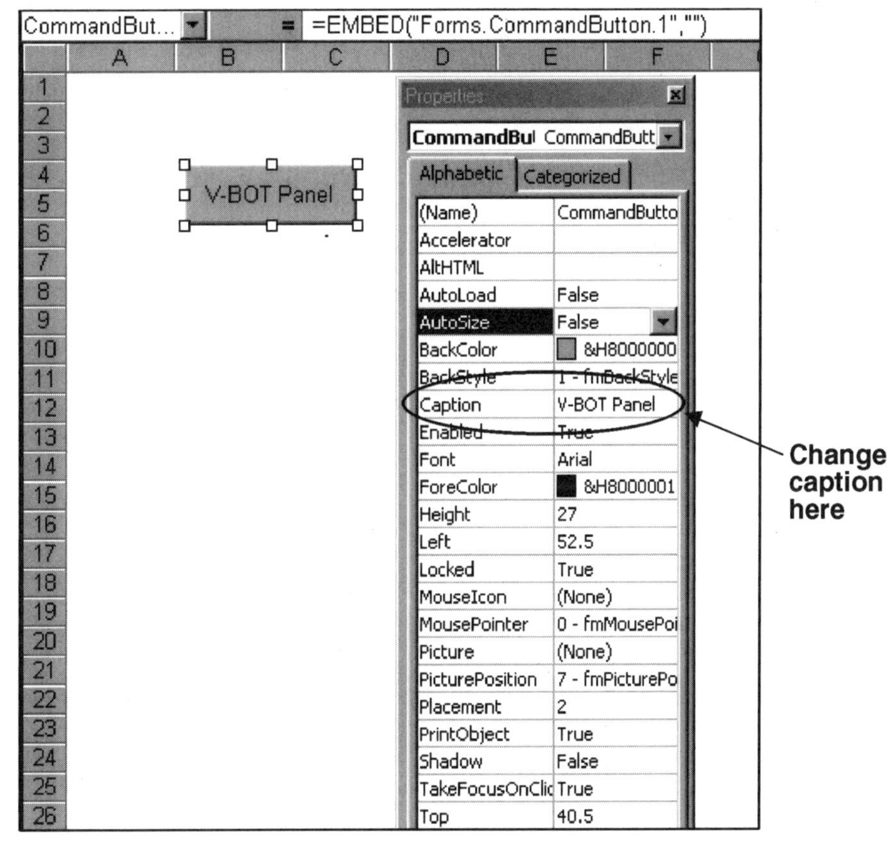

Single Click Event for the V-BOT Panel CommndButton

1. Move cursor to the V-BOT Panel button and right-click on object.
2. Click View Code from the pull-down menu. The VB Code Editor should be displayed. Under Private Sub CommandButton1_Click () heading, type the follow line of instruction: UserForm1.Show. Figure 2-28 shows the completed code for the V-BOT Panel CommandButton. Return to the Excel spreadsheet by clicking its icon on the menu bar.

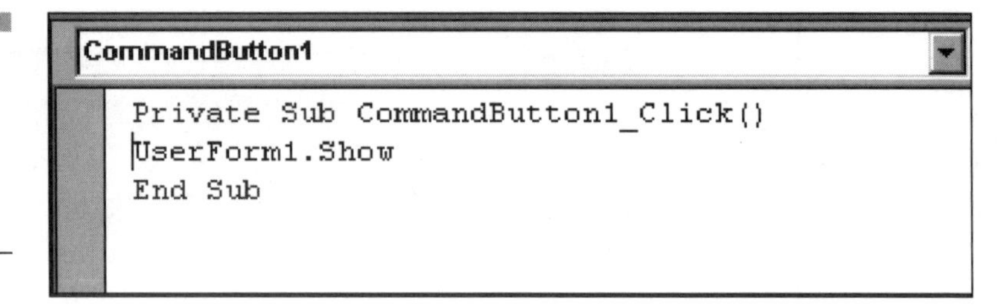

Figure 2-28 V-BOT Panel button embedded code for displaying the robot control panel on the Excel spreadsheet

3. Click the Design Mode icon to exit the designing session.
4. As a simple test, move the cursor to the ActiveX control. You will notice that the cross symbol changes to an arrow.
5. Click on the button to call up the control panel for V-BOT.

NOTE: *All other Excel functions will not operate while the GUI is running. This feature is inherent because VB is not a multitask programming language.*

6. Click the X in the upper-right corner of the GUI to exit the V-BOT control panel.

The robot control panel GUI is 90 percent complete. The layout of the controls and the capability to call up the GUI via an ActiveX button placed on the Excel spreadsheet works per the design requirements established earlier in the chapter. The final programming of the control panel's functions will be presented after the V-BOT's mechanical and electronics interfaces have been built and tested. Next, the concept of RDA will be explained using the system block diagram for V-Bot shown earlier in Figure 2-10.

What Is an RDA?

Traditional electronic circuits have simple output loads such as *Light-Emitting Diodes* (LEDs), motors, lamps, and solenoids to provide a visual indicator for electrical performance and operation. In order to make the behavior aspect of an electronic circuit more appealing to the experimenter, you can use a mechatronics device such as a robot. Therefore, the robot becomes a physical-digital assistant to the experimenter because of its capability to carry out instructional tasks embedded within the software. Just like the virtual office assistants that grace the Windows applications on the computer screen providing answers to keyboard-typed questions, the RDA can provide assistance to the experimenter by reacting to an input request via a switch or sensor and providing an output response based on an outside stimulus. The concept of the RDA is to use basic building components for reading input requests, processing the input data, and providing a correct output response in an *edutainment* (educational and entertaining) way.

Figure 2-10 shows the basic building blocks for V-Bot RDA. V-Bot is an animated robotic power supply that is capable of providing a regulated linear output voltage. Why create an animated robotic power supply? Instead of using simple electrical switch for powering up a voltage regulator circuit, an RDA can assist the experimenter in controlling the power supply in an entertaining and educational way. Also, the RDA can be thought of as interactive programmable tester for electronics *input/output* (I/O) interfacing circuits.

The focus of this book is to teach out-of-the-box design and development processes for creating robots. Traditional LEGO Mindstorms books have focused on robot-building mechanics or programming. This book's objective is to take a mechatronics approach to robot building by discussing electronics, mechanics, and software interfacing design, development, and test processes. In other words, the intent of this book is to teach a synthesis or whole-approach method to robotics building. Therefore, the RDA is an interactive method to which systems methodology can be put into practice by constructing LEGO robots. The RDA concept will be used throughout the remainder of the book to enhance the mechatronics development of robots. The following section outlines how to build the V-Bot so you can control the linear voltage power supply and monitor input sensors using the robot control panel built in Excel VBA.

Building a Mindstorms RDA

In product creation, the key is to shorten the development cycle by using successful bookshelf designs. The design selected should meet 90 percent of the requirements outlined in the product specification document. The remaining 10 percent enables design modification using low-cost, high-quality technologies and innovation. V-Bot is really Inventorbot from the 1.5 *Robotics Invention System* (RIS) Constructopedia with the exception of having a regulated output power supply. The reason for using Inventorbot as the based design for V-Bot is because its physical assistant behavior ties in with the RDA concept that V-Bot uses the main mechanics and sensors from the Inventorbot design, as shown in pages 58 through 82 of the

Constructopedia. Therefore, V-Bot is a slightly evolved Inventorbot that has the capability to provide laboratory assistance to simple electrical experiments requiring a small *direct current* (DC) voltage source. The innovation for this genre of Inventorbot relies on the robot's capability to control a power supply using a GUI. The following lab project outlines the procedure for building V-Bot and wiring a linear voltage regulator to one of its outputs. The rest of the code for completing the software control interface of the robot will also be provided in the project.

Bill of Materials (BOM)

Here is the list of materials required to build the 7805 5 Volt Linear Regulator circuit.

- RIS 1.0, 1.5, or 2.0 construction set
- One 7805 5 Volt Linear Voltage Regulator IC. Constructopedia Manual version 1.5.
- Two 0.01 capacitors
- One 10K resistor
- Constructopedia 1.5 Manual
- Solderless breadboard
- Alligator test leads
- LEGO handheld remote control
- LEGO electric wires

Objectives

The objective of this procedure is to build and test the V-Bot RDA built from the Inventorbot instructions of the Constructopedia 1.5 Manual. We also want to test the 5-volt linear voltage regulator circuit interface to output B of the robot.

Build and Test Procedure

1. Build the Inventorbot robot using pages 58 though 82 of the Constructopedia 1.5 Manual. If the Constructopedia 1.5 Manual is not available, use Figures 2-29 through 2-39 as a guide for building the robot.

2. Prepare the LEGO electric wire using Figure 2-38 or the wire harness assembly used in the beetle bot project in Chapter 1, "Wireless Basics." Set the assembly aside for later use.
3. Wire the 5-volt linear voltage regulator as shown in Figure 2-40.
4. Attach the LEGO electric wire output B of the Inventorbot robot and the 5-volt linear voltage regulator circuit. See Figures 2-41 and 2-42.
5. Turn on the RCX of the V-Bot (Inventorbot).
6. Press the B up-arrow button on the LEGO handheld remote control. Notice the LED and output voltage reading of the *digital multimeter* (DMM).
7. Press the B down-arrow button on the LEGO handheld remote control. Notice the LED and output voltage reading of the DMM.
8. Press the A up-arrow button on the LEGO handheld remote control. Notice that V-Bot's hat raises in the air. If the direction is opposite, change the orientation of the electric wire until the correct motion is observed.
9. Press the A down-arrow button on the LEGO handheld remote control. Notice that V-Bot's hat lowers.

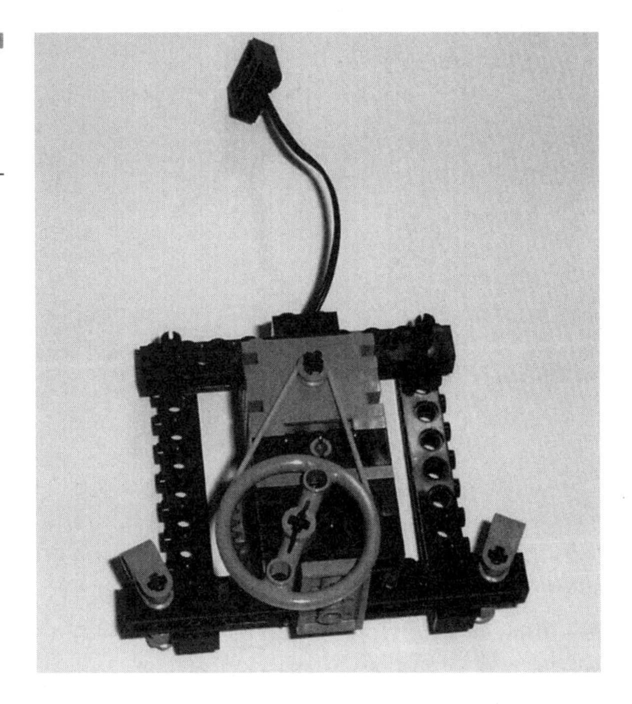

Figure 2-29
Stand base drive mechanism (bottom view)

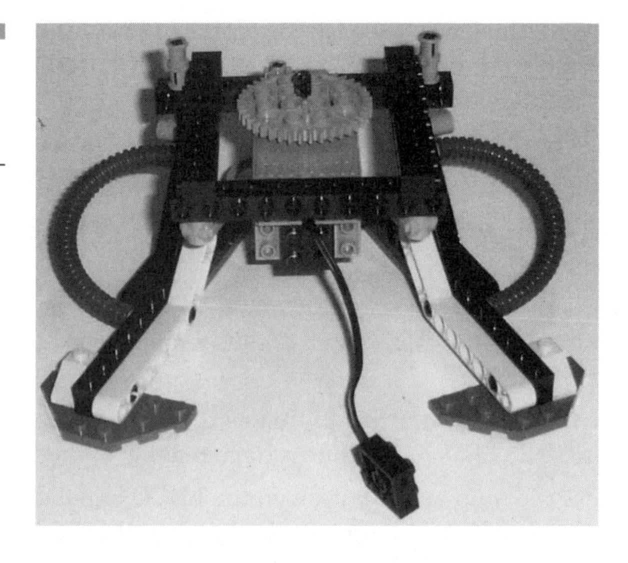

Figure 2-30
V-Bot's standing base with drive mechanism

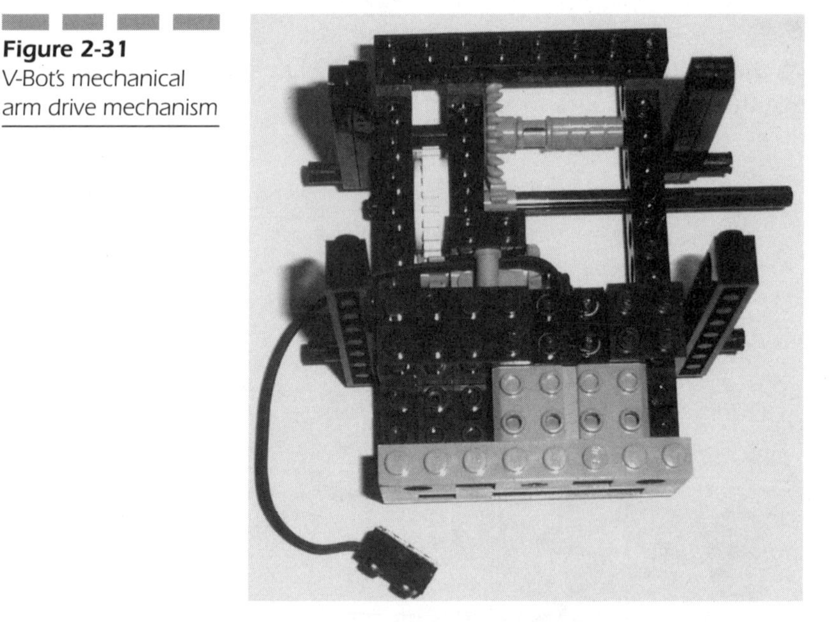

Figure 2-31
V-Bot's mechanical arm drive mechanism

10. Press the C up-arrow button on the LEGO handheld remote control. Notice that V-Bot's body turns left. If the direction is opposite, change the orientation of the electric wire until the correct motion is observed.
11. Press the C down-arrow button on the LEGO handheld remote control. Notice that V-Bot's body turns right.

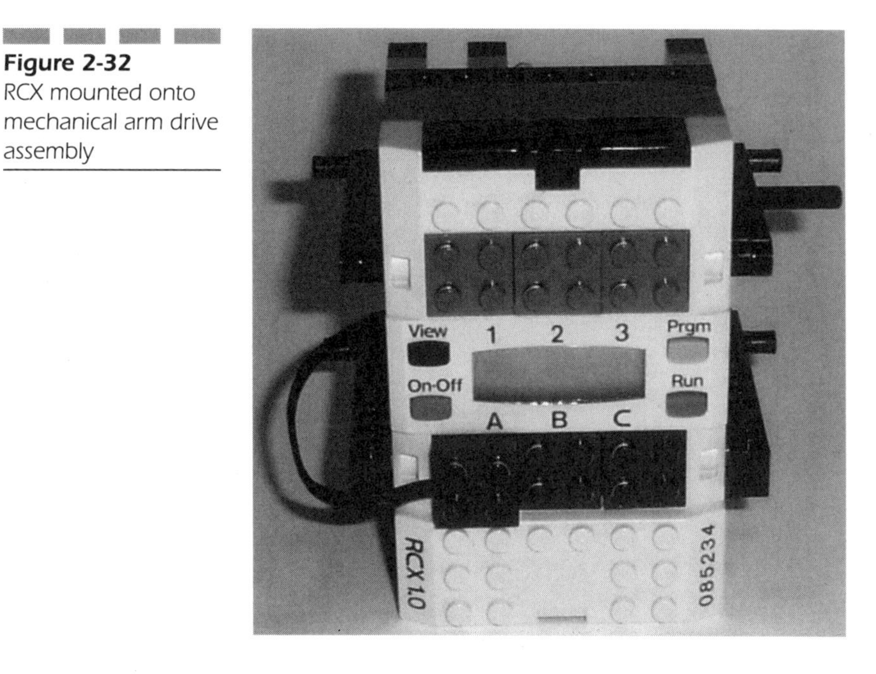

Figure 2-32
RCX mounted onto mechanical arm drive assembly

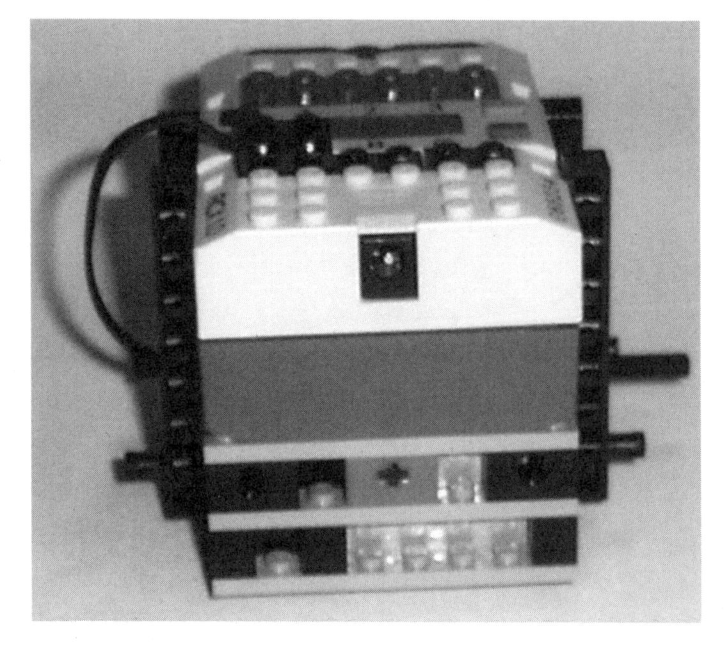

Figure 2-33
Bottom of RCX-mechanical arm drive assembly

Figure 2-34
V-Bot's head

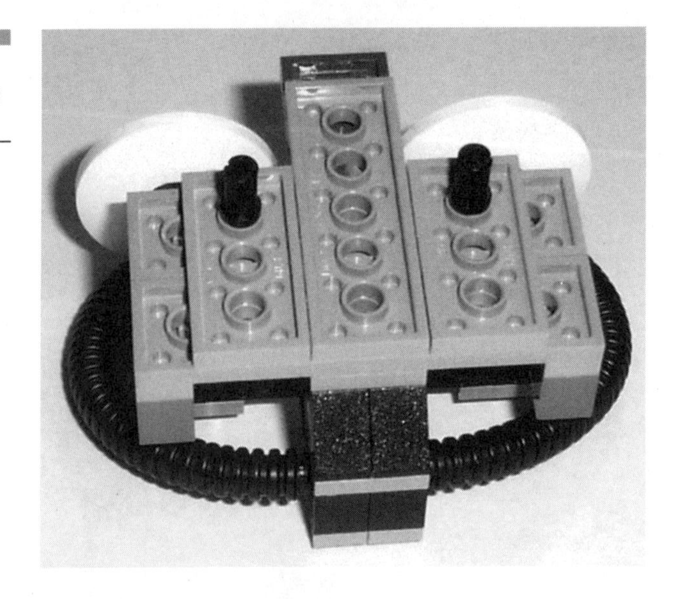

Figure 2-35
V-Bot's head (bottom view)

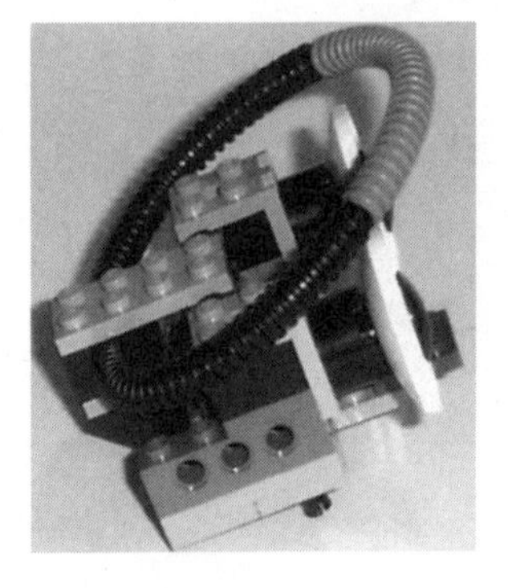

Figure 2-36
V-Bot's head (side view)

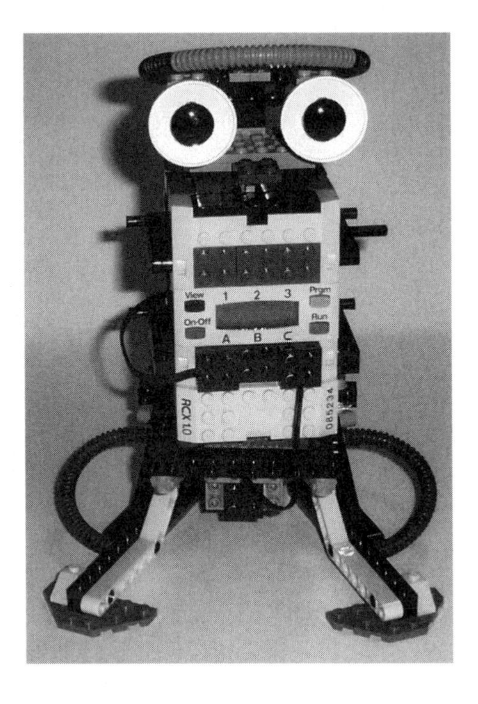

Figure 2-37
V-Bot with head and standing base assembled to mechanical arm drive

Figure 2-38
The completed V-Bot: Inventorbot

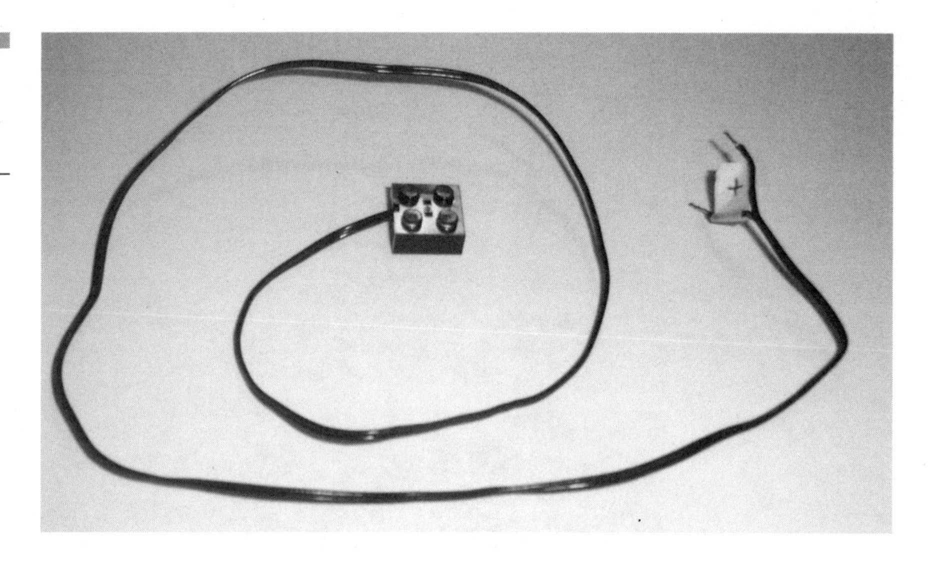

Figure 2-39
LEGO electric wire with tinned wires for breadboarding

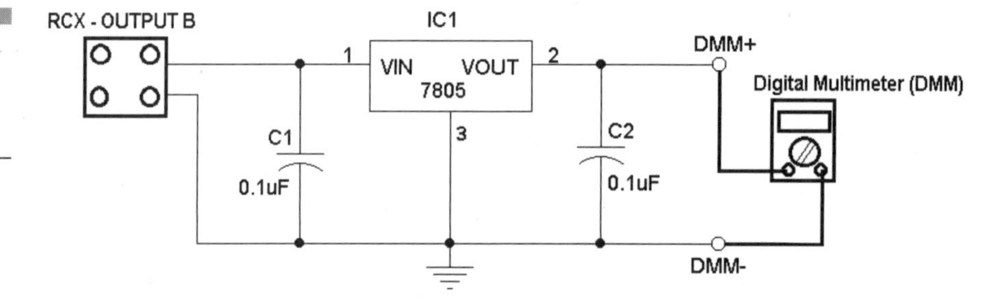

Figure 2-40
The 5-volt linear regulator circuit interface for V-Bot

Figure 2-41
The Michael Lachmann computer-aided design (MLCAD) model showing the orientation of the LEGO electric wire to the RCX

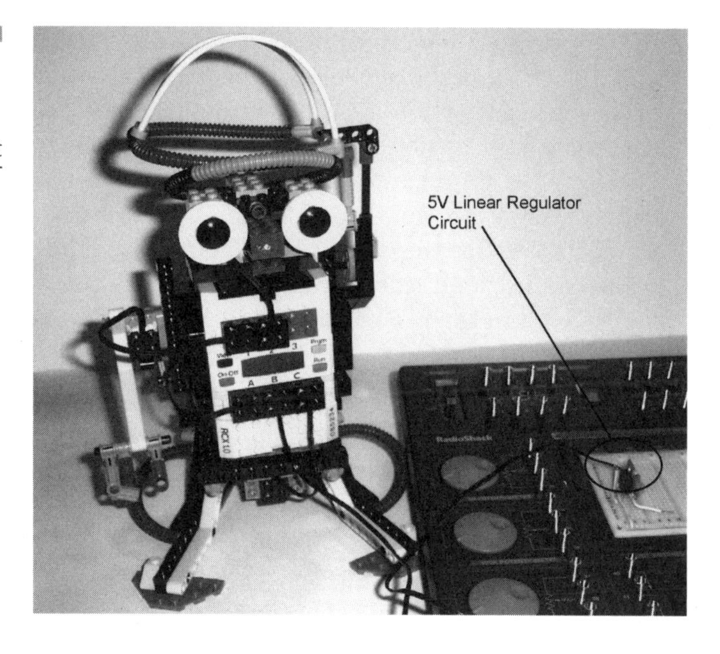

Figure 2-42
V-Bot electrically wired to the 5-volt linear regulator circuit

This completes the testing of V-Bot's mechanical drives and the 5-volt linear voltage regulator interface circuit. Now we will discuss state machine development along with writing code for the robot control panel. Hands-on lab projects will also be provided.

Using State Machines to Write Code for the RDA

Traditional flow charts are used to define the logic and organization of how a computer program will work within an embedded controller or desktop workstation. The concern that software developers grappled with for flow charts was the lack of hardware I/O definition and interfacing with the target system. State machines provide a graphical representation of how the hardware interface functions within the embedded environment's software layer. By creating states or hardware outputs, and transitions or input requests and events, the software developer can logically create a diagram that can be converted into code, thereby capturing the functional requirements of the embedded controller or electronic control unit.

To draw a state machine, a set of symbols is used to capture the states and transitions of the controller's embedded behavior or function. Circles or ovals represent states with arrows defining the transitions. Figure 2-43 shows the basic symbols for creating a state machine. To start the process of creating a state machine, the I/O of the embedded controller must be defined. A set of requirements defining the controller's function must be available to define the initial conditions of states and events. With the functional requirements on hand, the state machine can be created using circles and arrows. In the live example of V-Bot, the functional requirements are stated as follows:

- With the robot at rest, motors A, C, and B are off.
- By releasing the switch attached to one arm of the robot, motor C will turn on for 1 second.
- After 1 second elapses, motor A will turn on for 4 seconds.
- After 4 seconds elapse, motor C should reverse direction for 1 second.
- After 1 second elapses, motor B will turn on for 20 seconds.
- After 20 seconds elapse, motors A, C, and B are off.

Figure 2-44 shows the state machine for V-Bot's embedded code. The key item to note when creating a state machine is that all functions should start from an off state and transition to an on state back to an off state. This full-circle or closed-loop control ensures that the physical systems

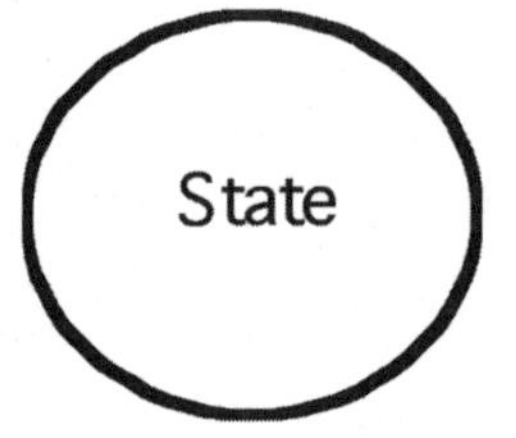

Figure 2-43 Basic symbols used in drawing state machines

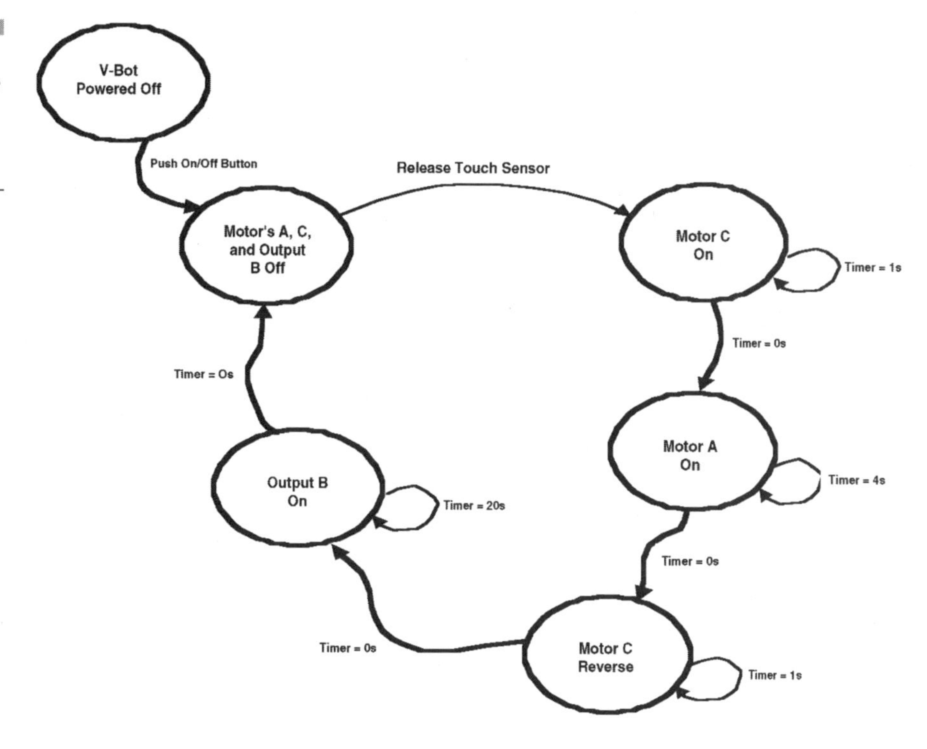

Figure 2-44 State machine for the V-Bot Inventorbot body—hat movement task

energy source will not be depleted through the continuous use of active I/O devices staying energized. After the state machine completes, writing code becomes quite easy because of the translation of a closed-loop to top-down code design convention. RCX code programming uses a top-down method because of the stacking of code instruction used to create the embedded function of the target controller. Therefore, tracing and debugging the code will be quite easy because there are no jumps or GOTOs used within the embedded software. Therefore, top-down programming makes the code more manageable for maintenance and reflects the system's hardware/software/mechanical interface architecture as well. Figure 2-45 shows the code written using the state machine in Figure 2-44 for V-Bot's embedded function.

The lab that follows the discussion on VBA and OOP basics outlines the procedures for completing the robot control panel software and testing the GUI using the RCX code of Figure 2-45.

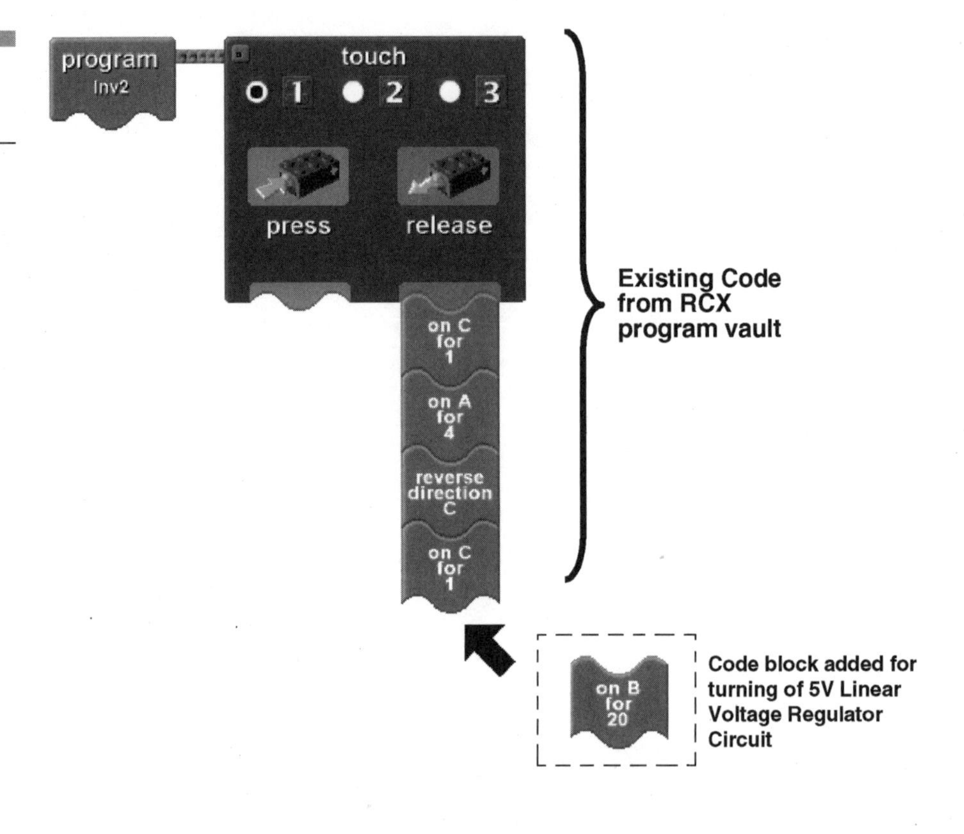

Figure 2-45 V-Bot Inventorbot RCX code

VBA and OOP Basics

As stated in the introduction of this chapter, VBA programming language uses an object-oriented method for creating Windows-application-specific GUIs. OOPs enable the software developer to encapsulate not only the function, but also the data associated with the behavior of the object. *Object* is the key word in object-oriented programming. It establishes the basis for the *visual* part of Visual Basic for Applications because of the use of objects such as buttons, text boxes, and sliders to embed functions and data within the programmable icons. The formats used in VBA for defining objects are as follows:

<Parent Object. Child Object. Property>

<Parent Object. Child Object. Method>

Parent Object is basically the main icon container providing a programming foundation of the smaller icon containers placed on top of it. For example, UserForm1 would be the Parent Object to text box because the ActiveX control is placed on top of the form, thereby providing a programming foundation for the graphical icon. The *Child Object* in this example is a text box because it depends on the UserForm providing a programming interfacing reference to other ActiveX controls placed on top of it. The properties of an object define the set attributes such as color, logical state, caption, and name. Changing a property for the ActiveX control CommndButton was illustrated earlier in this chapter. (See the section "Calling Up a VBA-Built Robot Control Panel GUI Using an Excel" for the procedure.) *Method* in OOP refers to the function of the VBA ActiveX control. Examples of a Method are paint, plot, scroll, and poll. The important item to keep in mind when programming VBA ActiveX controls is to use the <Parent. Child. Property/Method> construct for defining the operation of the user interface icon. By using this programming style, the Windows GUI application will be easy to manage, and upgrades to capture new features can be implemented within minutes.

The Robot Control Panel Programming Lab Project

The objective of this lab project is to provide functions to the robot control panel built in the section "V-Bot Control Panel Build Procedure 2" of this chapter. Upon completion of writing the VBA software for the panel, the GUI will be tested using the V-Bot Inventorbot.

VBA Software Embedding Procedure

1. Open the file V_Bot Control Panel.xls.
2. An Excel Dialog box will appear on the screen. Click the Enable Macros button.
3. Click on the VBE icon located on the menu bar. The VB IDE window will be displayed.
4. Double-click the Forms icon within the Project Explorer-VBA Project window. The robot control panel should be displayed along with the toolbox.

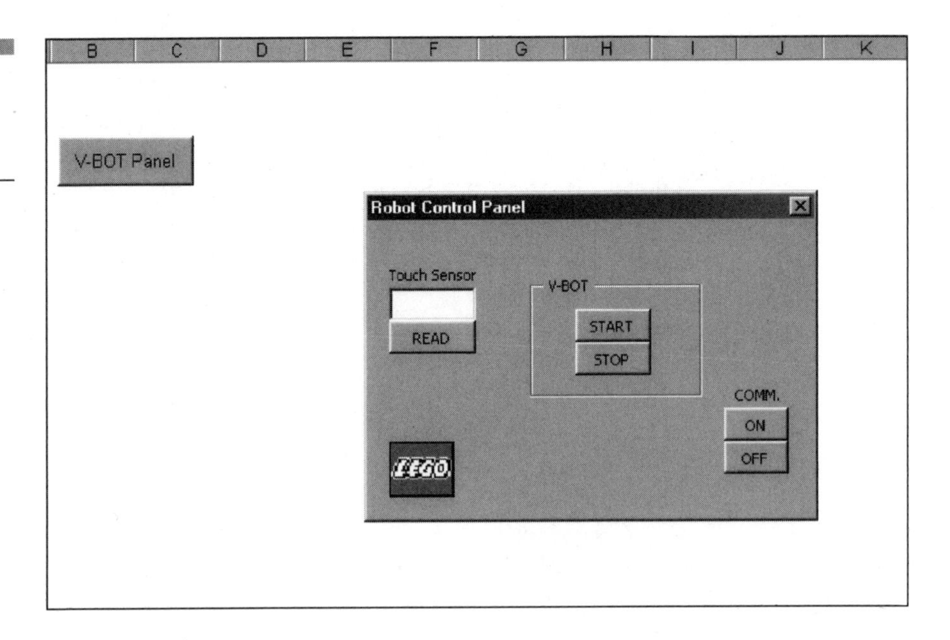

Figure 2-46 Adding LEGO Spirit ActiveX control to robot control panel

5. Select and drag the LEGO logo to the lower-left corner of the GUI, as shown in Figure 2-46.
6. Click the LEGO icon. Locate the Name property and type PBrickCtrl.
7. Within the Properties window of the UserForm, locate the Name property and type RDACTRL. Notice that RADCTRL is displayed in the Project Explorer window.
8. Select the READ CommandButton and right-click View Code. The VBE window will appear.
9. Under the Private Sub CommandButton_Click () heading, type the following code:

```
SENSOR1 = 0 'Read Input 1 of RCX
SENTOUCH = 1 'Reading Data from a Touch Sensor
SENVAL = 9 'Retrieve value from Touch Sensor at Input 1
PBrickCtrl.SetSensorType SENSOR1, SENTOUCH
RDACTRL.TextBox1 = PBrickCtrl.Poll (SENVAL, SENSOR1)
```

10. Click the CommandButton ActiveX control. Locate the Name property and type Read.
11. Above the Private Sub CommandButton_Click () heading, type the following code:

```
Dim SENSOR1 As Integer
Dim SENTOUCH As Integer
Dim SENVAL As Integer
```

12. To add code to the remainder ActiveX controls, repeat Steps 8 through 10 with exception of changing the property names.

    The following is the complete code for the V-Bot robot control panel:

```
Dim SENSOR1 As Integer
Dim SENTOUCH As Integer
Dim SENVAL As Integer
Private Sub Read Click ()
SENSOR1 = 0 'Read Input 1 of RCX
SENTOUCH = 1 'Reading Data from a Touch Sensor
SENVAL = 9 'Retrieve value from Touch Sensor at Input 1
PBrickCtrl.SetSensorType SENSOR1, SENTOUCH
RDACTRL.TextBox1 = PBrickCtrl.Poll (SENVAL, SENSOR1)
End Sub
Private Sub StartCmd_Click ()
PBrickCtrl.StartTask 1
End Sub
Private Sub StopCmd_Click ()
PBrickCtrl.StopTask 1
End Sub
Private Sub CommOn_Click ()
PBrickCtrl.InitComm
End Sub
Private Sub CommOff_Click ()
PBrickCtrl.CloseComm
End Sub
```

Note that in this code PBrickCtrl.InitComm is the ActiveX control for opening or initializing the serial port of the notebook or desktop PC. This communication interface enables the Infrared Tower interface to send and receive monitor and control instructions used within the LEGO Mindstorms Spirit object library. In Step 6 of the previous procedure, the original name for the LEGO ActiveX control was Spirit. It was renamed to follow the convention of names used in the *Software Development Kit* (SDK), which can be downloaded from the LEGO Mindstorms web site at http://mindstorms.lego.com/sdk/default.asp. PBrickCtrl.CloseComm will close the serial port, thereby stopping all software interaction with the robot. PBrickCtrl.SetSensorType is the LEGO Spirit object for defining the type of sensors connected to the RCX brick's inputs.

Once the sensors have been defined and mapped to the inputs, the data must be read. To accomplish this, the object used is PBrickCtrl.Poll (SENVAL, SENSOR1). The data value associated with the sensor will be read or polled upon the READ button being selected from the GUI. The information will be stored and displayed in RDACTRL.TextBox1 VBA ActiveX control object. The Spirit object PBrickCtrl.StartTask 1 enables the robot to perform the raising of the hat and body turning left and right upon the Start button being selected. To stop the V-Bot from carrying out this function, PBrickCtrl.StopTask 1 is used. The LEGO Mindstorms SDK is a very useful tool for developing VBA-based control panels. The 110-page document has snippets of code showing how to use the library for controlling and monitoring states and events of LEGO Mindstorms robots. This document will become a very useful reference for the electronics hobbyist or experimentalist exploring LEGO robotics.

The VBA program is a vital link to the hardware/software interface for V-Bot because the communication between the GUI and the robot's embedded code establishes an external behavior control for the electromechanical device. Therefore, the VBA GUI provides another state machine interface layer for the V-Bot Inventorbot. The method of having a state machine control another is called a *testbench* and is used to validate software for embedded controllers in industry.

Now that the VBA code is embedded, the robot control panel can be tested. The following test procedure outlines the steps for testing the external state machine controller (such as the robot control panel) and software interface for V-Bot.

The Software Interface Testing Procedure

The objective of this lab is to test the software interface between the robot control panel and V-Bot.

Testing Procedure

1. Write the RCX code (as shown in Figure 2-45) using the LEGO Mindstorms programming environment. Download the code into V-Bot.
2. Have the Infrared Tower pointed in the vicinity of the robot.
3. Turn on the LEGO RCX brick and call up the robot control panel using the Excel CommandButton displayed on the spreadsheet.
4. Click the On button for COMM. This establishes a communication link between the control panel and the LEGO RCX brick.
5. Click on the READ button under the Touch Sensor text box. A 1 should be displayed inside of the text box.
6. Press and hold the Slap Arm of V-Bot. Click the READ button again. A 0 should be displayed inside of the text box. Release the Slap Arm of the robot.

7. Click on the Start button and the robot should perform the task as defined in the RCX code, as shown in Figure 2-45.
8. With the robot running the RCX code, click the Stop button. The last movement executed should run continuously. Hit the Program button on the RCX brick to stop the movement.

The robot control panel has priority over the Slap Arm event because of the master-slave protocol embedded within the RCX brick's firmware. This firmware recognizes a manual command over an autonomous instruction; therefore, the robot responds immediately to the controlling device without hesitation. A critical feature such as this is very important in industrial environments because of the possibility of runaway machines endangering human operators. A manual override system must be available to bypass the autonomous functions embedded within the machine to prevent accidents within the industrial environment. The VBA-built control panel illustrates such a safety *Proof Of Concept* (POC) system by maintaining immediate control when the Stop button was initiated. Also, this lab provides insight into how software is a seamless control interface for managing the I/O of intelligent hardware machines demonstrated by V-Bot Inventorbot.

An Alternative Solution for Data Acquisition (DA) and Measurement

Building robots is a multidisciplinary field that requires electronics, mechanics, and software to work seamlessly within the mechatronics system. The testing and measurement of robots are other important elements of robotics because they define a baseline for quality and performance. By documenting the performance data of various robot designs, the electronics hobbyist or experimentalist can select the appropriate machine for events such as design competitions. Data collection such as torque, speed, voltage, and current are important parameters to document and analyze as they relate to the performance and reliability of the robot.

In Chapter 1, the speed of a beetle bot was measured and recorded using a LEGO Speed Computer. This simple measuring device is actually a small *Data Acquisition* (DA) system in a brick. Traditional DA systems have the capability to monitor, control, and log data from transducers or sensors. The type of system includes one or more sensors, transducers, interfacing circuits with data conversion (digital to analog [D/A]/analog to digital [A/D]), and a means for displaying the data using an LED, *Liquid Crystal Display* (LCD), or computer monitor.

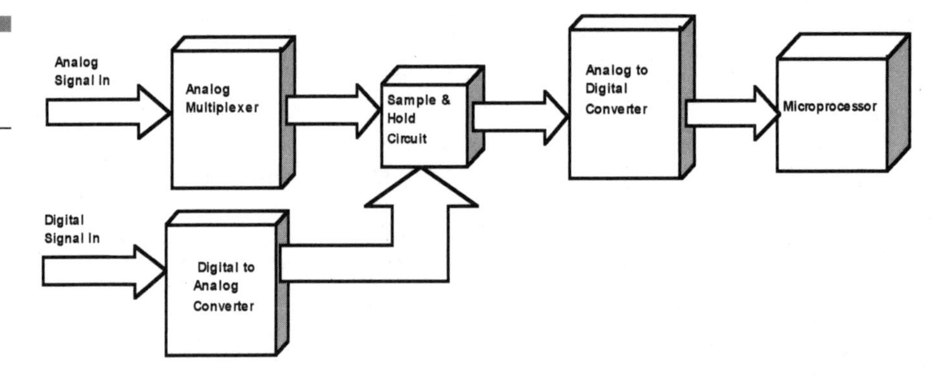

Figure 2-47 Block diagram of a traditional DA

Figure 2-47 shows a block diagram for a traditional DA system. The LEGO Speed Computer has a rotation sensor built into the small pickup attached to the measuring device via wires. The computer then processes the data received from the A/D converter and displays it on the LCD. Therefore, the LEGO Speed Computer can be considered an alternative DA system for collecting speed data from robots. To determine the type of data to collect, the hardware/software interface block diagram can be used. With the data flows shown on the diagram, the type of data measurement can be determined quite easily. Data collection for V-Bot will primarily be the input voltage received by the 5-volt linear voltage regulator and the output signal used to turn on a small electrical load such as an LED.

Although there are many DA systems out on the market, for the experimentalist on a shoestring budget, most of them are quite expensive. Therefore, an alternative solution of measuring and logging voltage data can be achieved by using a DMM with a PC interface. Radio Shack has a low-cost DMM with a PC interface (Cat. No. 22-805) that connects between the serial port of a notebook or desktop computer and one side of the digital meter. Inside the DMM with a PC interface, traditional measuring circuits for voltage, current, and resistance measurements are standard. A/D and D/A converters are external gateways for the microprocessor to connect to the robot's output voltages. The data is then displayed using an LCD that the experimentalists can read for interpreting the performance of the robot's controlled voltage interface circuit. Basically, the DMM can be categorized as a small alternative DA system because of its measuring and logging capabilities compared to traditional large-scale digital acquisition units out on the market.

DMMs can be used to measure input/output voltages of V-Bot and the data manually recorded, but plotting this information requires some man hours on the part of the experimentalists. Also, repeating the same testing conditions will prove challenging to experimentalists because of the large amount of data to be recorded. In the case of measuring voltage, the DMM with the PC interface is set to the appropriate volts scale and can be sent to a notebook or desktop computer using the 5-pin male connector/9-pin D-shell connector serial cable data. The software that comes with the DMM can display data continuously on an emulated LCD viewed on the computer monitor or on a single/repetitive strip chart recorder. The data can also be logged and stored on a 3-inch floppy diskette or hard drive using the recording feature of the software.

Figure 2-48 shows a block diagram of the DMM with a PC interface used as an alternative DA system for recording output voltage of the 5-volt regulator robot-controlled *Integrated Circuit* (IC). Data capture from this measurement setup is shown in Figure 2-49. The testing condition used to capture this data was used primarily to initiate the embedded task of V-Bot using the Start button on the robot control panel. The DMM was set to trigger at 5 volts on the rising edge of the output signal. Upon the signal being present, this peak value was recorded during its 20-second on state as programmed using the RCX code downloaded to V-Bot. After 20 seconds, the signal went to 0 volts and stayed there until another start event was present based on the CommndButton being clicked from the VBA GUI. Two cycles were recorded using this alternative DA system and the results were repeatable each time during the logging session.

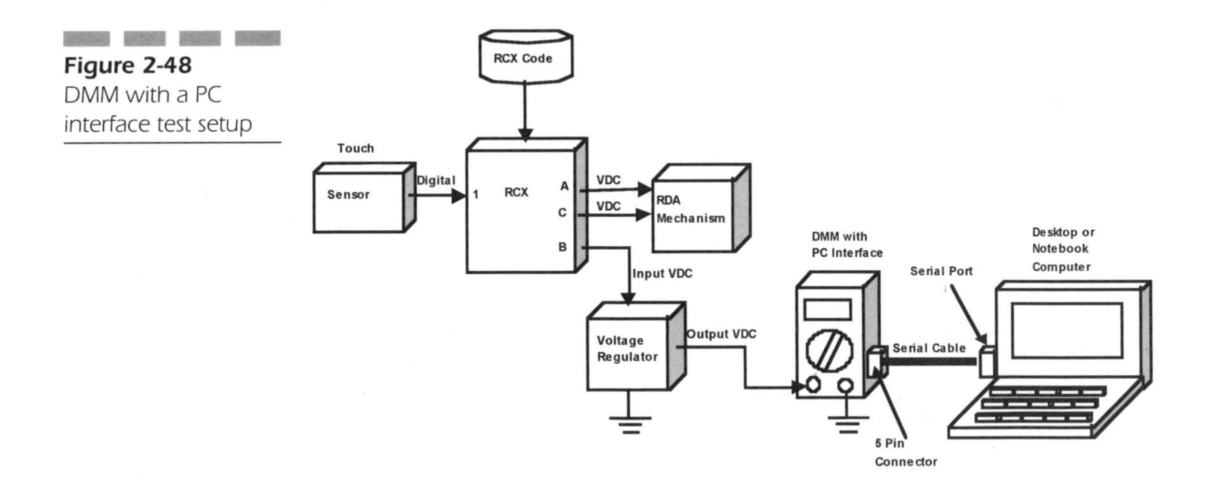

Figure 2-48 DMM with a PC interface test setup

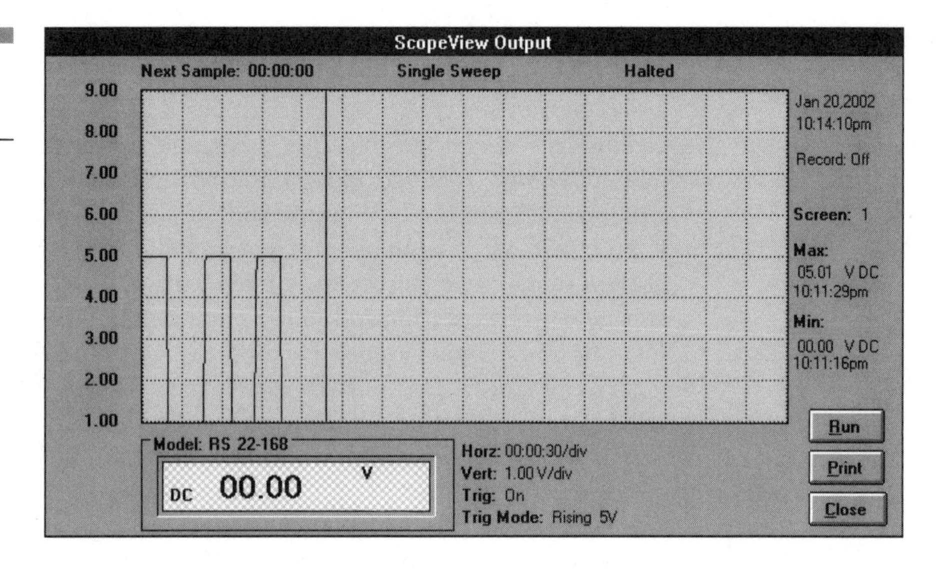

Figure 2-49 Data recorded from DMM Test Setup

In this chapter, the concept of using a software interface to control a robot was explored using Excel VBA and LEGO Mindstorms RCX software. By creating a hardware/software interface layer model, the robot's system I/O and mechanical architecture can be designed. Understanding the input and output signals generated by the RCX brick can be accomplished by assigning data flows to the base hardware/software interface layer model. Once data flows are defined, a state machine can be developed to create the embedded function or task the robot will perform. This state machine is then embedded into the target robot using Mindstorms RCX software. The robot control panel created in Excel VBA provided an external state machine or testbench used to test the embedded RCX code of V-Bot as well as a safety device for bypassing its internal software controls. Voltage data produced by the 5-volt linear voltage regulator interface circuit was collected using a DMM with PC interface; thus, an alternative DA test station was created.

In addition to building a VBA GUI for the software control of a robot, some of Excel's advanced features for spreadsheet automation were explored. By using OOP methods, a useful Windows application was created. With the buttons placed on the UserForm, their controlling interface functions to V-Bot were formed by changing the properties and methods, and adding additional object libraries using the Spirit ActiveX control (LEGO logo). The placement of controls on top of a UserForm follows the

OOPs programming conventions of <Parent Object. Child Object. Property> and <Parent Object. Child Object. Method>. Excel VBA enables you to change properties using the Property Explorer window within the VBA IDE. Once in the editor of VBA, methods are changed by selection from a drop-down list box that is displayed when a period is typed following a Child Object.

To understand the Spirit object library, the LEGO Mindstorms software development requires reference material for the electronics hobbyist or experimentalist. The 110-page document of the LEGO Mindstorms SDK has snippets of code showing how to use the library to control and monitor states and events of LEGO Mindstorms robots.

The next chapter will explore how electromechanical controls and software can be used to monitor input switching events of the Scout and LEGO RCX programmable bricks.

CHAPTER 3

Electromechanical Controls Interfacing

A typical electric switch falls under the category of electromechanical controls because it has the mechanical action of a lever that opens and closes its contacts and the electric current that flows through them. The touch sensor provided with the *Robotics Invention System* (RIS) and the *Robotic Discovery Kit* (RDK) is a simple electromechanical device for controlling robots. Upon depressing or releasing the yellow button, electric current is controlled using the opening and closing action of its contacts. Now the question might be asked, "Why is there a whole chapter on electromechanical controls?" The answer is that more robotic systems use electronics and mechanics to create a mechatronics interface, thereby enabling a microcontroller to make the machine more adaptable to its environment based on software logic controls. Therefore, this chapter illustrates how the simple electric switch can be transformed into a mighty sensory device that an *Robot Command Explorer* (RCX)- or Scout-based *programmable brick* (P-Brick) can use for maneuvering around its environment. The following topics will be discussed in this chapter:

- Understanding the touch sensor RCX/Scout interface
- How to build the homebrew touch sensor
- Resistive switches
- Electromechanical relays for RCX/Scout input control
- How to build a smart switch using the RCX P-Brick
- How to build a Switch Monitor Detector

In order to create electromechanical controls for the LEGO Scout and RCX P-Bricks, you must understand the touch sensor and its electrical interface to these devices. This will be discussed in the following section.

Understanding the Touch Sensor RCX/Scout Interface

The touch sensor is basically an electrical switch with a resistor wired in series. This simple series circuit provides an electrical interface to the RCX and Scout P-Bricks' three input ports. The RCX has three input ports and three output ports that are connected to a microcontroller and a solid-state driver circuit, respectively. After further dissection, the sensor and resistor circuit connects with an internal 10-kilo-ohm component. The input port of

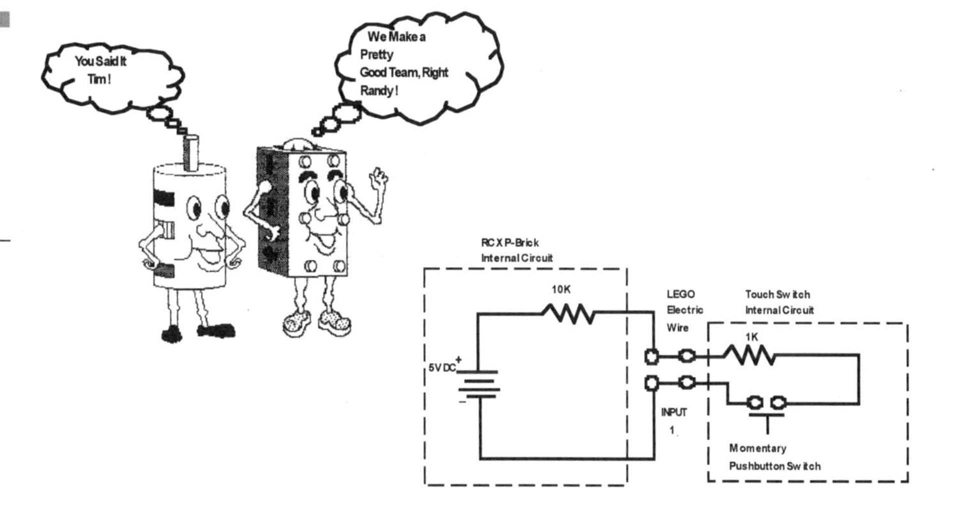

Figure 3-1 Tim Touch Sensor and Randy Resistor discuss the internal circuit of the RCX P-Brick and the sensor.

the RCX is configured as a voltage divider circuit, as shown in Figure 3-1. By pressing the touch sensor, the internal 1-kilo-ohm resistor becomes electrically connected to the 10-kilo-ohm device. Voltage drops are established across the two resistors produced by the current flowing from the internal 5-volt *direct current* (DC) source. The voltage across the internal series resistor of the touch sensor is one-tenth of the drop for the 10-kilo-ohm element. The two resistive elements shown in Figure 3-1 provide two voltage levels that the microcontroller inside of the Scout or RCX P-Brick reads when the touch sensor is pressed or released. The voltage divider circuit is connected to an *analog-to-digital* (A/D) converter that converts the two input voltage levels or sensor input events to an equivalent digital value that the microcontroller uses for data processing. To calculate the two voltages, the voltage divider equation can be used.

DC Circuit Analysis

The voltage across the internal 1-kilo-ohm resistor of the touch sensor can be found using the voltage divider equation. To find this voltage,

$$V_{1K} = (V_{S-INT} \times R_{TSensor})/R_{RSensor} + R_{RCX}$$

where

$$V_{S-Int} = 5\text{V}$$
$$R_{Tsensor} = 1\text{K}$$
$$R_{RCX} = 10\text{K}$$

Substituting the numeric values into the following equation,

$$V_{1K} = (5V \times 1K) / 1K + 10K$$
$$V_{1K} = 5{,}000/11{,}000$$
$$V_{1K} = 0.45V$$

The 0.45 volts represent the voltage drop across the 1-kilo-ohm resistor when the touch sensor is pressed, as shown in Figure 3-2. The voltage is then applied to the microcontroller using an A/D converter inside of the Scout or RCX P-Brick. When the touch sensor is released, the voltage across it can be calculated by the following equation:

$$V_{TSensor} = V_{RRCX} - V_{1K}$$

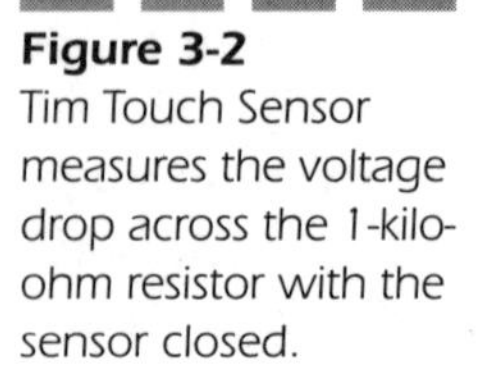

Figure 3-2
Tim Touch Sensor measures the voltage drop across the 1-kilo-ohm resistor with the sensor closed.

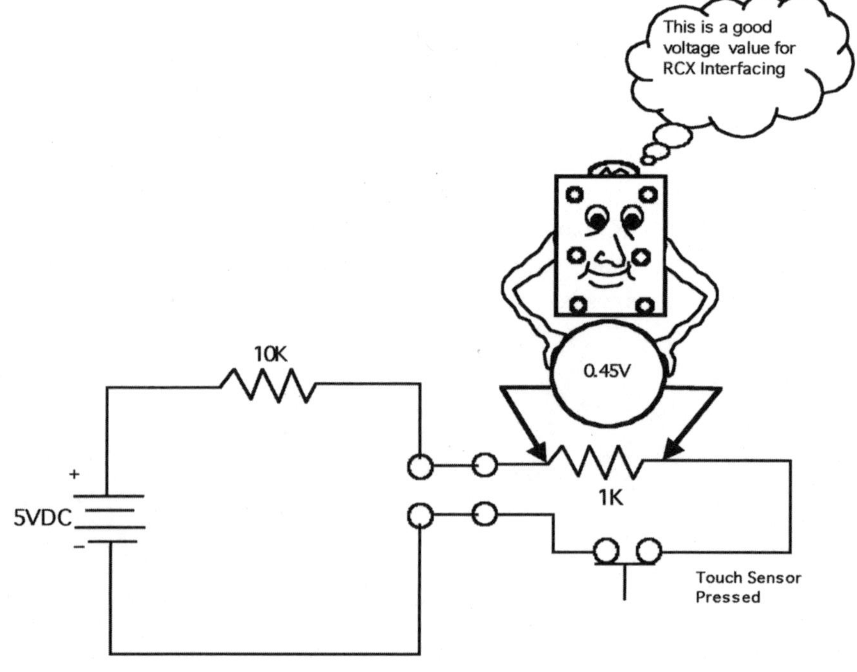

Current flowing through the series network with the sensor pressed equals the following:

$$I_{Total} = V_{S-Int}/R_{total}$$

where

$$V_{S-Int} = 5V$$
$$R_{total} = 11K$$

Therefore,

$$I_{Total} = 5V/11K$$
$$I_{Total} = 455\mu A$$

The voltage drops across each series element should equal 5 volts. If the touch sensor is opened, $I_{Total} = 0$ A instead of the 455 μA. Therefore, $V_{Tsensor}$ equals

$$V_{S-Int} = V_{1K} + V_{RRCX} + V_{Tsensor}$$
$$5V = (0 \times 1K) + (0 \times 10K) + V_{Tsensor}$$
$$5V = V_{Tsensor}$$

The microcontroller for both the Scout and RCX P-Bricks would handle the 0.45 volts as a logic 1 and the 5 volts as a logic 0. If there is a touch sensor connector to any input on the LEGO RCX P-Brick, the logic values can be seen on the *liquid crystal display* (LCD) quite easily. Press and release the yellow button on the touch sensor and set the P-Brick in view mode. By pressing the touch sensor connected to inputs 1 or 2 on the Scout, the *light-emitting diodes* (LEDs) will turn on, thereby representing a logic 1. To see the logic values in an analog form of the Scout P-Brick, you must use the Scout tool as discussed in Chapter 1, "Wireless Basics," which is set in the Advanced Monitoring box. The values will be displaced as RAW A/D count values with 255 equal to logic 1 and 0 equal to logic 0. An important item note about the internal resistor is that a small amount of current must be present in the electric switches in order to prevent oxidation from forming on the metal contacts. Another term used to describe this electrical cleaning process is known as a *whetting* current. The 455 uA current is sufficient for cleaning the contacts of the touch sensor because of its small packaging and the amount of metal material it uses. In industrial applications, this whetting current can be as high as 10 *milliampere* (mA). It all depends on the size and switching application of the electric switch.

The following lab shows several methods of reading the touch sensor's logic values using software and analog tools. The procedures outlined will serve as valuable testing tools for checking out other electromechanical controls discussed in this chapter.

Reading Logic Values from a Touch Sensor

The objective of this lab is to describe testing methods for reading logic values from a touch sensor attached to the input of a Scout or RCX P-Brick. This lab also demonstrates how circuit analysis can be validated through a testing procedure.

Bill of Materials (BOM)

- Scout P-Brick
- *Robotics Invention System* (RIS) Software version 1.5 or 2.0
- RCX P-Brick
- Radio Shack Electronic Learning Lab breadboarding system or equivalent
- Touch sensor
- LEGO electric wires
- *Digital multimeter* (DMM)
- RIS construction set
- Additional LEGO bricks
- *Double Pole-Double Throw* (DPDT) electric switch
- Momentary pushbutton electric switch
- An experimenter's board (Radio Shack Cat. No. 276-147)

Touch Sensor Lab Procedure

1. Modify LEGO electric wires as shown in Figures 3-3 and 3-4.
2. Place the modified LEGO electric wire into the breadboard system, as shown in Figure 3-5.

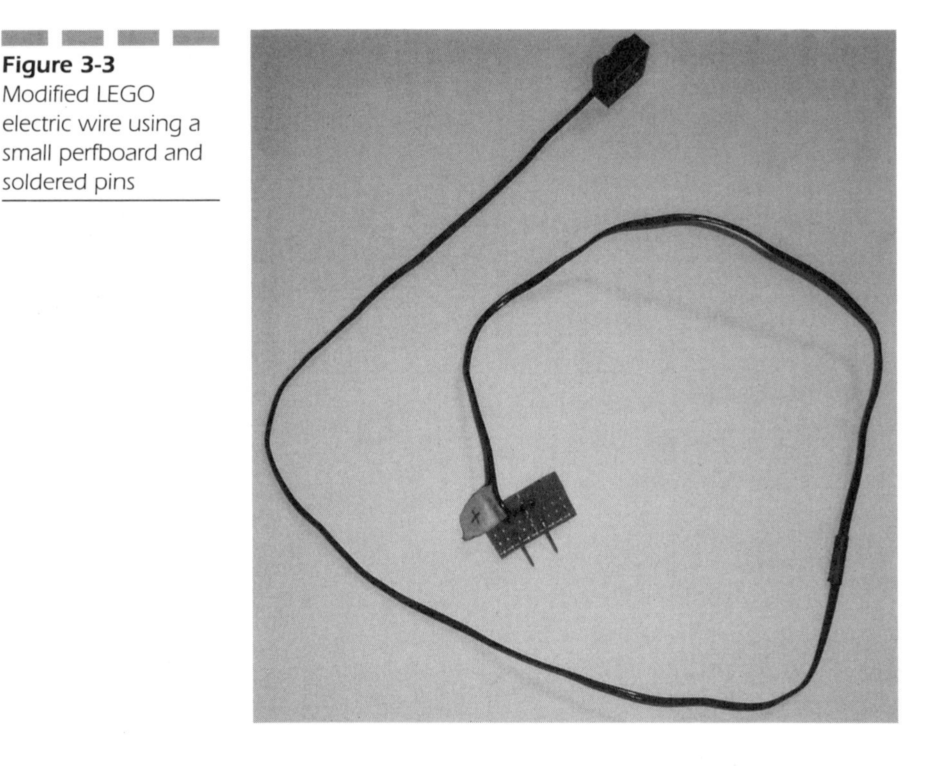

Figure 3-3 Modified LEGO electric wire using a small perfboard and soldered pins

Figure 3-4 The back view of a perfboard that shows soldered connections

3. Add two wires to the breadboard system, as shown in Figure 3-5.
4. Attach the other end of the LEGO electric wire to input 1 of the RCX P-Brick.
5. Connect the DMM to the wires added in Step 3. Turn on the DMM and set the appropriate voltage scale on the measurement device.
6. Turn on the RCX P-Brick and push the View button once. The LCD should display a 0, a standing man, and a 5.

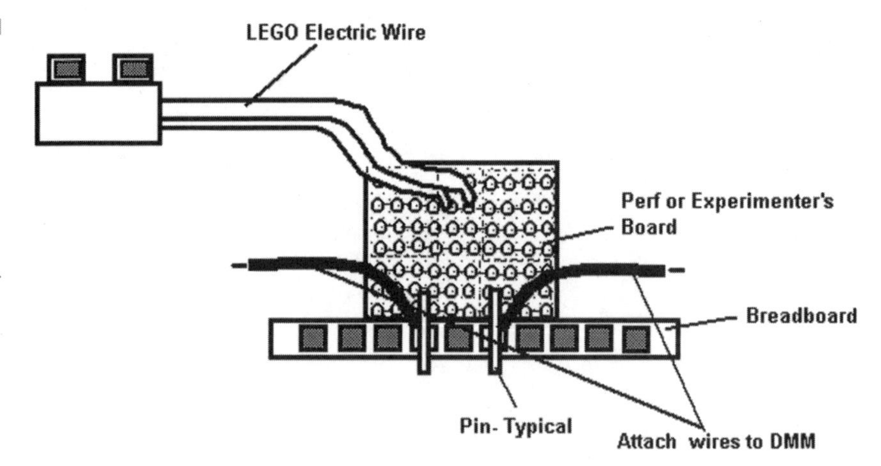

Figure 3-5 Modified LEGO electric wires and DMM test wires being inserted into a solderless breadboard (sectional view)

7. Press and hold the touch sensor and record the voltage on the DMM. Record the digital value shown on the LCD of the RCX P-Brick (V = 0.49, logic = 1).
8. Release the touch sensor and record the voltage on the DMM. Record the digital value shown on the LCD of the RCX P-Brick (V = 4.98, logic = 0).

RCX Code—Procedure for Testing the Touch Sensor

1. Open the Mindstorms software. Go to the RCX code section of the software.
2. Turn on the RCX P-Brick and press the View button. A 0 should be displayed on the LCD.
3. Click on Sensor Watchers and select the Touch icon. Move the Sensor Watcher to Program Untitled, as shown in Figure 3-6.
3. Click on the Tryout Tool and move to the Touch Sensor Watcher. Click the Press icon. A small box with a 0 should be displayed.
4. Press and hold the yellow button on the touch sensor. A 1 should be displayed on the Touch Sensor Watcher. A 1 should be displayed on the LCD as well.

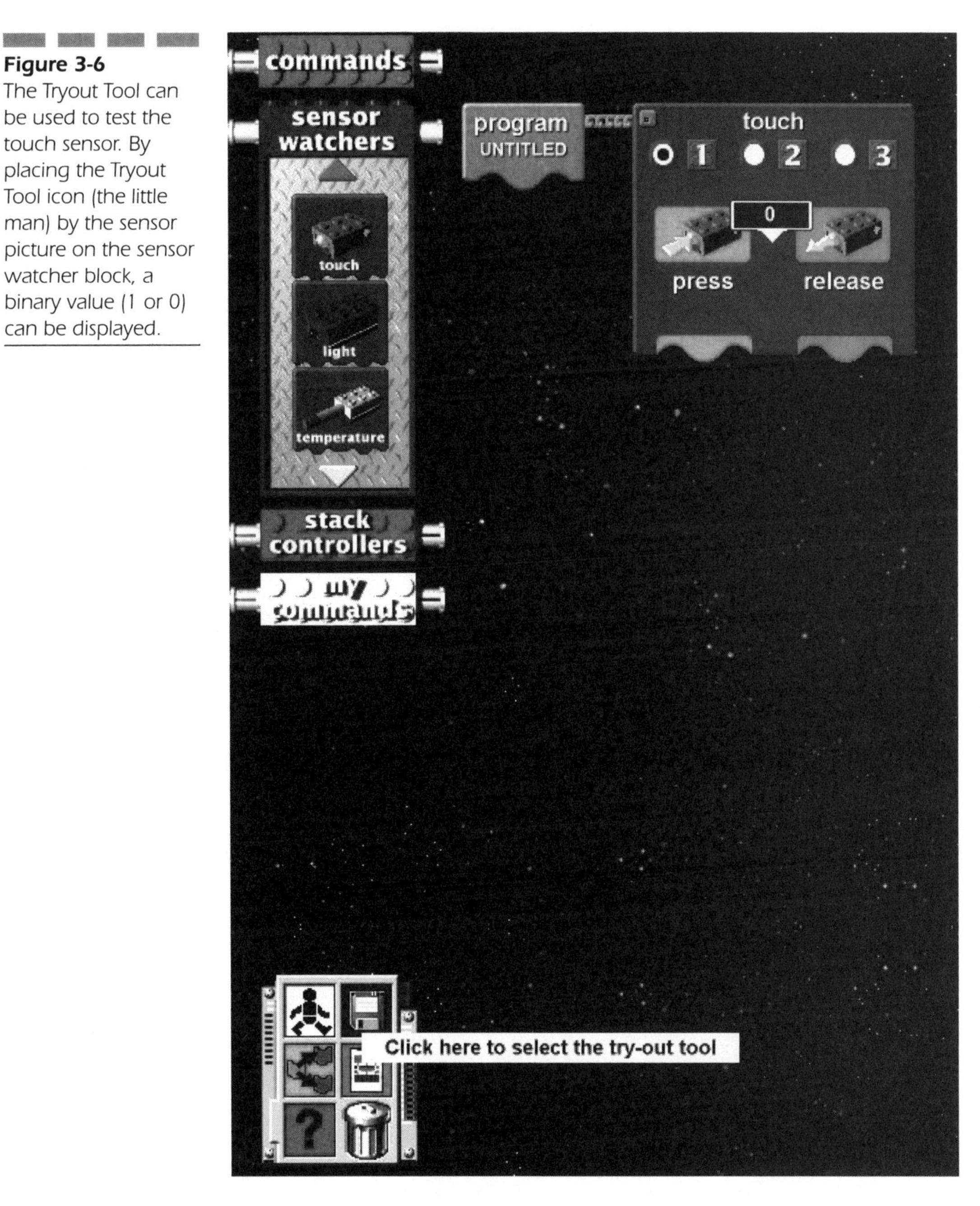

Figure 3-6 The Tryout Tool can be used to test the touch sensor. By placing the Tryout Tool icon (the little man) by the sensor picture on the sensor watcher block, a binary value (1 or 0) can be displayed.

5. Close out of the RCX software. Remove the LEGO electric wires from the RCX P-Brick. Turn off the RCX P-Brick.

Scout Tool—Procedure for Testing the Touch Sensor

1. Add the electric wires used in the RCX touch sensor lab to input 1 of the Scout P-Brick.
2. Turn on the Scout P-Brick.
3. Open the Scout tool. Click on the Init Ghost button displayed on the *graphical user interface* (GUI).
4. Click the Advanced button of the Scout tool. The Advanced Monitoring window should be displayed.
5. Click the Enable Monitoring button. The light bar graph and motors A and B LED indicators should be active and on, respectively.
6. The touch sensor Touch 1 should have an A/D count value of 255. This value represents a logic value of 0. The Scout's LED will be turned off.
7. Press and hold the touch sensor. The reading should be 25, representing a logic 1. The Scout's LED will be turned on.
8. Close out of the Scout tool. Turn off the Scout and remove the LEGO electric wires from the P-Brick.

How to Build a Homebrew Touch Sensor

The information provided in the previous discussion will help the electronics hobbyists or experimentalists to build a homebrew touch sensor. The key to building the sensor is to add the 1-kilo-ohm resistor to the chosen switch to keep the contacts clean and provide the correct input event for proper output control. The switch can be either a momentary or latching device. Figure 3-7 shows one way of building a touch sensor using a switch from Radio Shack's Electronics Learning Lab (Cat. No. 28-280). The 1-kilo-ohm resistor is wired in series with the momentary or pushbutton switch. The switch assembly is then wired to the modified LEGO electric wire to connect to the input of the Scout or RCX P-Brick. The operating conditions will be the same as a standard LEGO touch sensor. A timed latching function can be created for a pushbutton using software. The output can stay on for a predetermined time once the switch has been released. The experimental touch sensor lab outlines the procedure for creating a timed latch-

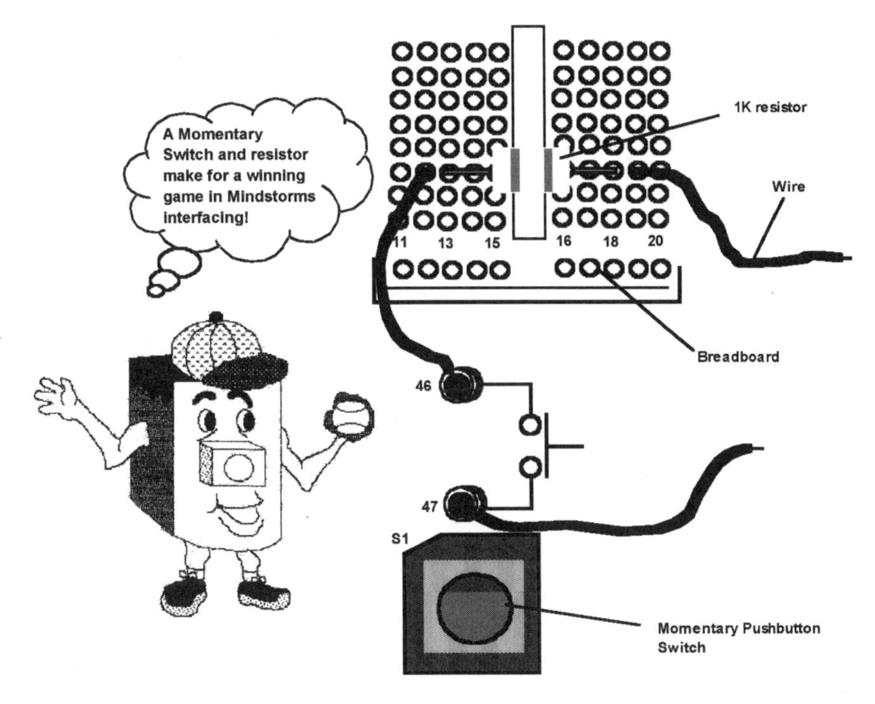

Figure 3-7 A pictorial diagram showing how to build an experimental touch sensor using the Radio Shack Electronics Learning Lab as explained by Mike Momentary Switch

ing function. If the switch doesn't work like the original sensor, check the wiring and test it again.

A latching circuit using the DPDT switch can be built quite easily. This switch has a centered off position where the component will not be electrically connected to the series resistor. The neat thing about this switch is that either the up or down positions can be used to provide unique input switching for the Scout or RCX P-Brick. By using software and the wired switch position, the LEGO robot can perform some interesting and unusual tasks. For example, an inverted logic function can be created where the switch in the down position can make the robot move forward.

In the workbook that comes with the Radio Shack Electronics Lab Kit, on pages 20 through 22 Forrest Mims[1] shows several switching circuit applications. The LEDs and buzzers can be replaced with a Scout or RCX P-Brick and programmed to emulate the output electrical devices. A *Robot Digital Assistant* (RDA) can be built in a way where its output can turn on the LED or buzzer and remain on until the touch sensor or experimental switch is pressed. The electrical output signal can be plotted using a DMM

[1]Forrest Mims is an amateur scientist and electronics engineer who writes the Mini Engineering Notebook series for Radio Shack.

with a PC interface as discussed in Chapter 2, "Developing GUIs: Software Control Basics."

Buggel—The eInsect

We will use an experimental robotic called *eInsect* to explore homebrew touch sensors. eInsect is a variety of insect that I call *Buggel*. This robot bug was inspired by the centipede design in the book *Joe Nagata's LEGO Mindstorms Idea Book*[2] by Joe Nagata. On pages 15 through 20, he outlines the assembly procedure for building his autonomous centipede. Buggel is a stripped-down version of the centipede because the RCX is not mounted on top of the insect's mechanical drive. I decided to build Buggel differently because I wanted the insect to have a small-streamline design. One thing I found interesting about the centipede was its motor attachment to the worm gear assembly. The small hinges enable the motor to be adjusted to the appropriate angle for the axle joiner to attach the shafts of the motor and the worm gear assembly together.

Another interesting mechanical design feature of the centipede is the drive train. A *24-tooth* (24t) gear is used as the driver gear and an 8t gear is used as the driven gear. If you take these two gears and multiply them over the insect's body length, a compound gearing system is created for moving the bug. Figure 3-8 shows the basic drive train design using the 24t and 8t gears. To calculate a number for describing a mechanical system's capability to increase or decrease speed, the gear ratio can be used. The gear ratio for this mechanical system is 3:1. In Chapter 1, the gear ratio formula was expressed mathematically as a fraction.

$$\textit{Gear ratio} = \textit{driver gear}/\textit{driven gear}$$

Substituting the 8t and 24t gears into the equation,

$$\textit{Gear Ratio} = 24\text{t} / 8\text{t}$$

$$\textit{Gear Ratio} = 3$$

[2]The LEGO Idea book, ISBN 1-886411-40-9, is published by No Starch Press, http://www.nostarch.com.

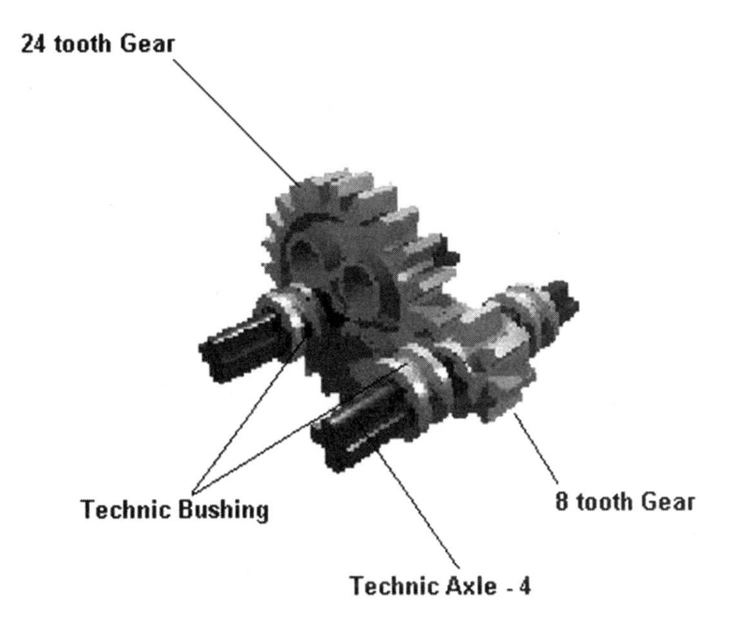

Figure 3-8
A Michael Lachmann computer-aided design (MLCAD) model of the 24t and 8t drive train for Buggel eInsect

For every one turn the 24t gear rotates, the 8t gear makes three revolutions. This speed is then decreased by three based on another 24t gear being applied to the 8t component. Therefore, the drive train balances the speed of the mechanical system. This balanced velocity enables all eight legs to move at the same speed, thereby producing an animated motion of a slow-moving insect foraging through the grass. Offset the legs to prevent any two of them from colliding with each other during insect movement. The LEGO remote control can be used to check the offset of Buggel's legs by pressing the appropriate forward and reverse buttons. The speed of Buggel can be measured by attaching the LEGO Speed Computer to the last 24t gear axle of the insect. Figure 3-13 shows the LEGO Speed Computer attached to the last 24t gear of Buggel. The speed measured with the LEGO Speed Computer was 20 *revolutions per minute* (rpm). Buggel uses the same drive train with the exception of the number of legs used to provide forward motion for the insect. Joe's design uses 16 legs total (8 per side) for his insect, whereas Buggel has 8 legs total (4 per side).

Figures 3-9 through 3-11 show the mechanical design of Buggel. The wonderful thing about building Mindstorms robots is that you have the flexibility to change designs. Buggel can be controlled using either the RCX or a Scout P-Brick. The ability to exchange controllers enables you to be more creative and experimental in mechatronics exploration. With this flexibility in construction, experimental play is key to the *proof of concept* (POC) studies in mechanical design.

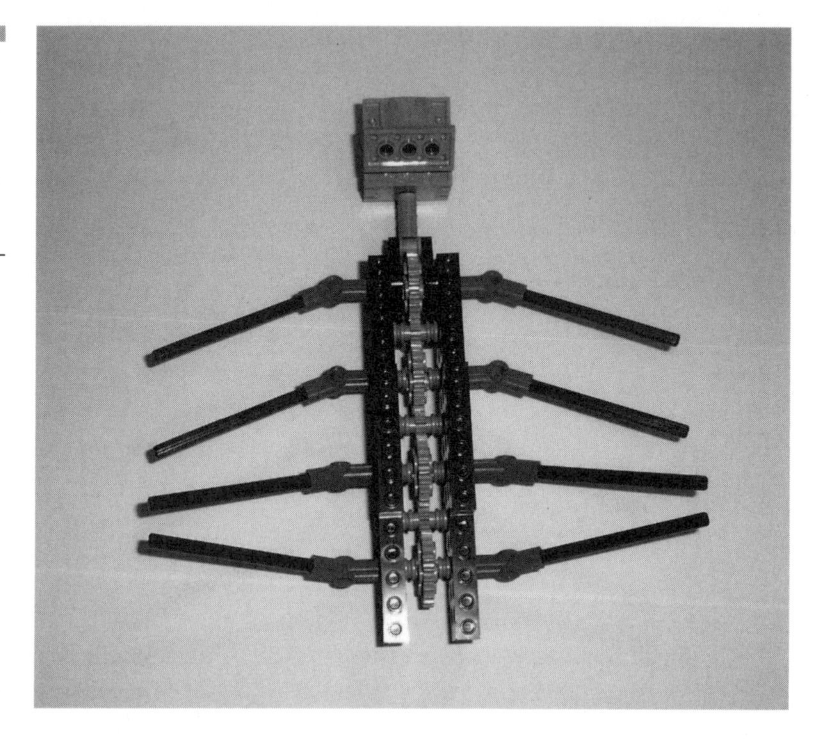

Figure 3-9
The top view of the drive train for Buggel eInsect. Notice the 24t and 8t gear pairs within the body of the robot.

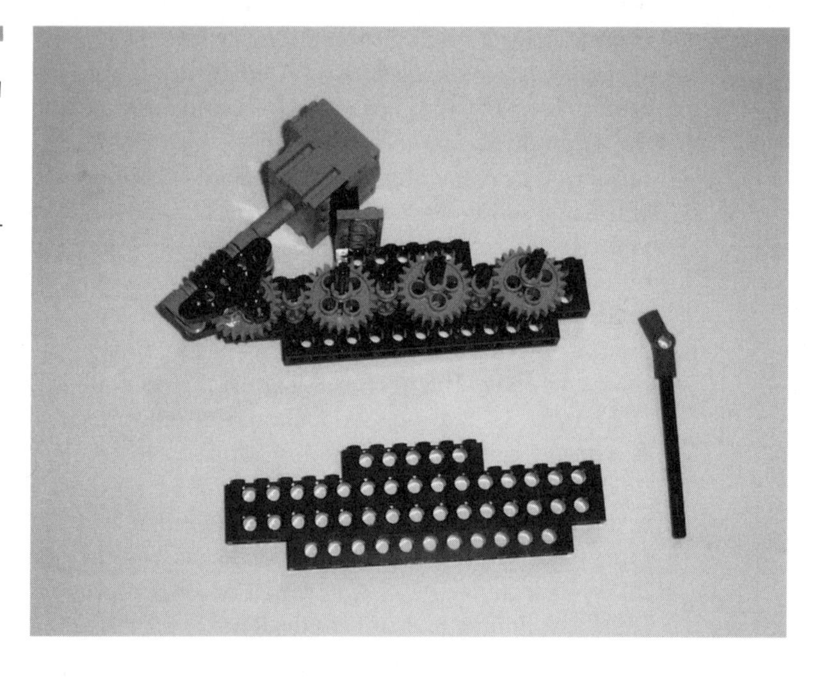

Figure 3-10
Details of the 24t and 8t gear pairs, the outer walls, and a typical leg assembly of Buggel

Figure 3-11
A close view of the drive train and the mounting of the LEGO motor using small hinges

Controlling Buggel with a Homebrew Touch Sensor—Lab Procedure

This lab explores how to build and use a homebrew touch sensor to control Buggel. The challenge of controlling Buggel is writing software that recognizes that when the button on the touch sensor is pressed, the insect should move forward. When the button is released, Buggel should move backwards. This inverse switching function simulates the insect detecting an object in its path. The insect then proceeds to avoid the obstacle by going in the other direction.

The procedure used to control Buggel with a homebrew touch sensor is as follows:

1. Build Buggel using Figures 3-9 through 3-13 as an assembly guide.
2. Wire the circuit on the Radio Shack Electronics Learning Lab breadboard or equivalent using the schematic shown in Figure 3-14.
3. Attach Buggel to the circuit using the modified LEGO electric wire discussed in Step 2 of the touch sensor lab.
4. Attach the complete circuit/robot assembly to input 1 of the RCX P-Brick.
5. Open the LEGO Mindstorms RCX software and build and download the program shown in Figure 3-15.

Figure 3-12
The completed Buggel eInsect robot

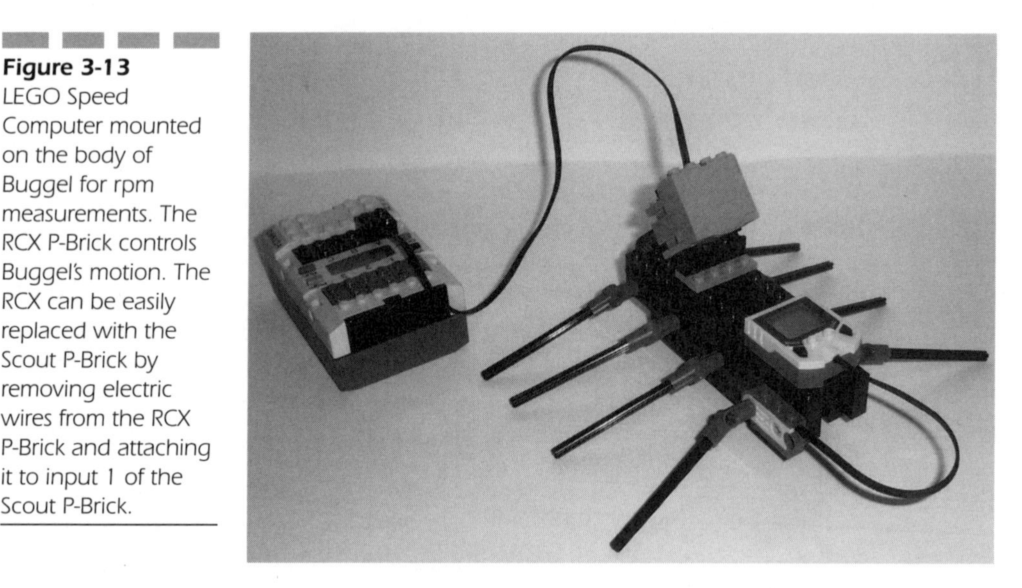

Figure 3-13
LEGO Speed Computer mounted on the body of Buggel for rpm measurements. The RCX P-Brick controls Buggel's motion. The RCX can be easily replaced with the Scout P-Brick by removing electric wires from the RCX P-Brick and attaching it to input 1 of the Scout P-Brick.

6. Turn on the RCX P-Brick, press Run, and observe Buggel. Did Buggel move forward? If not, check the RCX code and repeat this step.
7. Press and hold the homebrew touch sensor. Did Buggel move backward? If not, check the wiring and RCX code and repeat this step.

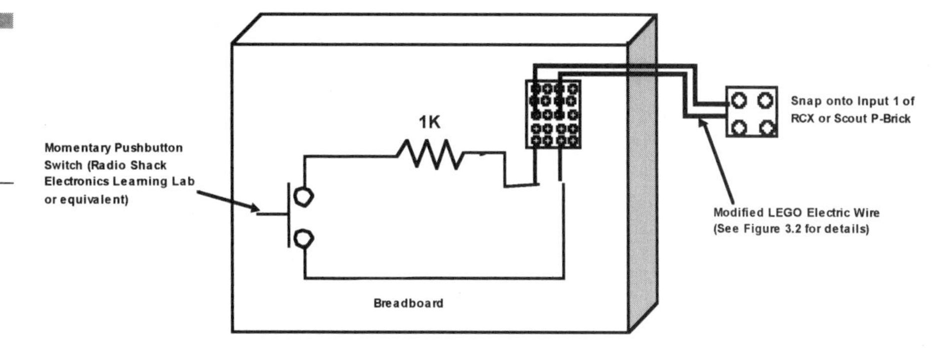

Figure 3-14
A wiring schematic for building an experimental touch sensor

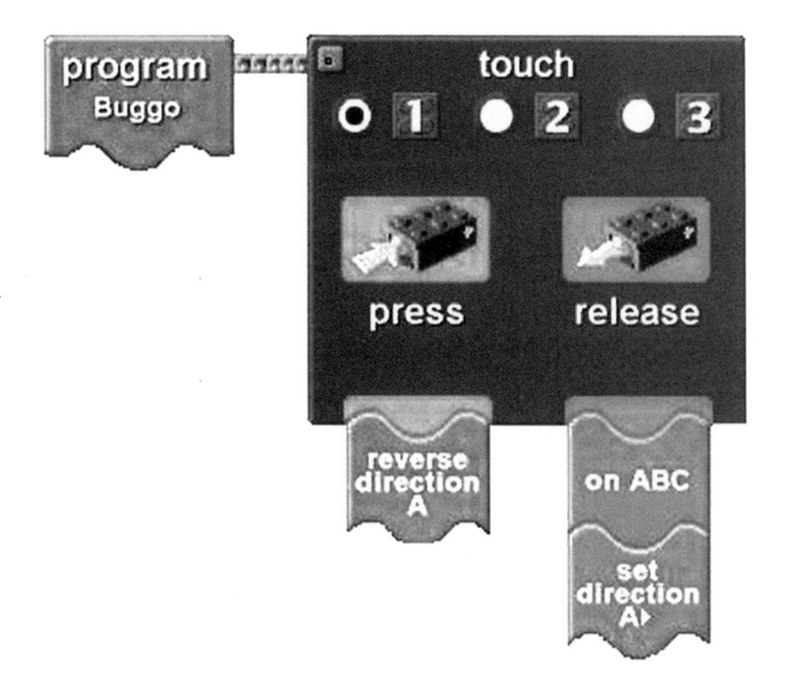

Figure 3-15
The RCX code for Buggel. With the touch sensor released, the eInsect moves forward. Pressing the touch sensor makes Buggel go in reverse.

8. Stop Buggel and replace the momentary switch with a DPDT switch. Use the circuit schematic shown in Figure 3-16 for this lab activity.
9. Turn on the RCX, press Run, and observe Buggel. Did Buggel move forward?
10. Slide the DPDT switch up. Did Buggel move backward? If not, check the wiring and repeat this step.

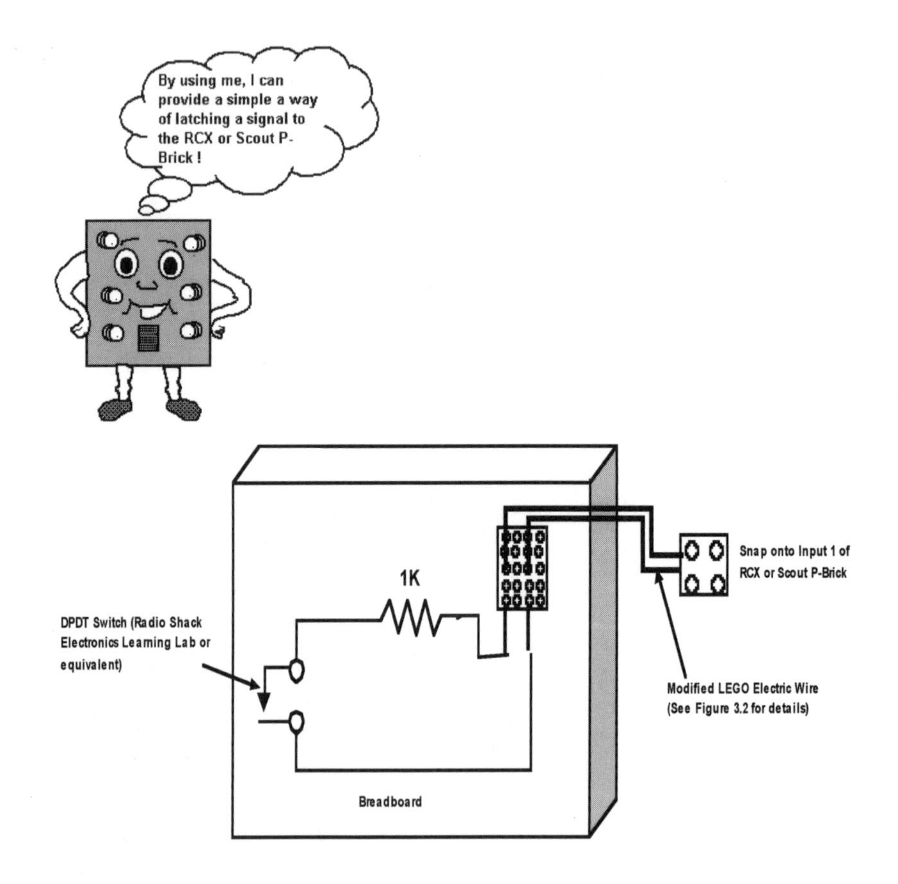

Figure 3-16 Dutton Double Throw discusses the wiring schematic for building a simple latching circuit using a DPDT switch.

Enabling a Switch for a Homebrew Touch Sensor—Lab Procedure 2

This lab illustrates how to handle a second switch for allowing, or enabling, the homebrew touch sensor to activate Buggel. The circuit is wired as an AND logic gate, where the DPDT switch AND the homebrew touch sensor will change the direction of the robotic eInsect.

The procedure to enable a switch for a homebrew touch sensor is as follows:

1. Wire the AND logic gate on the Radio Shack Electronics Learning Lab breadboard or equivalent using the circuit schematic in Figure 3-17.
2. Turn on the RCX P-Brick, press Run, and observe Buggel. Did Buggel move forward?

3. Slide the DPDT switch up and press and hold the momentary pushbutton. Did Buggel move backward? If not, check the wiring and repeat this step.
4. Slide the DPDT switch down and press and hold the momentary pushbutton. Did Buggel move backward? If not, explain why the robotic eInsect did not change directions. Complete the truth table depicted in Figure 3-18.
5. Remove the LEGO electric wires from the RCX P-Brick and set them aside. This wiring assembly and electromechanical control circuit will be used in the Scout P-Brick enable switch lab.

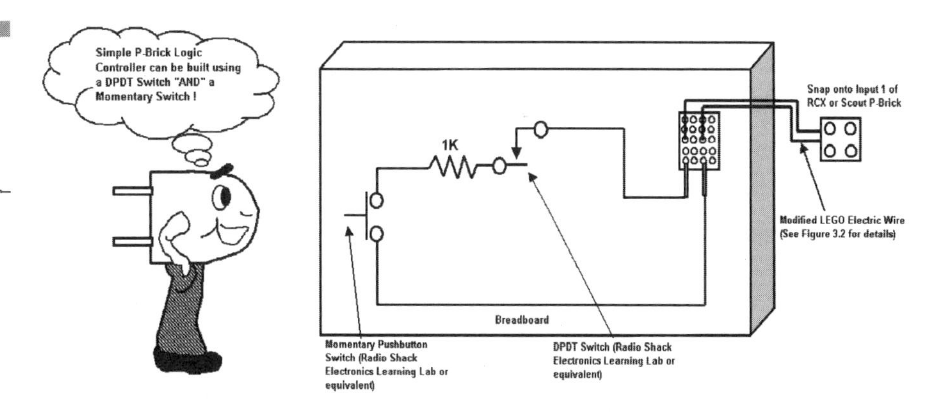

Figure 3-17 Andy on the wiring schematic for a simple logic controller

Figure 3-18 A blank truth table to be used with the enable switch lab project

Input A	Input B	Output C
DPDT Switch	Momentary Pushbutton	eInsect Reverse Direction

Analysis of the Enable Switch Lab This lab illustrates how simple electric switches can be wired to perform basic computer logic. If the DPDT switch contacts are closed or the slider is in the up position and the home-brew touch sensor is pressed, the 1-kilo-ohm resistor was electrically connected to the internal 10 kilo-ohm of the RCX P-Brick. The software embedded within the P-Bricks' microcontroller was able to make a logical decision on what action to perform based on the input event received. This processed information was then converted to an output voltage, thereby providing the appropriate state of motion for Buggel. If the DPDT switch contacts are open, the opposite input event was processed. There was no state of motion for Buggel; therefore, it continued to move forward. Buggel would not move unless the DPDT switch "AND" the touch sensor electrical contacts are closed simultaneously. The two electromechanical switches therefore create a "master" control signal for the RCX P-Brick allowing Buggel to move forward. The completed truth table along with the Boolean expression is shown in Figure 3-19.

The following lab procedure outlines how to use the Scout P-Brick to control Buggel the eInsect with an equivalent *Stand-Alone Code* (SAC) for forward/reverse robotic control.

Input A	Input B	Output C
DPDT Switch	Momentary Pushbutton	eInsect Reverse Direction
Open	Open	No
Open	Close	No
Close	Open	No
Close	Close	Yes

Figure 3-19 Professor Andy on the results of the enable switch lab project

Scout P-Brick Enable Switch Lab Procedure

This lab demonstrates how the Scout P-Brick SAC can be used as a software interface for the enable switch feature described in the RCX experiment. The same circuit will be used as demonstrated in the previous lab exercise except the preprogrammed behaviors embedded within the Scout P-Brick will be used. Also, the function performed by the SAC will be slightly different than the RCX code. See if you can notice the difference between the two software functions.

The procedure used for the Scout P-Brick enable switch lab is as follows:

1. Attach the LEGO electric wires and electromechanical control circuit used in the RCX P-Brick lab to input 1 of the Scout P-Brick.
2. Turn on the Scout P-Brick and program it using the SAC shown in Figure 3-20.
3. Press Run. Which direction does Buggel move?
4. Slide the DPDT switch up to close its contacts. Press and release the homebrew touch sensor. Buggel should move in the opposite direction.
5. Momentarily press the homebrew touch sensor. Again, Buggel should move in the opposite direction.
6. Slide the DPDT switch down to open its contacts. Press and release the homebrew touch sensor. Buggel should continue to move in the same direction.
7. Press Run on the Scout P-Brick and turn off the power using the ON/OFF button on the P-Brick.

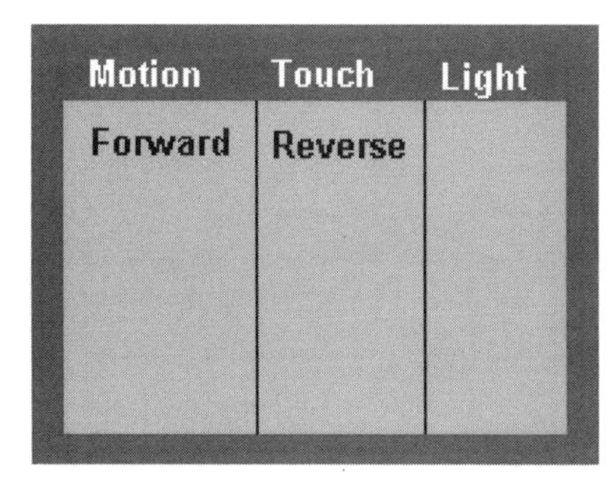

Figure 3-20
SAC for the Scout P-Brick

Analysis of the Scout P-Brick Enable Switch Lab Just like the RCX P-Brick enable switch lab, Buggel was able to change direction only if the DPDT switch and homebrew touch sensor contacts were closed. The significant difference between the two software implementations is the latching function used in the Scout SAC. The pre-programmed code embedded within the Scout P-Brick has a latching feature where by releasing the touch sensor, the selected output behavior remains until another input event occurs. This memory state can be programmed for the RCX P-Brick and will be explored on page 101, "How to Build a Smart Switch Using the RCX P-Brick," within this chapter.

Resistive Switches

A dual function controller can be built by using a DPDT switch and two resistors. Buggel has two distinctive output states based on two discrete input events. If an electric switch's contacts are open, the eInsect moves forward; if the contacts are closed, the robot bug moves backward. The solution of using one switch to do two control functions is an elegant design because expensive electronics switching is eliminated and it's simple. The resistors used will have to act like a touch sensor being pressed and released. Therefore, the resistors selected would be a 1-kilo-ohm component and have a value of resistance that has a voltage drop across it greater than 0.45 volts. To find the resistor value, create two ratios set equal to each other:

$$\frac{\text{Resistor 1}}{0.45\text{V}} = \frac{\text{Resistor 2}}{5\text{V}}$$

where:

$$Resistor\ 1 = 1\ kilo{-}ohm$$

$$Resistor\ 2 = unknown\ resistor$$

Solving for Resistor 2,

$$5V(Resistor\ 1) = 0.45V(Resistor\ 2)$$

$$\frac{5V(Resistor\ 1)}{0.45V} = \frac{0.45V(Resistor\ 2)}{0.45V}$$

$$\frac{5V(Resistor\ 1)}{0.45V} = Resistor\ 2$$

Substituting 1 kilo-ohm for Resistor 1,

$$\frac{5V(1K)}{0.45V} = Resistor\ 2$$

$$11.11\ kilo\text{-}ohms = Resistor\ 2$$

To calculate the voltage drop using this resistor,

$$V_{11.11K} = \frac{5V(11.11K)}{(11.11K + 10K)}$$

$$V_{11.11K} = 2.631V$$

Therefore, the 11-kilo-ohm resistor will emulate or act as the input action of a touch sensor being pressed. This input event enables the Buggel to move backwards, thereby meeting the design requirements discussed in the section "Controlling Buggel with a homebrew touch sensor—Lab Procedure" of this chapter. Figure 3-21 shows the circuit schematic for the resistive switching scheme.

The following lab explores the controlling function of this elegant electromechanical control circuit.

Dual Function Control Using the Resistive Switching Lab Procedure

This lab demonstrates how Buggel's direction can be changed using a simple DPDT electric switch and two resistors. Resistive switching, or multiplexing, is a convenient method of providing multiple signals using one switch.

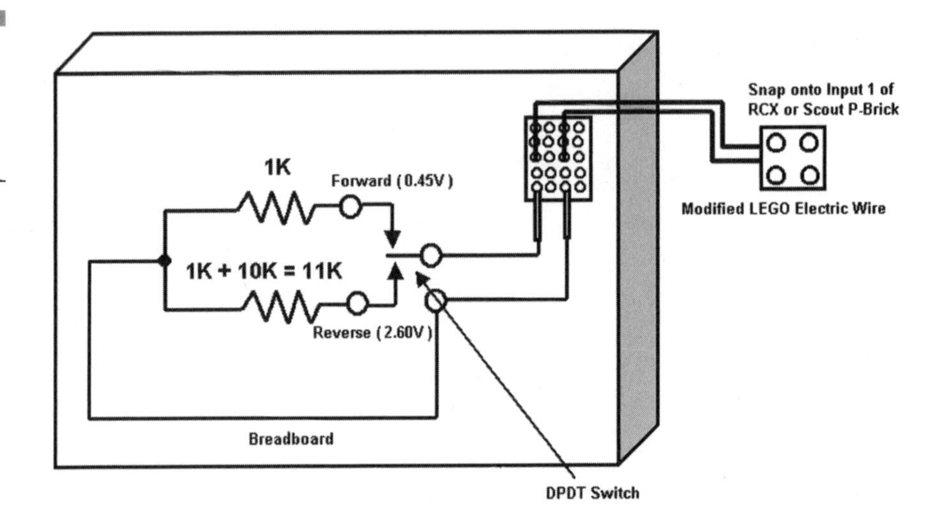

Figure 3-21 A resistive switch wiring diagram for the dual function control lab project

BOM

- Two 1-kilo-ohm resistors, 1/4 watt
- One 10-kilo-ohm resistor, 1/4 watt
- One DPDT switch

The procedure for enabling dual function control using resistive switching is as follows:

1. Wire the circuit shown in Figure 3-21 using the Radio Shack Electronics Learning Lab breadboarding system or equivalent.
2. Attach the LEGO electric wires to input 1 of the LEGO RCX P-Brick.
3. Attach a DMM to the specified points shown in Figures 3-22 and 3-23.
4. Turn on the RCX P-Brick and run the Buggel program used in the previous lab experiments.
5. Slide the DPDT switch down to reverse Buggel. Did the eInsect move backward? If not, check the wiring and repeat this step. What voltage is read on the DMM with the switch in this position? Does the measured value equal to 2.60 volts? (See $V_{11.11K}$ for circuit analysis details.)
6. Slide the DPDT switch up to move Buggel forward. Did the eInsect move in the specified direction? If not, check the wiring and repeat this step. What voltage is read on the DMM with the switch in this position? Does the measured value equal to 0.45 volts? (See V_{1K} for circuit analysis details.)

Figure 3-22
Complete measurement setup with the wired DPDT latching circuit with attached DMM

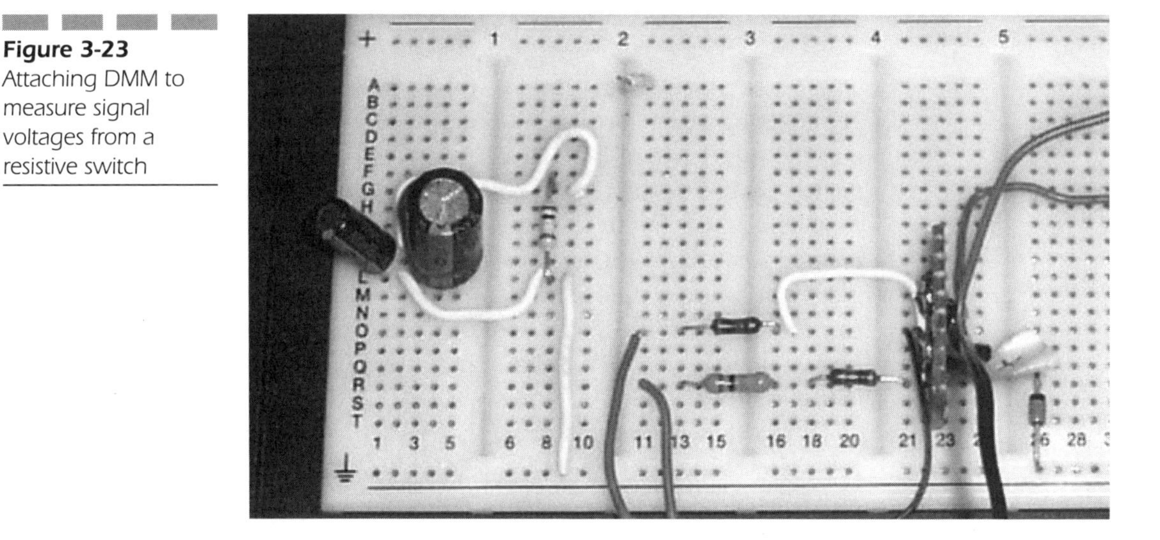

Figure 3-23
Attaching DMM to measure signal voltages from a resistive switch

7. Stop Buggel program running in RCX P-Brick. Turn off the P-Brick and remove the LEGO electric wire from it.
8. Connect the resistive switch to the Scout P-Brick's input 1 using the LEGO electric wire removed from the RCX P-Brick in Step 7.
9. Attach a DMM to the specified points shown in Figure 3-23.

10. Turn on the Scout P-Brick and run the SAC used in Step 2 of the Scout P-Brick enable switch lab.
11. Slide the DPDT switch up to move Buggel forward. Did the eInsect move in the specified direction? What voltage is read on the DMM with the switch in this position? Does the measured value equal 0.45 volts?
12. Slide the DPDT switch down, up, and then down to move the Buggel backward. Did the eInsect move in the specified direction? What voltage is read on the DMM with the switch in this position? Does the measured value equal to 2.60 volts?
13. Repeat Step 11 with the switch sequence of up, down, and then up. What direction does Buggel move: forward or backward?
14. Stop the program and turn off the Scout P-Brick.

Electromechanical Relays for RCX/Scout Input Control

In a nutshell, electromechanical relays are electrically actuated switches. By applying a control signal current through the coil, a magnetic field is created that attracts a flexible spring-loaded metal strip from one relay switch contact to another. Electromechanical relays are used in high-current application [typically 2 to 100A (amperes)] for the purpose of turning on/off an external electrical load. The control signal used to turn on or energize the coil is a couple of milliamperes (typically 10 to 50 mA). The switch contact section of the device comes in many of the common manual switch packages of *Single Pole-Single Throw* (SPST), *Single Pole-Double Throw* (SPDT), and DPDT.

Figure 3-24 shows a wiring schematic of a typical electromechanical relay circuit interface with the Scout or RCX input circuit. The circuit operates by current flowing through the coil of the relay initiated by closing the electric switch. The electric switch, therefore, provides a control signal to the relay's coil. When the coil receives this signal, the flexible switch contact makes with the stationary metal strip thus creating a closed circuit. The 1-kilo-ohm resistor is then electrically connected to the RCX or Scout's internal resistor and voltage source. The embedded RCX code or SAC will then perform the task specified in the mini computer program. If the control signal is absent when the electric switch is opened, the relay contacts break the external RCX interface, thereby stopping the mini computer pro-

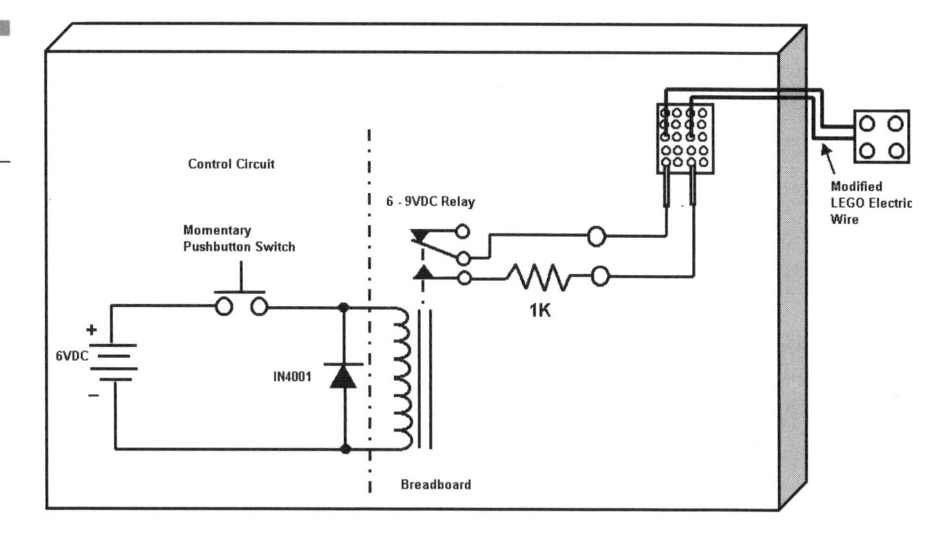

Figure 3-24
Wiring schematic for the electromagnetic relay lab project

gram of the P-Brick. The diode wired across the electromechanical relay's coil suppresses the inductive flyback voltage stored in its windings when the magnetic component was energized from the control signal. The following lab demonstrates the basic operation of using an electromechanical relay for controlling the Buggel eInsect robot.

Controlling Buggel with an Electromechanical Relay Control Circuit Lab Procedure

This lab demonstrates the basic operation of a simple electromechanical control circuit using a relay, an electric switch, and a 6-volt DC source. The circuit will be used to run the Buggel program of the RCX P-Brick and the SAC of the Scout P-Brick.

BOM

- One 6- to 9-volt electromechanical relay
- One electric switch
- One 1-kilo-ohm resistor
- One IN4001 silicon diode
- One 6-volt DC battery or power supply
- Wires
- LEGO electric wire assembly

The procedure for controlling Buggel with an electromechanical relay control circuit is as follows:

1. Wire the circuit shown in Figure 3-24.
2. Connect the Buggel eInsect to the circuit using the LEGO electric wire assembly used in the previous lab projects for the Scout P-Brick. Attach the electric wire to input 1 of the Scout P-Brick.
3. Turn on the Scout P-Brick and run the SAC program used in the previous lab projects.
4. Press and release the momentary pushbutton. The eInsect should advance in the forward direction.
5. Press and release the momentary pushbutton. The eInsect should now go in the reverse direction.
6. With a voltmeter attached across the input connector, play with the control circuit and note the voltage readings as you press and release the momentary pushbutton.
7. Stop the Scout P-Brick program and turn off the P-Brick.
8. Remove the LEGO electric wire from the input of the Scout P-Brick and attach it to input 1 of the RCX P-Brick.
9. Turn on the RCX P-Brick and run the Buggel program.
10. With the eInsect advancing in the forward direction, press and hold the momentary pushbutton. Buggel should now move in the reverse direction.
11. Release the momentary pushbutton and notice the direction of the eInsect.
12. Repeat Step 6 and notice the voltage readings when you press and release the momentary pushbutton.
13. Stop the Buggel program and turn off the RCX P-Brick.

This ends the controlling Buggel lab project. The following section discusses a unique electromechanical switch application using software and hardware to provide an output timed-delay off feature for controlling Buggel.

How to Build a Smart Switch Using the RCX P-Brick

The RIS is a great design tool for prototyping *smart devices*. With the RCX as the *microbrain*, hobbyists and experimentalists can develop a vast array of thinking things that can make decisions based on input data received using sensors and switches. Once the information is received, the P-Brick will make a decision and the appropriate output response will be presented.

As an Electronics Systems Engineer in the auto industry, I have the fun job of developing smart devices that provide convenience to the driver of the vehicle. The smart devices that I've worked with have ranged from smart door modules, automatic headlamps, and body computers to power distribution centers. All of these devices require an input signal from a switch or sensor so the microcontroller inside of the *electronics control unit* (ECU) can process the data and make the appropriate electromechanical load turn on or off. The *smart switch* is a mechatronics device that can delay the turning off of an output device based on a predetermined time. This programmable time-delay feature, therefore, remembers to turn off the device for you. By pressing a momentary pushbutton, the smart switch processes the release event action with a timed-delay off output state response. The function provides a built-in energy saver because the smart switch turns off the electrical load when the programmed on time has expired. Smart switches for robots give the mechatronics system flexibility because you can change how the mechanics are actuated on the device using embedded software. The software layer of the smart switch, therefore, enables the hobbyist and experimentalist to change the behavior of the electromechanics without drastically changing the hardware interface of the device.

The electromechanical control circuits described in this chapter can be changed into a smart switch quite easily. As an example, let's design a smart switch for Buggel where the release of a momentary switch or touch sensor will cause the output of the P-Brick to stay on, enabling Buggel to move for 30 seconds. After 30 seconds have elapsed, the smart switch will turn off the output, stopping Buggel in its tracks. If the switch is pressed continuously, Buggel will move until the smart function is initiated. Basically, a latching timed-delay output on function is embedded within the smart switch control circuit. The logic of the smart switch is shown in the state machine diagram of Figure 3-25. As seen in this diagram, once the control signal has been initiated with the touch sensor, the timed output states cycle their way on until 30 seconds have elapsed. After the time has expired, motor A will be off.

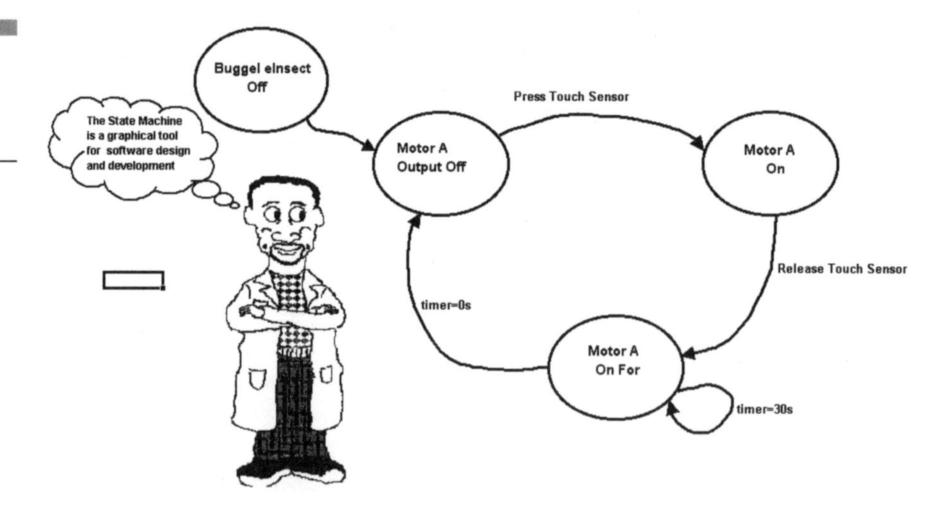

Figure 3-25 Mr. Don and the smart switch state machine diagram

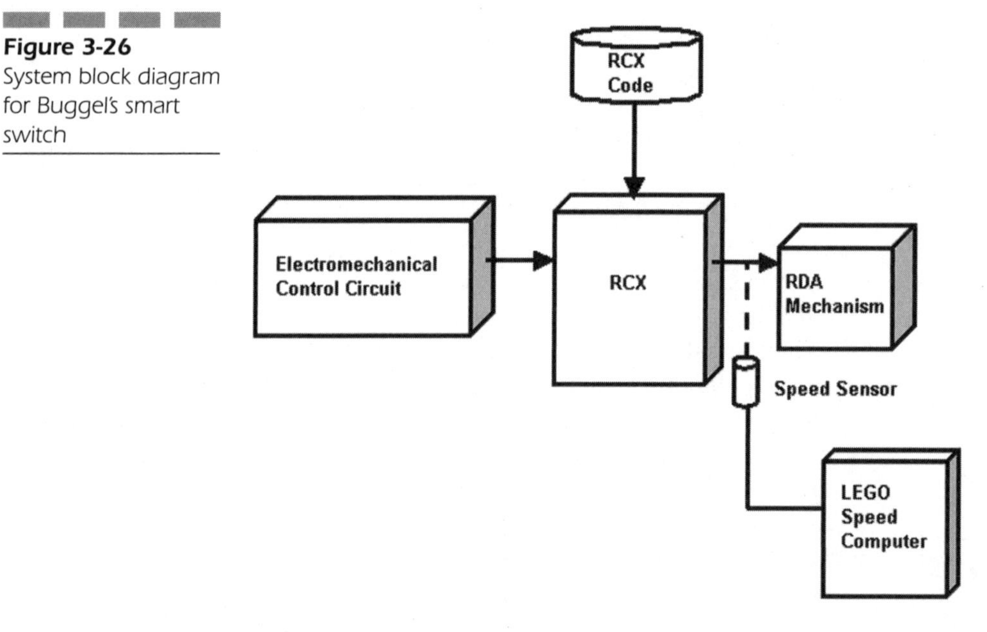

Figure 3-26 System block diagram for Buggel's smart switch

Figure 3-26 shows the system block diagram and how the state machine interfaces with the mechatronics device. The LEGO Speed Computer is shown as an integrated instrument for measuring the compound gearing drive rpms as well. The system block diagram should be complete in its graphical representation of the target unit it is intended for. Therefore, all instrumentation, output indicators, input switches, and electrical loads should be shown on the system block diagram. Ideally, when developing a POC model for product development feasibility studies, the system block diagram and the state machine should be drawn prior to the actual building of the smart device.

The following lab project outlines the procedure for developing the smart switch for Buggel.

Smart Switch Lab Procedure

The objective of this lab is to illustrate how software can be used to extend a typical electromechanical control circuit switching function for LEGO Mindstorms robots. By embedding software, the output state of the electromechanical control circuit can be delayed upon the release of the momentary pushbutton or sensor.

BOM

- One prebuilt electromechanical control circuit
- One prebuilt robot
- LEGO Mindstorms RCX software

The procedure used for the smart switch lab is as follows:

1. Open the LEGO Mindstorms RCX software and program the code using Figure 3-27.
2. Turn on the RCX P-Brick and download the smart switch code.
3. Use the electromechanical control circuit built in Step 1 of the controlling Buggel lab.
4. Attach the RCX P-Brick to the electromechanical control circuit and Buggel using input 1 for the circuit interface and output A for Buggel's worm gear drive.
5. Run the smart switch program downloaded into Buggel in Step 2.
6. Buggel should start moving and remain in motion for 30 seconds. After 30 seconds, Buggel will stop moving. This initial movement calibrates the software with hardware of the eInsect, thereby enabling the embedded function to be repeatable upon an input event or request signal.
7. Press and release the momentary pushbutton. Buggel should start moving and remain in motion for 30 seconds. After 30 seconds, Buggel

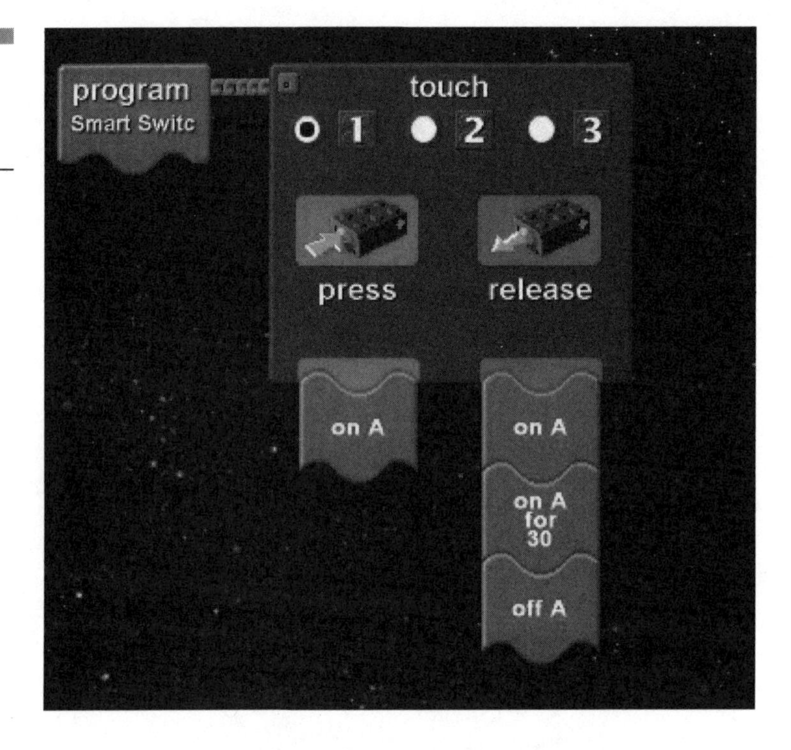

Figure 3-27 RCX code for the smart switch lab project

will stop moving. If this function doesn't work, check the interface wiring of the LEGO electric wire to input 1 of RCX P-Brick and review the software.

8. Press and hold the momentary pushbutton switch. Buggel should start moving and remain in motion until the control signal is removed.
9. Stop the smart switch program and turn off the RCX P-Brick. This ends the smart switch lab project.

As seen in the lab, software enhanced the electromechanical control circuit of Buggel. The lab also demonstrated the philosophy of hardware and software interface layers and the seamless integration they require when defining robotic functions. The hardware and software interface model discussed in Chapter 2 plays an important part when developing robots. Once the function has been defined and embedded within the robot's hardware interface layer, modifying the overall feature of the intelligent machine becomes less demanding on the developer. For example, if the requirement for Buggel has changed from directional movement based on object detection using a sensor to movement when no object is present, a slight modification in the RCX code would accomplish this new function. Therefore, hardware and software are inseparable elements for robotic systems, and hobbyists and experimentalists should understand both equally.

As a final wrap-up to this chapter, the last section will discuss modifying the V-Bot robot panel for recording the number of times the momentary pushbutton has been pressed on the Excel spreadsheet. The data can then be statistically analyzed using any of the built-in Excel stat functions packaged with the software.

How to Build a Switch Monitor Detector

When building robots, a hobbyist or experimentalist needs a set of testing tools to validate the hardware, particularly the *inputs/outputs* (I/Os), of the machine. As demonstrated in Chapter 2, Excel *Visual Basic for Applications* (VBA) enables the developer to build no-cost effective testing tools for LEGO RCX-based robots. This section is a continuation of the discussion on software controls for monitoring input events occurring from switches and sensors. The function of the testing tool application that will be built in the following lab project is to display input events from an electromechanical control circuit on a panel as well as log all subsequent switching activities generated from the switch. The tool should enable a table to be displayed on the spreadsheet that shows the recorded data from the control circuit logging session.

Reusable Code Concept

The software solution to the design challenge can be found quite easily by using an existing application's framework as the foundation for code development. In practice, a roboticist should have packets of functional-reusable code available for future robot projects. The purpose behind having this software library is to reduce the code development cycle time. When a design challenge presents itself and a software solution is required, having a library of functional code that can be reused prevents a lot of development frustration. The software used in the lab projects up to this point is available on the CD packaged with this book. Depending on the application, the code will have to be modified to perform the specified robot function or task. Although some minor modifications may be necessary, the amount of time spent coding and testing is kept to a minimum.

Building the Switch Monitor Detector in VBA

The robot control panel used in Chapter 2, "Developing GUIs: Software Control Basics," will serve as the framework for building the Switch Monitor Detector. Figure 3-28 shows the code building blocks required for the testing tool. With this *Software Context Diagram* (SCD), an architectural framework for *partioning*, or creating sections of code, can be used as a guide in writing the application. The software blocks that have been added are the LogCmd, PrintCmd, ResetCmd, and ClearCmd VBA events. The LogCmd block enables the logging session to occur by initializing or setting a counter's variable to zero. The PrintCmd block enables all of the recorded switching data obtained from the monitor detector to be displayed on the Excel spreadsheet. The ResetCmd block will clear the touch sensor text box and the ClearCmd VBA event will clear all of the data displayed on the specified spreadsheet cells. The complete code for the Switch Monitor Detector is shown in the following software listing:

```
Dim SENSOR1 As Integer
Dim SENTOUCH As Integer
Dim SENVAL As Integer
Dim n As Integer
Dim b As Integer
Dim Reading As Single
Dim MyReading(30) As Single
Private Sub Read_Click()
SENSOR1 = 0 'Read Input 1 of RCX
SENTOUCH = 1 'Reading Data from a Touch Sensor
SENVAL = 9 'Retrieve value from Touch Sensor at Input 1
PBrickCtrl.SetSensorType SENSOR1, SENTOUCH
SWMON.TextBox1 = PBrickCtrl.Poll(SENVAL, SENSOR1)
Reading = SWMON.TextBox1.Value
MyReading(n) = Reading
If n = 30 Then
MsgBox "Maximum Data Points Have Been Reach," vbExclamation, "Warning"
Else
n = n 1 1
End If
End Sub
Private Sub CommOn_Click()
PBrickCtrl.InitComm
End Sub
Private Sub CommOff_Click()
PBrickCtrl.CloseComm
End Sub
Private Sub LogCmd_Click()
n = 0
End Sub
Private Sub PrintCmd_Click()
For i = 0 To n
Cells(i 1 3, 1).Value = i
Cells(i 1 3, 2).Value = MyReading(i)
Next i
End Sub
```

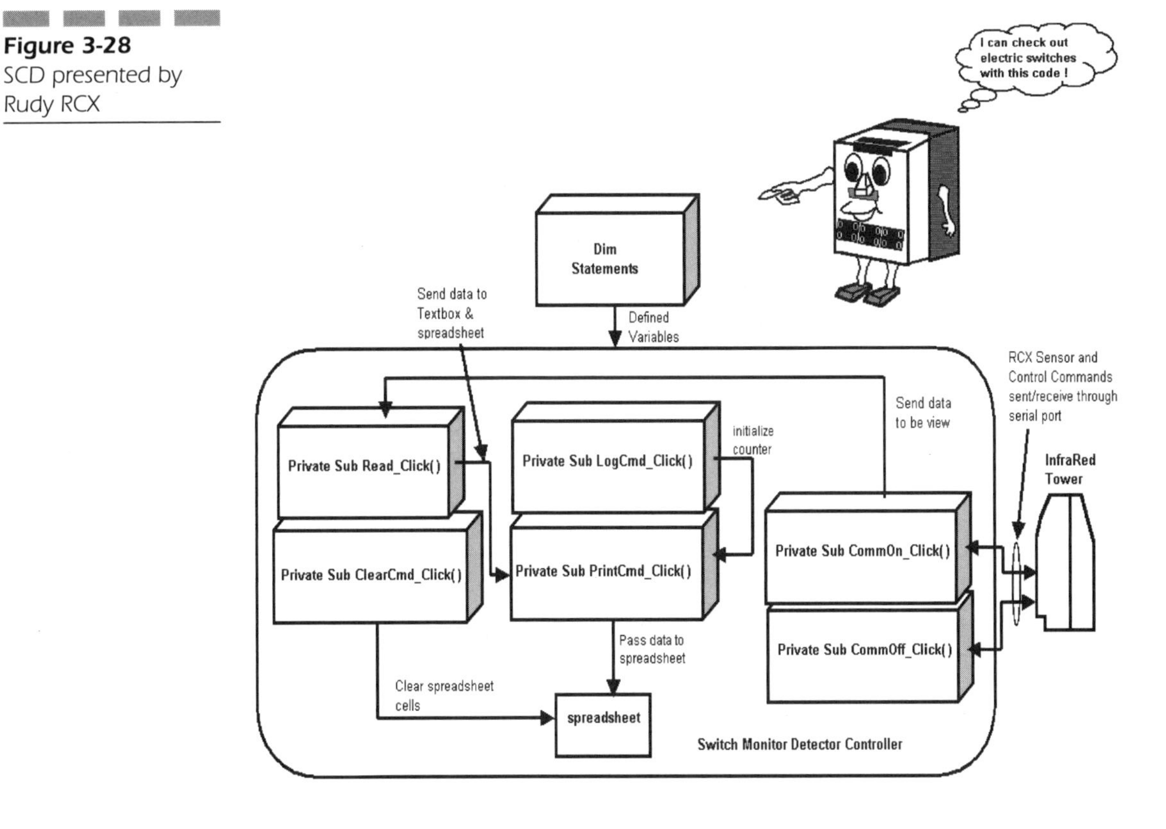

Figure 3-28 SCD presented by Rudy RCX

```
Private Sub ClearCmd_Click()
For b = 0 To 30
Cells(b 1 3, 1).Clear
Cells(b 1 3, 2).Clear
Next b
End Sub
Private Sub ResetCmd_Click()
SWMON.TextBox1 = " "
End Sub
```

Software Listing for Switch Monitor Detector

After running the application, the Switch Monitor Detector will display the electromechanical control circuit digital data (1 or 0) within the text box when the Read button is pressed. With additional presses/releases of the pushbutton, this data will be logged under the record variable name MyReading(n). The subscript (n) is the location for where the data is stored or read into memory. To see the data, it must be released from the stored location using the VBA instruction "Cells (rows+ 1, columns).value". A counter is used to write the data to the specified by cell using the "For I = 0 to n, Next I" loop instruction. With the iteration of the counter, data is placed onto the specified Excel cell location using the Cells instruction. The maximum size of data that can be recorded for this application is 30 data points. If more data bandwidth is required, the MyReading(30) within the Dim statements can be changed accordingly. One last item to point out is that the software blocks explained are available for reuse whenever a data table is from a lab experiment. The Switch Monitor Detector is a flexible tool that will grow based on the testing requirements of the LEGO Mindstorms robot application.

Switch Monitor Detector Software Lab Procedure

This lab project demonstrates how an existing VBA application can be used to create an electromechanical control circuit testing tool. The project also illustrates the reusable code concept by using existing code from a previous VBA application to help build the testing tool.

BOM

- One electromechanical control circuit
- One RCX P-Brick
- One Buggel eInsect robot (optional)
- V-Bot VBA control panel GUI software

The procedure for the Switch Monitor Detector software lab is as follows:

1. Use the electromechanical control circuit built in Step 1 of the controlling Buggel lab.
2. Attach the electromechanical control circuit to input 1 of the RCX P-Brick using the modified LEGO electric wires.
3. Turn on the RCX P-Brick.
4. Open the V_Bot Control Panel.xls built in Chapter 2. Using Figure 3-29, modify the UserForm accordingly.
5. Save the file as Switch Monitor Tool.xls on your hard drive.
6. Type the code in *Visual Basic Editor* (VBE) using the software listing for the Switch Monitor Detector.

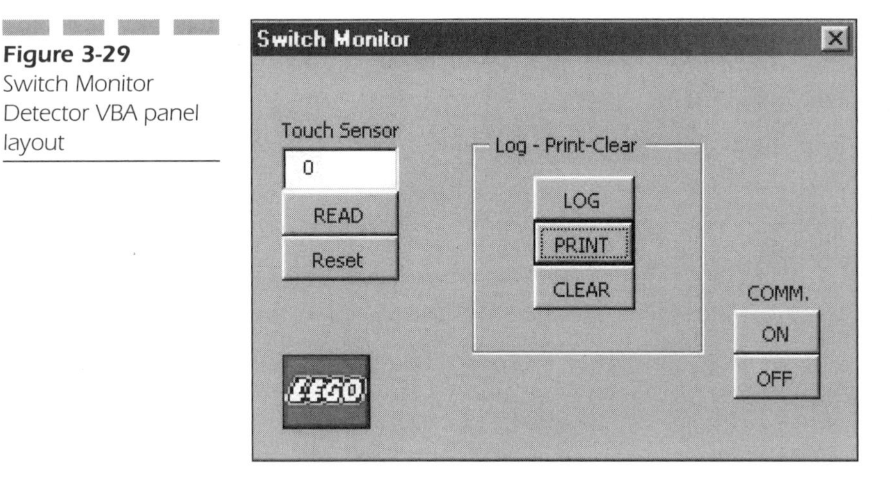

Figure 3-29 Switch Monitor Detector VBA panel layout

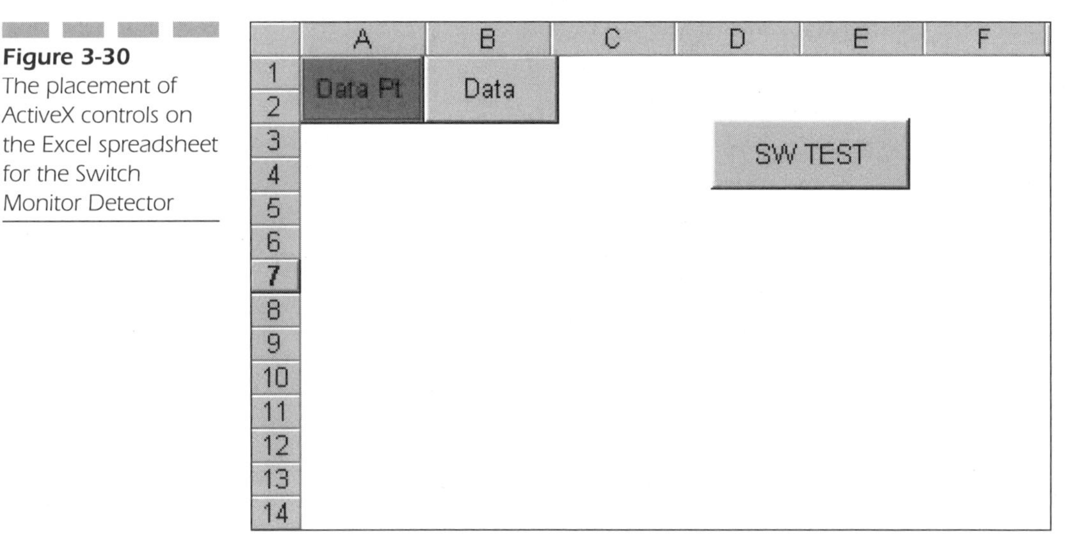

Figure 3-30 The placement of ActiveX controls on the Excel spreadsheet for the Switch Monitor Detector

7. Change the appropriate properties of the ActiveX controls to the names used for each button event shown in the listing.
8. Build the spreadsheet as shown in Figure 3-30. The Data Pt and Data labels where created using CommandButtons from the Toolbox menu.

9. Modify the SW TEST ActiveX control by right-clicking the button. Note that in order to modify code for this button, you must be in design mode.
10. Type in the following code within the VBE window:

```
Private Sub SWTESTCmd_Click()
SWMON.Show
End Sub
```

11. Click the Design Mode icon within the Toolbox menu.

Now that the software is built, the Switch Monitor Detector can be tested.

Procedure for Testing the Switch Monitor Detector

1. Turn on the RCX P-Brick.
2. Press View to display the digital data for input 1 of the RCX P-Brick. The LCD should display logic 0.
3. Click on the SW TEST button displayed on the spreadsheet. The Switch Monitor Panel should be displayed on the spreadsheet.
4. Click ON to open the communication port for the Serial Infrared Tower.
5. Click LOG to initialize the counter for recording data.
6. Click Read to display the logic value of the P-Brick. The text box on the test panel should display logic 0.
7. Press and hold the momentary pushbutton from the electromechanical control circuit.
8. Click Read to display the logic value of the P-Brick. The LCD of the P-Brick should display logic 1. The text box should have the same logic value as the P-Brick LCD.
9. Click Reset. The text box should be cleared.
10. Repeat Steps 6 through 7 several times varying the push/hold and release switching events.
11. Click Print to display the recorded data from the electromechanical control circuit.
12. To clear the data from the spreadsheet, click Clear on the Switch Monitor GUI.

13. Click Read 30 times. A "Maximum Data Points Have Been Reached" warning message should be displayed on the screen. Click OK to remove the message from the screen.
14. To reset the counter, click LOG.
15. Click OFF to close the serial communication port of the computer.
16. Click the X in the upper-right-hand corner of the GUI to exit the testing tool.
17. Turn off the RCX P-Brick. This ends the testing procedure for the Switch Monitor Detector.

Switch Monitor Detector Analysis This lab illustrates the concept of reusable code for developing the Switch Monitor Detector. This POC project demonstrates how the hardware and software interface layer can be used to build test panels in Excel VBA. The functions for logging and printing switch data from the electromechanical control circuit are key code blocks and can be reused for future applications quite easily. This lab project is a good foundation for the remainder of the book because test-monitoring and control applications will be built using the code blocks written from this GUI. In addition, the methods used to record and print data collected from a Mindstorms robot sensor can be improved and adapted for future robotic applications systems. This lab project also demonstrates how sophisticated test and control tools can be built using a Windows spreadsheet for collecting data from electromechanical control circuits and Mindstorms robots.

In this chapter several electromechanical controls were investigated for interfacing with the Scout and RCX P-Bricks. A basic AND computer logic gate, resistive switching, an electromechanical relay control and a smart switch were some of the circuits investigated using outlined test procedures. Depending on the type of electromechanical control circuit used and eInsect known as Buggel provided forward and backward motions using a special gearing mechanism. By adding a similar VBA software application, Buggel's motion data was logged and later displayed on a spreadsheet for further analysis. Electromechanical Controls are just but one way of managing a LEGO Mindstorms robot movement as well as providing a hardwire method of robotics control interfacing.

In the next chapter, electonic circuits using transistors, FETs, and a graphics calculator will demonstrate how solid-state devices can provide a method of controlling a LEGO Mindstorms robot.

CHAPTER 4

Electronic Switching Circuits

In Chapter 3, "Electromechanical Controls Interfacing," the electromechanical relay was the primary external switching actuator that enabled the RCX or Scout *programmable brick* (P-Brick) input circuit to be electrically connected to a 1-kilo-ohm resistor. This simple *direct current* (DC) circuit is critical because the firmware programmed into the *Robot Command Explorer* (RCX) code's operating system uses the voltage levels for digital switching, and detection is established by the resistor network for robotic applications. As demonstrated in the resistive switching circuit, the input interface has a maximum resistance value that it uses to determine if the signal is a binary 0 or 1.

Another item to note about the electromechanical switching circuits is that current requirements have to be about 1 mA or greater in order for the P-Brick to process the input data for an appropriate output response. The input circuits discussed in this chapter can have current levels that are less than 1 *milliamperes* (mA). These circuits have an active component that enables a small signal to be amplified to drive an electromechanical actuator for input detection to a robot controlled by either an RCX or Scout P-Brick. Therefore, an external power supply is required for signal amplification to occur with the electronic switching circuit. The electromechanical controls discussed in Chapter 3 were passive because an external voltage was not required for switching the 1-kilo-ohm resistor.

The RCX or Scout P-Brick is also capable of driving external circuits for audible and visual indication. By using circuits that isolate the P-Brick's output driver circuits from external devices that may produce small *transients*, or noise, elaborate audible and visual indicators can be built with the RCX or Scout P-Bricks. This chapter focuses on the following topics:

- A transistor relay driver input circuit
- Transistor Relay Monitor: *Visual Basic for Applications* (VBA)–C-Bot application
- A *Power Metal Oxide Semiconductor Field Effect Transistor* (PMOSFET) relay driver input circuit
- A wireless relay driver input circuit
- Electronic output switching
- *Calculator-Based Laboratory* (CBL)/*Digital Control Unit* (DCU) for robot diagnostics

A Transistor Relay Driver Input Circuit

As discussed in Chapter 3, the electromechanical controls for Mindstorms interfacing provide a variety of ways to control robots using simple electric switch and relay topologies. The RCX code and Excel VBA provide feature programmability into changing the robot's behavior based on the embedded task written in software. The electromechanical controls discussed in Chapter 3 require currents that are higher than 10 mA in order to energize the relay's coil so it can switch its contacts. Also, the electric switch contacts used to provide a control signal must be large enough to handle the current flowing through the relay coil circuit.

The transistor relay driver circuit provides low-current control switching of the external 1-kilo-ohm resistor using an electromechanical relay. The transistor amplifies the current provided by the closed contacts of the electric switch based on its h_{fe}, (β) or *beta*. Beta is the transistor amplification gain factor and is typically around 50 to 100 for NPN or PNP bipolar transistors. If there is a beta of 100, the transistor saturates, or turns on hard, enabling the maximum current to flow through its collector-emitter junction leads.

The current flowing through this junction will be enough to turn on the electromechanical relay. Figure 4-1 shows a transistor relay driver circuit taken from Forrest Mims' "Basic Electronics: Workbook 1" on page 55 of the Radio Shack Electronics Learning Lab (Cat. #: 28-280). The original circuit had a buzzer wired in series with its *Normally Opened* (NO) contacts. The output circuit has been modified to enable the 1-kilo-ohm resistor to interface to the inputs of an RCX or Scout P-Brick. Although this circuit is simple, the potential applications are only limited by the hobbyist or experimentalist's imagination.

The following section outlines the process for developing a *Proof Of Concept* (POC) study to explore interfacing applications using the transistor relay driver circuit. *Conveyor Robot* (C-Bot) will be used as the *Robot Digital Assistant* (RDA) in the lab project for testing the hardware circuit using VBA and RCX code controls for the software interface layer.

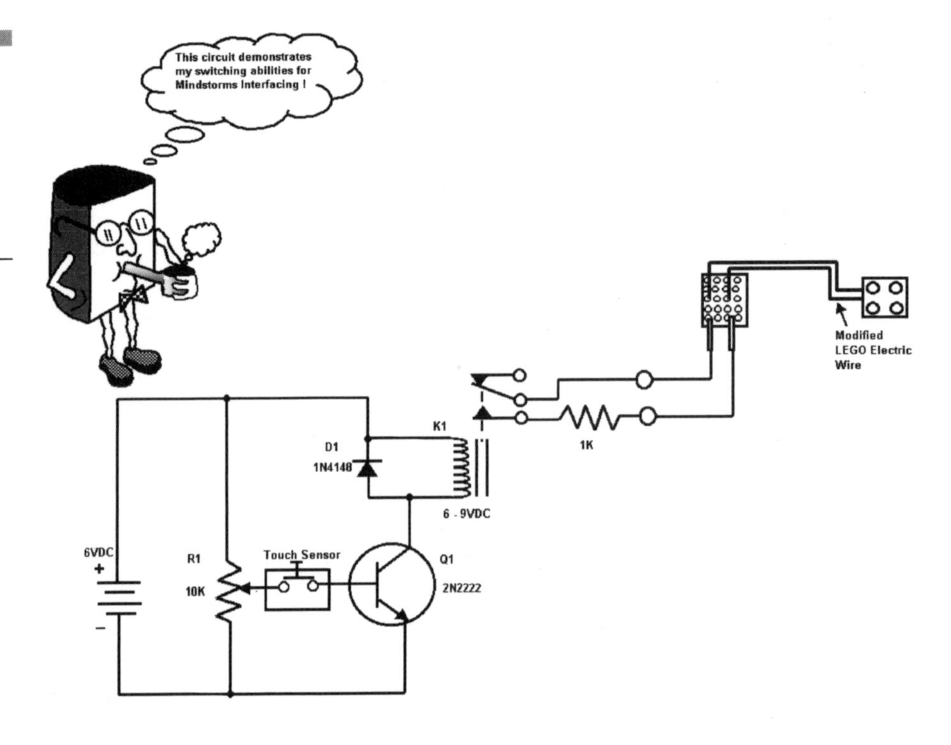

Figure 4-1 Tom Transistor introduces the transistor relay input circuit for the simple controller lab experiment.

Developing C-Bot

As discussed in Chapter 2, "Developing GUIs: Software Control Basics," a simple sketch is a good way to capture the basic POC idea for robotics design. C-Bot is an RDA that uses a touch sensor to generate a control signal for the transistor relay drive circuit. The 1-kilo-ohm resistor electrically interfaces with the RCX or Scout P-Brick input. The embedded code in the P-Brick moves the conveyor for 30 seconds, stops, and then reverses direction for the same duration. After 30 seconds, the conveyor stops.

Figure 4-2a shows a sketch of C-Bot with the POC design challenge and solution. Figure 4-2b shows the built robot that resulted from the sketch. As shown, the hand sketch is a good tool for capturing requirements for developing POC Mindstorms robots. The POC design challenge statement is important because it explains the problem in a simple single-line sentence. The possible design solution can be explained using the same approach of a block diagram, as shown in this example. The system block diagram shown

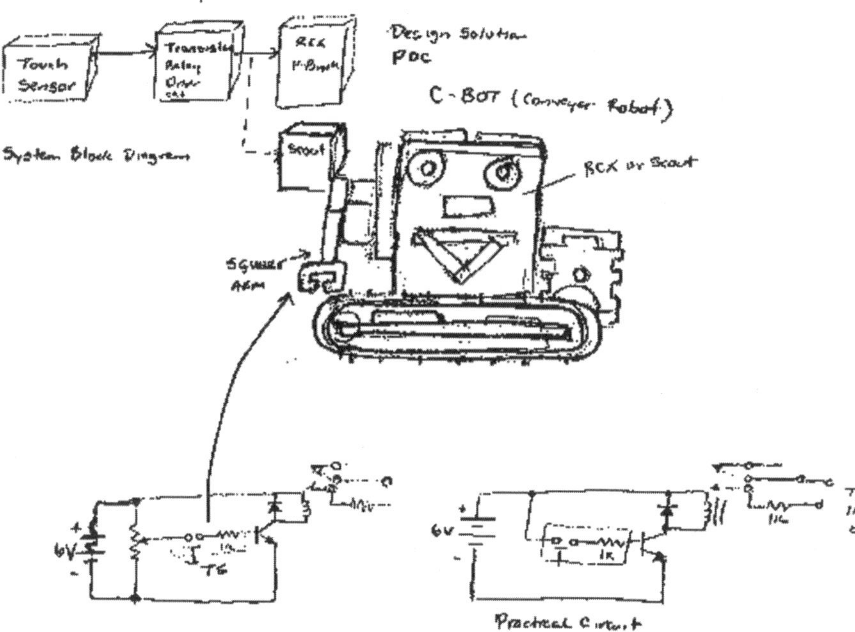

Figure 4-2a
A simple hand sketch can be used to solve POC design challenges. C-Bot was created to solve how to use a transistor relay circuit controlled by a touch sensor design challenge.

Figure 4-2b
The LEGO Mindstorms version of C-Bot

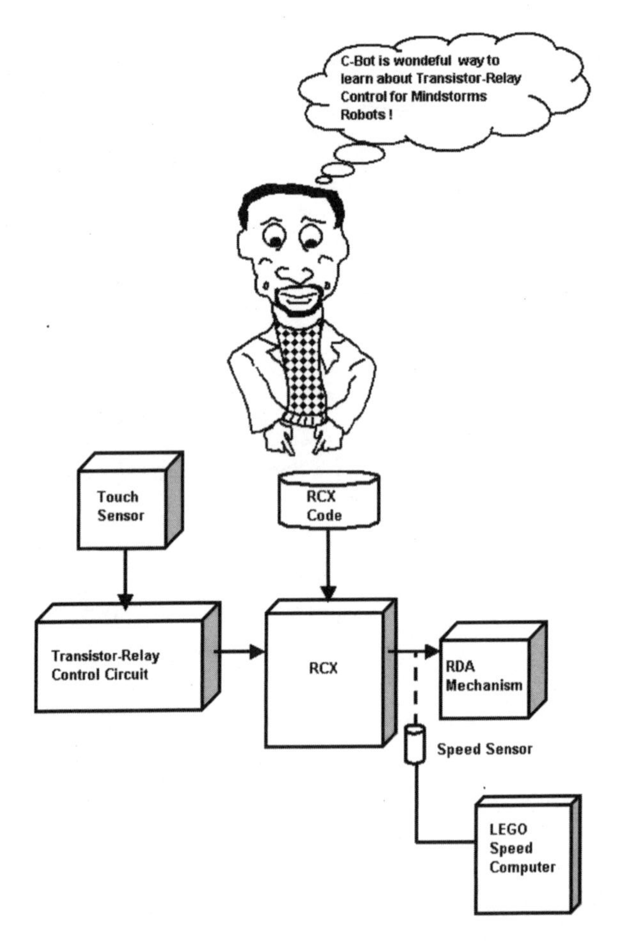

Figure 4-3
Mr.Don discusses the system block diagram for C-Bot.

in Figure 4-3 is also important because it shows the preliminary hardware architecture for the interface circuit and P-Brick.

C-Bot's function is to control a small conveyor belt upon receiving an input request by squeezing its hand. A control signal is produced when the transistor relay driver circuit enables base current to flow through the touch sensor. The collector-emitter junction current flows through the electromechanical relay, closing its switching contacts. The RCX code software embedded within C-Bot drives the conveyor forward. If the robot's hand is released, the conveyor reverses direction and remains in motion for about 20 seconds. Figure 4-4 shows the state machine for C-Bot's operation. The following lab projects explore using the transistor relay driver to control C-Bot. A VBA *Graphical User Interface* (GUI) will be used to log the switching circuit logic state of the relay contacts.

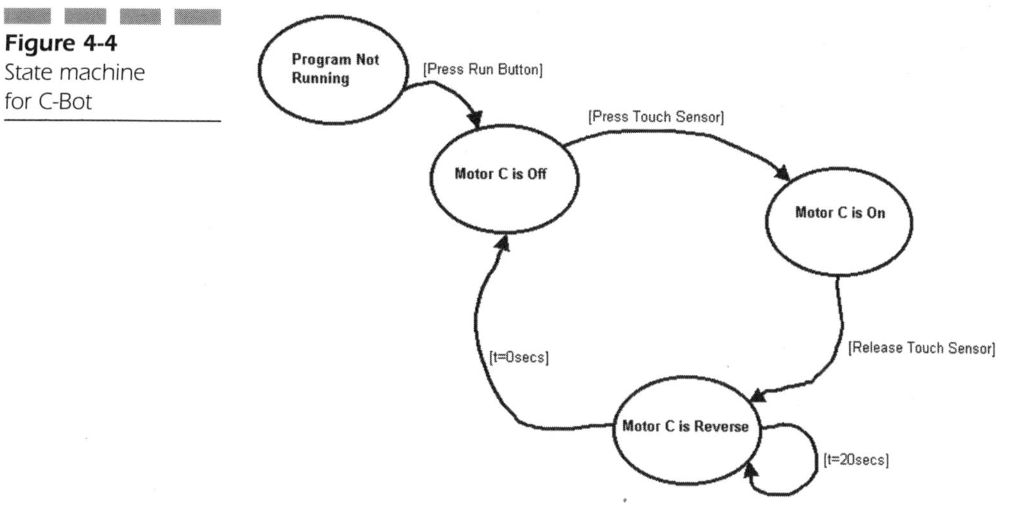

Figure 4-4 State machine for C-Bot

A Simple Controller for C-Bot

The objective of this lab project is to evaluate a simple controller built from a transistor relay driver used to operate C-Bot. The RCX code will use the state machine shown in Figure 4-4 for reading the appropriate digital logic signal from the transistor relay driver circuit.

Bill of Materials (BOM)

- One 10-kilo-ohm potentiometer with knob
- One 1-kilo-ohm resistor
- One 2N2222 NPN transistor or equivalent
- Two modified LEGO electric wires
- One 6- to 9-volt DC electromechanical relay
- One 1N4818 silicon diode or equivalent
- One LEGO Speed Computer (optional)
- Scout P-Brick
- RCX P-Brick
- Touch sensor
- *Robotics Invention System* (RIS) construction set
- Radio Shack Electronic Learning Lab breadboarding system or equivalent

- 6-volt lantern battery or DC power supply
- *Digital Multimeter* (DMM)
- Wires

Transistor Relay Driver Lab Procedure

1. Build C-Bot using Figures 4-5 through 4-8 as a guideline.
2. Build C-Bot's state machine shown in Figure 4-4 using RCX code. The completed code is shown in Figure 4-9.
3. Download the code to program location 1.
4. Modify another LEGO electric wire using the instructions from Chapter 3.
5. Build the transistor relay driver circuit shown in Figure 4-1 using the Electronic Learning Lab breadboarding system or equivalent.

Figure 4-5
Top view of C-Bot. Note the modified LEGO electric wires.

Figure 4-6
Back view of C-Bot. The 9×5 stud plate provides good structural support for C-Bot's legs and arm.

Figure 4-7
Side view of C-Bot. The legs are supported by 2×2 stud plates, bricks, and right-angle stud brackets. The conveyor is spaced where there is not interference with C-Bot's leg supports, enabling the motor drive to rotate freely.

Figure 4-8
The completed Mindstorms C-Bot is ready for electronic switching interface action.

Figure 4-9
RCX code for C-Bot built from the state machine in Figure 4-4

6. Insert the two modified LEGO electric wires into the specific locations shown in Figure 4-10.
7. Using the connector end of the modified wire coming from the 1-kilo-ohm relay contact circuit, attach it to input 1 of C-Bot. (See Figure 4-10.)
8. The other connector end will go to the touch sensor of C-Bot's squeeze arm.

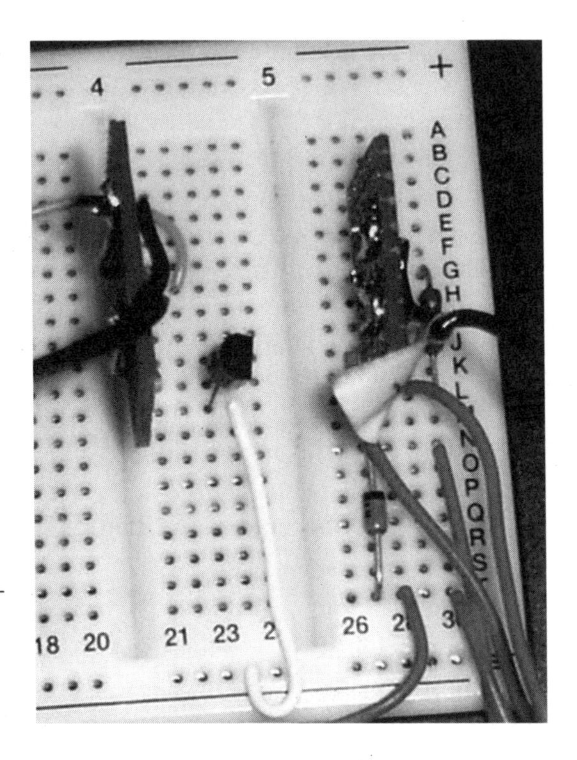

Figure 4-10
Location for the modified LEGO electric wires. The left side of the breadboard shows the modified LEGO electric wire touch sensor tied to the base circuit of the 2N2222 NPN transistor. The right side of the breadboard shows the modified LEGO electric wire 1-kilo-ohm resistor physical connection.

Circuit Calibration Procedure

Once C-Bot is wired to the electronics controller, the system is ready to be calibrated.

9. Turn on C-Bot.
10. Press View once. A binary logic value of 1 should be shown in the *Liquid Crystal Display* (LCD).
11. Rotate the 10-kilo-ohm potentiometer's knob fully counterclockwise. Turn on the 6-volt DC power supply or lantern battery.
12. Rotate the 10-kilo-ohm potentiometer's knob clockwise.
13. The electromechanical relay's contact should click once. If it does not, check the wiring and repeat Steps 12 through 13 and check the code and proper program location.
14. Run C-Bot's program stored in location 1 of the RCX P-Brick. The conveyor should be running.

15. Squeeze and release the robot's hand. The conveyor should run for 20 seconds in the opposite direction and then stop. Subsequent squeeze and release inputs from C-Bot's hands will always reverse the direction of conveyor from its previous forward motion.
16. If the LEGO Speed Computer is available and mounted on the 9-volt motor (as shown in Figure 4-8), measure the velocity of the conveyor. The speed measured from my machine was 80 *revolutions per minute* (rpm).
17. Turn the 6-volt DC power supply off. The C-Bot's conveyor should run continuously in one direction.
18. Squeeze C-Bot's hand. The conveyor should not be running.

DC Measurements Lab Procedure

In the next steps, take some DC measurements from the transistor relay driver circuit:

1. With the circuit wired, measure and record the transistor's biasing voltage, V_B, using a DMM, as shown in Figure 4-11. The value should be approximately 0.87 volts. Use Table 4-1 to record the value.

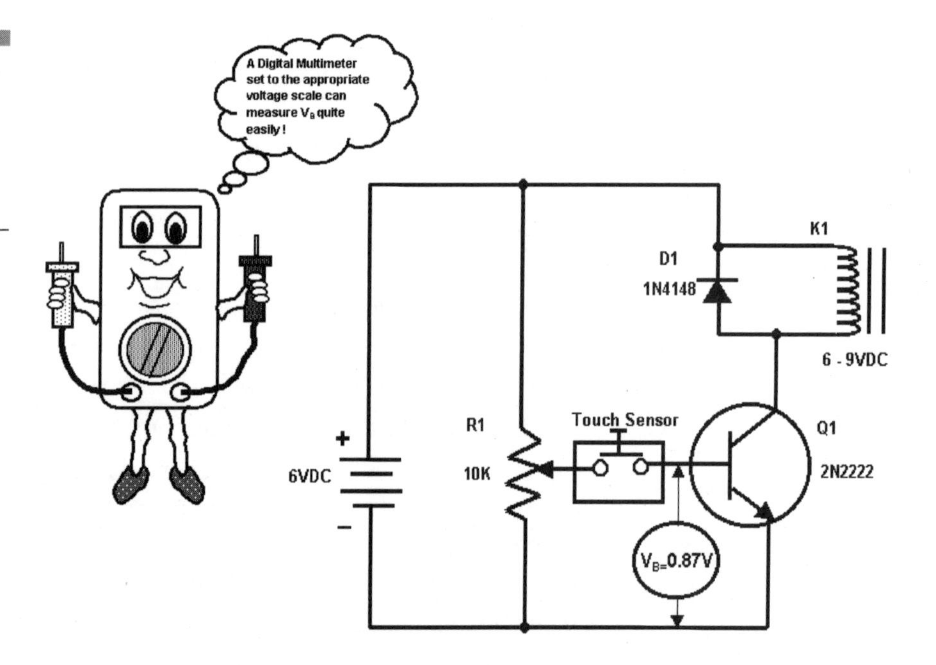

Figure 4-11 Multiman discusses setting the DMM to measure the V_B of the transistor relay input circuit.

Table 4-1

Data table for recording transistor relay input circuit measurements

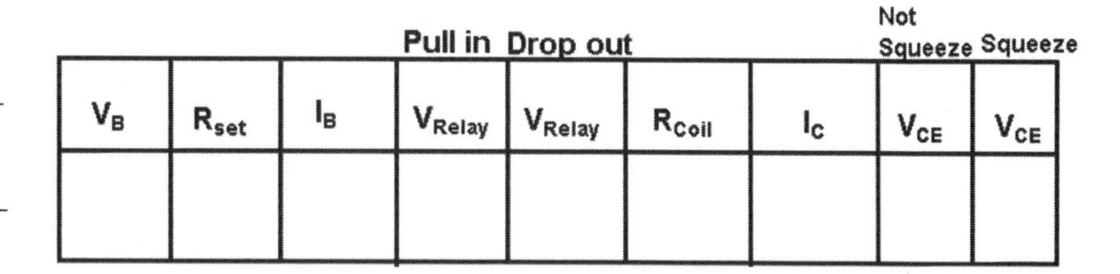

			Pull in	Drop out			Not Squeeze	Squeeze
V_B	R_{set}	I_B	V_{Relay}	V_{Relay}	R_{Coil}	I_C	V_{CE}	V_{CE}

2. Turn off the DC supply to the circuit. Measure the resistance setting of the 10-kilo-ohm potentiometer using an ohmmeter. Record this value onto Table 4-1.
3. Turn on the circuit and measure the voltage across the electromechanical relay's coil onto Table 4-1.
4. Slowly turn the 10-kilo-ohm potentiometer's knob to the left until you hear a small clicking sound of the contacts opening. Record the voltage reading off the DMM onto Table 4-1.
5. Slowly turn the 10-kilo-ohm potentiometer's knob to the right until you hear a small clicking sound of the contacts closing. This should be the value recorded in Step 2.
6. Turn off the power and remove a wire from the electromechanical relay's coil. Measure the resistance of the component and record the reading onto the data table. After taking the measurement, reconnect the wire to the coil.
7. Break the base circuit of the 2N2222 transistor and place a milliammeter in series with the open circuit. Figure 4-12 shows the

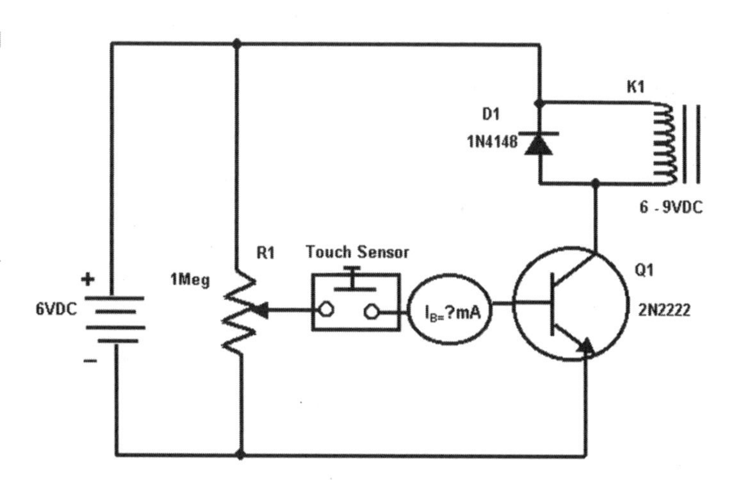

Figure 4-12
Insert a milliammeter between the 10-kilo-ohm potentiometer and base circuit of the 2N2222 NPN transistor.

principle of measuring base circuit current using a milliammeter. Set the DMM for a current measurement.

8. Turn on the circuit power and measure and record the current onto Table 4-1.
9. Squeeze C-Bot's hand. The current value will go to zero.
10. Turn off the power and replace the base circuit wires of the 2N2222 NPN transistor.
11. Break the electromechanical relay circuit and insert the milliammeter in series. Record the current value onto the data table.
12. Squeeze C-Bot's hand. The current value will go to zero.
13. Turn off the power and replace the collector circuit wires of the 2N2222 NPN transistor.
14. Turn the DC power supply on and measure the collector-emitter voltage (V_{CE}) of the transistor relay circuit and record the value onto the data table.
15. Squeeze C-Bot's hand. V_{CE} will increase by a factor of 12.
16. Turn off C-Bot and the transistor relay circuit.

My data is shown in Table 4-2. Note that the values shown depend on factors such as battery usage versus DC power supply, the type DMM used to capture data, and the ambient temperature of the room. A percent error of +/− 10 percent is acceptable

C-Bot Analysis

Test procedures 1 through 18 of the Transistor Relay Driver and Calibration Lab procedures (pages 120-124) enable hobbyists and experimentalists to build and test an RDA capable of operating a small conveyor system. The steps outlined were designed with the robot's state machine being validated

Table 4-2

My data recorded from the transistor relay input circuit

			Pull in	Drop out			Not Squeeze	Squeeze
V_B	R_{set}	I_B	V_{Relay}	V_{Relay}	R_{Coil}	I_C	V_{CE}	V_{CE}
0.87 V	2.51K	273uA	3.29 V	1.25 V	130.6 K	38.4 mA	0.47 V	5.78 V

by following the transition and state of the paper behavioral model. The squeeze hand provides interaction between the hobbyist and the experimentalist, maintaining the RDA concept discussed in Chapter 2.

The DC measurements lab provided a hands-on experience of capturing and recording important transistor parameters. In doing a manual circuit analysis or running a computer simulation using either Electronics Workbench or OrCad *Electronics Design Automation* (EDA) tools, the results generated can be used as baseline data for performing a physical test and determining the measurement of the breadboard circuit. Figure 4-13 shows the simulation results generated using Electronics Workbench (version 4.0) of the transistor relay driver circuit. The EDA tool is not meant to replace the traditional paper-and-pencil calculation; rather, it complements it. Also, the EDA tool's simulation results can be used to troubleshoot when the actual circuit is being breadboarded. Physical measurements should never be replaced with simulation studies because certain design assumptions have been used to model the behavior of the active or passive components within the EDA tool. The RDA concept discussed in Chapter 2 in the section "What Is an RDA?" assists with Steps 12 and 15 of the DC measurement lab.

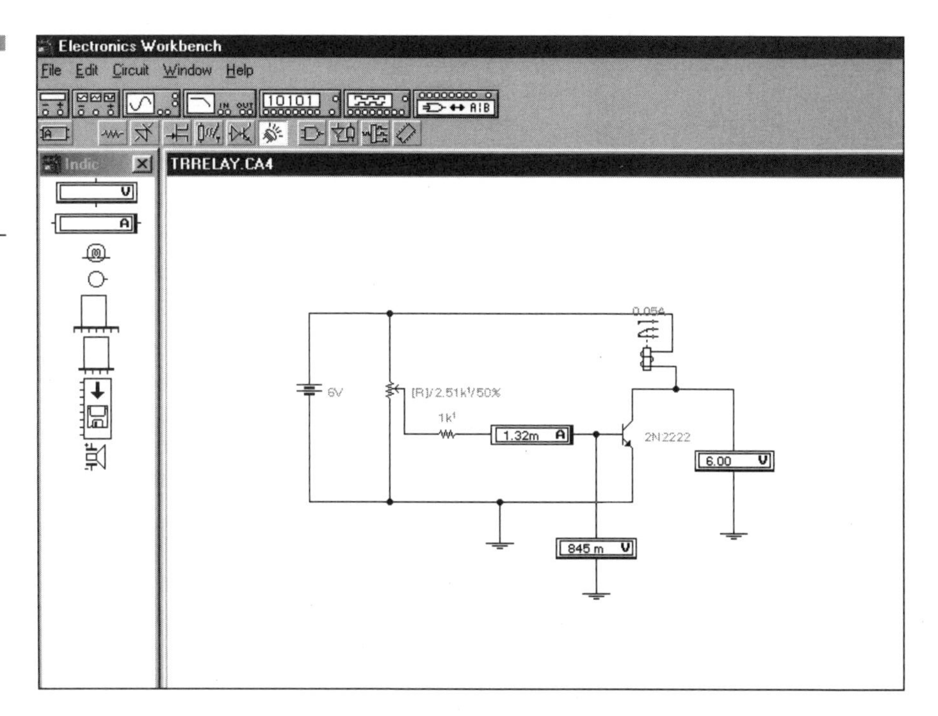

Figure 4-13 EDA example: Electronics Workbench simulation results of transistor relay circuit

Design Challenge

In the previous chapters, the Scout-equivalent robot was always demonstrated with a detailed test procedure. In this section, building and testing a Scout-equivalent C-Bot is left as a design challenge to the reader. Steps 1 through 18 in the test procedure, shown on pages 120-124, can be used as a guide with the exception of programming in RCX code. Use the appropriate *Stand-Alone Code* (SAC) elements to accomplish the same behavioral task programmed for the RCX C-Bot. The Scout tool makes an excellent software controller when performing the DC measurements lab for the circuit. Experiment, investigate, and by all means have fun!

Transistor Relay Monitor: VBA–C-Bot Application

Having a robot that is interactive provides a convenient application for exploring further developments in VBA programming. The basic circuit that was investigated in the previous lab project enables the hobbyist or experimentalist to develop a VBA applet that monitors the status of the relay contacts of the transistorized electromechanical switch for controlling C-Bot. The project uses the software development practice of reusable code for expediting GUI programming. Also, this software tool is very handy in checking out any transistor regardless if it's a bipolar or PMOSFET (which will be discussed in the following section of this chapter).

The basic operation behind this software control GUI is to read the state of the transistor relay circuit's contacts and display them either as open or close in an ActiveX control text box. Then the logged data can be printed on a spreadsheet and analyzed using Excel's stat function library. The code is the same as what was developed in Chapter 3 except for a few VBA programming enhancement features.

Figure 4-14 shows the GUI for the Transistor Relay Monitor tool. When pressed, the COMM ON/OFF buttons will show ON or OFF using the Label ActiveX control. Also, instead of showing a 0 or 1 inside of the text box, Close or Open will be displayed. The contact status data can also be logged onto the spreadsheet using an array, as discussed in Chapter 3. The key concept to note is the continuous improvement activity of product development. Products have development cycles for improving features based on customer wants. The design team must meet the functional objectives derived

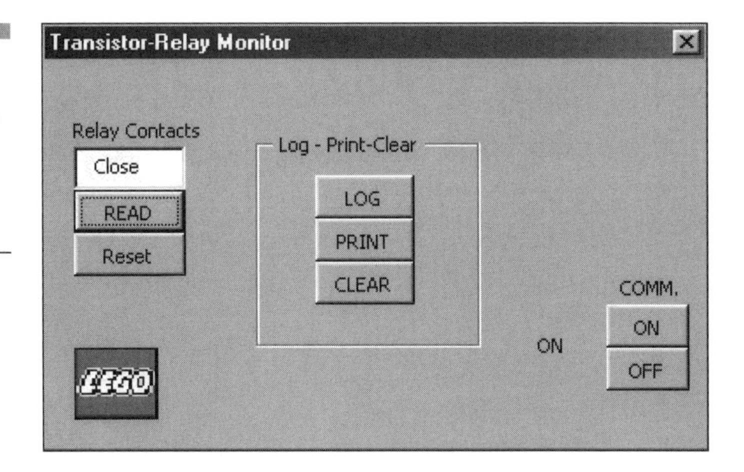

Figure 4-14 Transistor Relay VBA Monitor tool used to collect switched input data from the electronic driver circuit

from customer wants. The requirements specification is then written to capture the objectives of the product.

In this project, the need to display contact closure information is required in order to build a tool that can show the input event that C-Bot receives from the transistor relay circuit. The software GUI along with the breadboard electronics hardware provide a good development base for the POC study that can help enhance an understanding of the operation of the tool and the integration between mechanical, electronics, and software interface controls for Mindstorms robotics.

Transistor Relay Monitor Lab Procedure

The objective of this lab is to illustrate further practices in reusing code and to improve upon an existing VBA GUI for monitoring and recording contact closure of the transistor relay input circuit built for controlling C-Bot.

BOM

- One C-Bot built from the previous lab project
- One transistor relay input control circuit
- One Switch Monitor tool developed from Chapter 3

The first step in reusing code is to find software that meets most of the requirements of the given project. The code is then imported into the new design for modification and evaluation.

Procedure for Developing a New GUI Using an Existing UserForm

The following steps outline the procedure for building a new GUI using an existing UserForm design:

1. Open the spreadsheet with the appropriate UserForm.
2. Go to the *Visual Basic Editor* (VBE) by clicking on the icon in the toolbar.
3. Under the VBA Project window find the target UserForm.
4. Select and right-click it.
5. Under the pull-down menu select Export File . . . .
6. Put the UserForm (name.frm) into the appropriate directory.
7. Close out of the worksheet and open a new .xls file.
8. Go to the VBE.
9. Select File and click Import File . . . .
10. Go to the directory where the .frm will be placed.
11. Double-click the UserForm. A folder named Forms should be visible within the VBA project.
12. Open the folder and the UserForm should be there.
13. Change the name of the UserForm to TRMON under the Properties window.
14. Highlight the name TRMON and right-click on View Code.
15. The VBA code for this applet should be displayed in the VBE.

The method cuts down on starting from scratch and expedites GUI and software development tremendously. The rest of the lab project focuses on modifying the GUI and making minor software changes to the monitor code.

16. Change the UserForm to look like Figure 4-14. The COMM ON/OFF status was built using the Label ActiveX control.
17. The following code changes are required for the tool to work properly:

```
Dim Status As Integer
Dim Reading As String
Dim MyReading(30) As String
(Under Private Sub Read_Click(): Status = PBrickCtrl.Poll)
If Status = 1 Then
TRMON.TextBox1.Value = "Close"
Else
```

```
TRMON.TextBox1.Value = "Open"
Reading = TRMON.TextBox1.Value
End If
(Under Private Sub CommOn_Click())
TRMON.COMMSTATUS.Caption = "ON"
(Under Private Sub CommOff_Click())
TRMON.COMMSTATUS.Caption = "OFF"
```

With the following changes made to the software, the spreadsheet layout can be built using Figure 4-15 as a guide. The code for the TR Monitor CommandButton is as follows:

```
Private Sub CommandButton3_Click()
TRMON.Show
End Sub
```

Make sure design mode is active when building the code. With the software done, exit out of design mode to complete the spreadsheet layout. The code/VBA tool is now ready to be validated.

Validating the Code/VBA Tool

1. Click on the TR Monitor Command button.
2. The Transistor Monitor tool should be displayed on the spreadsheet.
3. Turn on C-Bot and the transistor relay input circuit.
4. Click the ON button under the COMM section of the tool.
5. The ON status label should be displayed on the UserForm.
6. Run C-Bot's RCX code in program location 1.

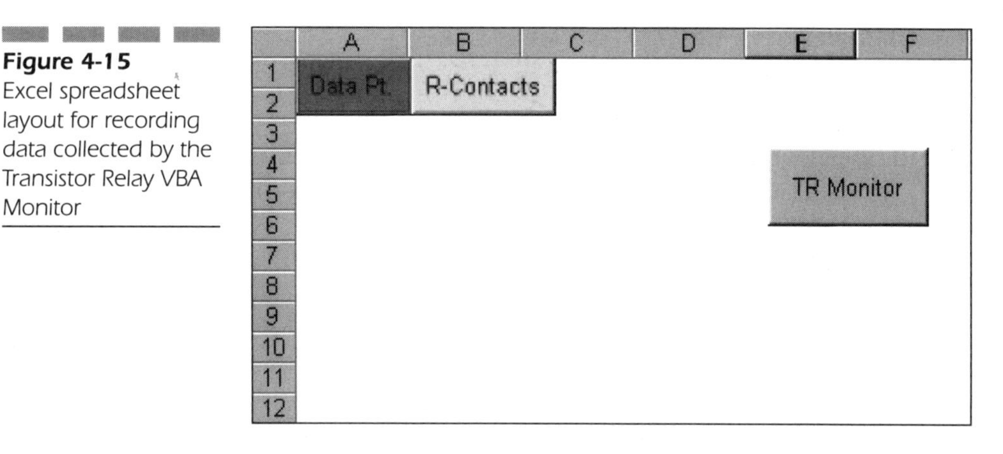

Figure 4-15 Excel spreadsheet layout for recording data collected by the Transistor Relay VBA Monitor

7. Press Read under the Reading Contacts heading on the tool. Close should be displayed in the text box.
8. Squeeze C-Bot's hand and click Read. Open should be displayed in the text box.
9. Repeat Step 7 and note that the Close message should be displayed in the text box.
10. Close out of the Transistor Relay Monitor tool. Power off the electronic switching input circuit and C-Bot.

VBA Analysis

This lab project illustrates the concept of code reuse. An existing application was modified to meet the new requirement for the robot. By using the Export/Import feature within Excel VBA, the Switch Monitor application used in Chapter 3 could be easily ported into the spreadsheet VB environment. The code was modified to meet the new requirements and tested. This VBA application along with C-Bot shows a semiautomated process for testing NPN transistors. The same tool can be used to test PNP components by simply changing the power supply leads to accommodate the semiconductor's lead arrangement and current flow requirements. The next section discusses how a PMOSFET can be used to switch an electromechanical relay used for C-Bot interface control.

A PMOSFET Relay Driver Input Circuit

Electronic switching of the Scout or RCX P-Brick input can be accomplished using a PMOSFET relay driver circuit. PMOSFETs are perhaps the most popular transistors used today; they draw very little input current, are easy to make (require few semiconductor ingredients), can be made extremely small, and consume very little power. In terms of applications, PMOSFETs are used in ultrahigh input impedance amplifier circuits, voltage-controlled resistor circuits, and switching circuits, and are found with large-scale integrated digital *Integrated Circuits* (ICs).

No electrical bias or signal is applied to the gate in PMOSFET. As a result, no current can flow because there will always be a blocking *positive-negative* (pn) junction. When the gate is forward biased with respect to the source and has an applied drain-source voltage, the free-hole carriers in the p-epitaxial layer are repelled away from the gate area creating a channel, which enables electrons to flow from the drain to the source. The device is in the normally off mode in this case—that is, the switch blocks the current until it receives a signal to turn on. The opposite is depletion mode, which is a normally on device. Therefore, the PMOSFET is a voltage control device instead of a current control component such as the bipolar transistor.

By having the ability to switch loads with no current flowing through the gate circuit, how can the PMOSFET be used as the input switching circuit? The following lab project demonstrates how a PMOSFET along with an electromechanical relay can be used to control C-Bot.

PMOSFET Controller for C-Bot

This lab demonstrates the switching capability of a PMOSFET relay driver that can be used to control C-Bot. The circuit topology (wiring configuration) is similar to the transistor circuit discussed earlier. Instead of having base, emitter, and collector circuits, gate, drain, and source are the equivalent leads for the PMOSFET.

BOM

- One 1-mega-ohm potentiometer with knob
- One touch sensor
- One IRF630A N-Channel PMOSFET or equivalent
- Two modified LEGO electric wires
- One 6- to 9-volt DC electromechanical relay
- One 1N4818 silicon diode or equivalent
- 6-volt lantern battery or DC power supply
- Radio Shack Electronic Learning Lab breadboarding system or equivalent

PMOSFET Relay Input Circuit Lab Procedure

1. Build the PMOSFET relay driver circuit shown in Figure 4-16 using the Electronic Leaning Lab breadboarding system or equivalent.
2. Insert the two modified LEGO electric wires into the specified locations shown in Figure 4-17.
3. Turn on C-Bot. Push the View button once.
4. Turn the PMOSFET relay driver circuit.

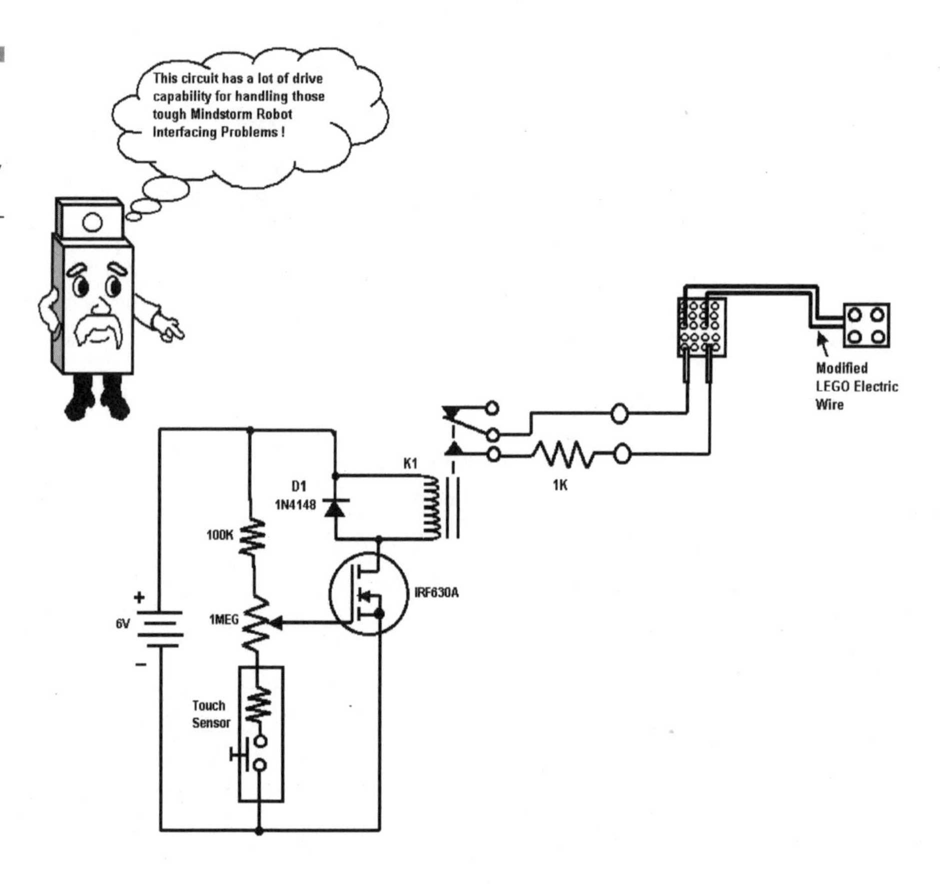

Figure 4-16 A PMOSFET input circuit for controlling C-Bot as explained by Fritz FET

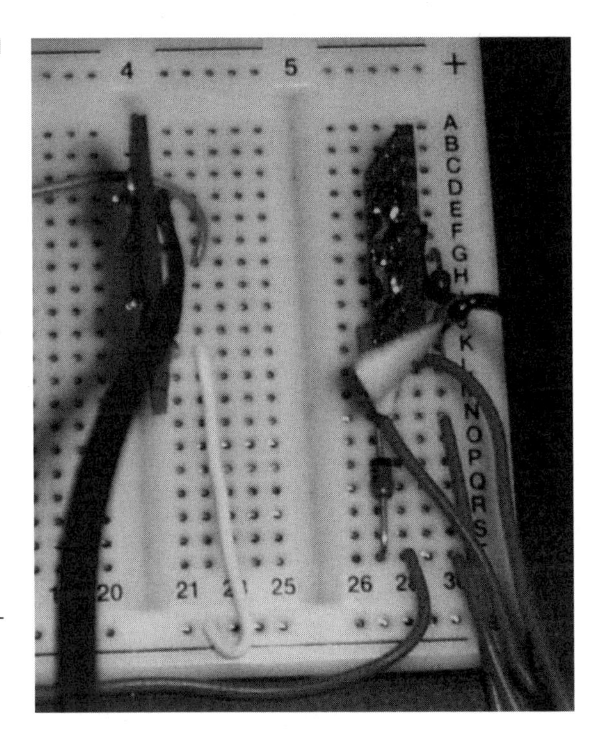

Figure 4-17 The location for the modified LEGO electric wires. The left side of the breadboard shows the modified LEGO electric wire touch sensor tied to the 1-meg potentiometer for gate control of the IFR630 N-Channel PMOSFET. The right side of the breadboard shows the modified LEGO electric wire 1-kilo-ohm resistor physical connection.

Calibration of the PMOSFET Input Driver Circuit

The following four-step procedure calibrates the FET relay circuit for the proper control operation of C-Bot.

1. Adjust the 1-mega-ohm potentiometer until the LCD of the RCX P-Brick displays the logic 0.
2. Squeeze and release C-Bot's hand and observe the LCD of the RCX P-Brick. A logic 1 should be present on the mini display.
3. If a logic 1 is not displayed in the LCD, power off the circuit and check the wiring. If the wiring is correct, the electromechanical relay clicks upon squeezing C-Bot's hand.

4. Squeeze and release C-Bot's hand several times and notice the logic transition between logic 0 and 1. The relay contacts should click on and off during the squeeze and release of the robot's hand.

This completes the calibration of the PMOSFET input driver circuit. The remainder of the test procedure validates the circuit's interface control operation of C-Bot.

1. Start C-Bot's code stored in program location 1 by pressing Run on the RCX P-Brick.
2. The conveyor should start running immediately. Note the direction of the conveyor. Is it forward or reverse? After 20 seconds, it should stop.
3. Squeeze and release C-Bot's hand and the conveyor should move in the opposite direction of test procedure 2.
4. Squeeze and release C-Bot's hand and the conveyor should move in the opposite direction of test procedure 3.
5. Adjust the 1-mega-ohm potentiometer until the LCD displays a logic 1.
6. Squeeze and release C-Bot's hand. The conveyor should not move.
7. Measure the V_{GATE} using a DMM set for VDC and the appropriate volt scale range. Use Figure 4-18 as a guide for measuring V_{GATE}. Record the

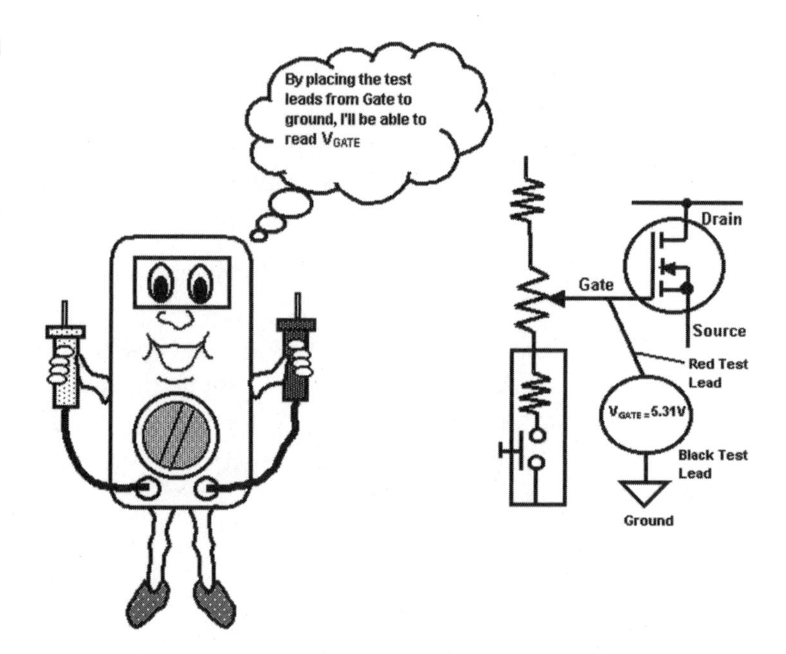

Figure 4-18 Multiman explains basic measurement technique for obtaining V_{GATE}.

value for V_{GATE}: ______ volts. Note that upon attaching the DMM test leads, C-Bot's conveyor starts moving.

8. Squeeze and hold C-Bot's hand and measure V_{GATE} with the DMM. Record the value for V_{GATE}: ______ volts. Did the reading go up or down?
9. Turn C-Bot's program off by pressing the Run button. The conveyor should be running continuously. Note that squeezing C-Bot's hand has no effect on the conveyor's operation.
10. Press View once. Adjust the 1-mega-ohm potentiometer until the LCD shows a logic 0. The conveyor should move in the opposite direction and will stop after 20 seconds.
11. Squeeze and release C-Bot's hand. The conveyor should move in the opposite direction. Note that the DMM reading should have jumped from the recorded value in Step 8.
12. Repeat Step 11 several times noticing the instantaneous increase of V_{GATE}.
13. Measure the $V_{DRAIN\text{-}SOURCE}$ voltage using the DMM and Figure 4-19 as a guide. Record the value for $V_{DRAIN\text{-}SOURCE}$: ______volts.
14. Squeeze and release C-Bot's hand. The conveyor will start to move. Record the value for $V_{DRAIN\text{-}SOURCE}$: ______volts.
15. Turn off the power to the circuit and C-Bot.

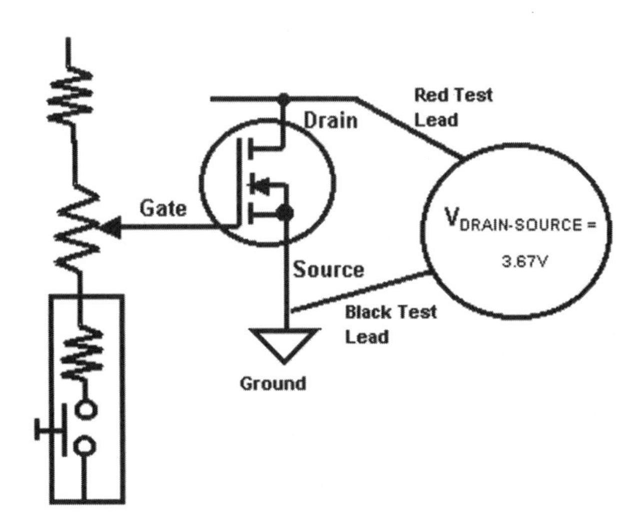

Figure 4-19 How to measure $V_{DRAIN\text{-}SOURCE}$ with a DMM

PMOSFET Relay Input Circuit Lab Analysis

This lab project explores the switching capabilities of the PMOSFET used to control an electromechanical relay. The relay's contacts provide the 1-kilo-ohm resistor required by the input circuit of the P-Brick for proper robotics operation. The PMOSFET required 1.61 volts (5.31[squeeze] through 3.70[no squeeze]) to turn it on. The increase of 5.31 volts when squeezed is due to the intrinsic gate capacitance of the PMOSFET charging. This quick-turn spike can induce noise on electronic devices that are susceptible to *Electromagnetic Interference* (EMI). Therefore, a soft start is more desirable and can be accomplished by adding a resistor between the gate circuit and ground. The 1-mega-ohm potentiometer provides a variable control that enables you to adjust the turn on threshold for V_{GATE}. The drain-to-source voltage ($V_{DRAIN\text{-}SOURCE}$) of 3.67 volts is present due to the impedance of the junction being higher than the relay's resistance. This value exists when the PMOSFET is not turned on; therefore, an open circuit exists at this junction. The voltage drops to 0 volts when the circuit FET is turned on, enabling the electromechanical relay to energize because of the drain-to-source (I_{DS}) current flowing through this part of the circuit.

The circuit performed the same way as the bipolar device; therefore, the PMOSFET is capable of providing an alternative switching circuit for Mindstorms robots. The Transistor Relay Monitor tool can be used to check the switching operation of the FET input circuit. As an exercise for the hobbyist or experimentalist, use the procedure outlined under the section "Validating the Code/VBA Tool" in this chapter. The results should be the same as those received when testing the transistor relay driver circuit discussed in this section. The next section uses the switching topologies of the transistor (bipolar or PMOSFET) discussed to build a wireless switching input circuit for C-Bot.

A Wireless Relay Driver Input Circuit

This lab project is a continuation of the discussion in Chapter 1, "Wireless Basics," on building a wireless controller for the Scout P-Brick. The spin to this P-Brick application is that the RDA will provide a signal to the *Radio Control* (RC) transmitter instead of the hobbyist or experimentalist and the receiver will provide the logic input to the P-Brick. This lab project provides a core for wireless sensing/detection robots because the RC transmitter-receiver blocks can be changed to either send/receive wireless signals for Mindstorms robotic control. The following procedure provides a fundamental method for understanding how a wireless device using a transistor relay driver circuit can control C-Bot.

Wireless Relay Driver Circuit Lab Procedure

The objective of this lab project is to wirelessly control C-Bot using a salvaged RC transmitter-receiver pair. A transistor relay input circuit will be driven by the wireless device to control C-Bot's conveyor upon squeezing its hand. C-Bot's squeeze arm is wired to the transmitter used to transmit the *radio frequency* (RF) signal to the receiver/transistor relay input circuit.

BOM

- One salvaged RC transmitter-receiver pair
- One dual voltage DC power supply
- One 100-ohm resistor
- One 2N3904 NPN transistor or equivalent
- One 1N4818 silicon diode or equivalent
- Two modified LEGO electric wires
- C-Bot
- DMM
- Radio Shack Electronics Learning Lab breadboarding system or equivalent
- One 1-kilo-ohm resistor

The procedure used to have wireless control of C-Bot using a salvaged RC transmitter-receiver pair is as follows:

1. Build the circuit in Figure 4-20 using the Electronics Learning Lab breadboard or equivalent.
2. Attach C-Bot's squeeze arm touch sensor to the transmitter using the modified LEGO electric wires. Use Figure 4-21 as a guide.
3. Attach C-Bot's input 1 to the transistor relay circuit. Use Figure 4-22 as a guide.

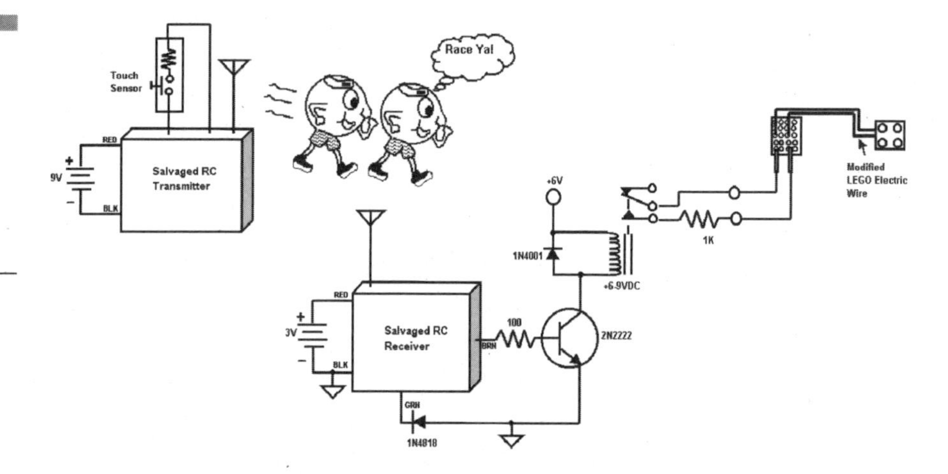

Figure 4-20 Wireless control of C-Bot using a salvaged RC transmitter/receiver, a touch sensor, and a transistor relay input circuit

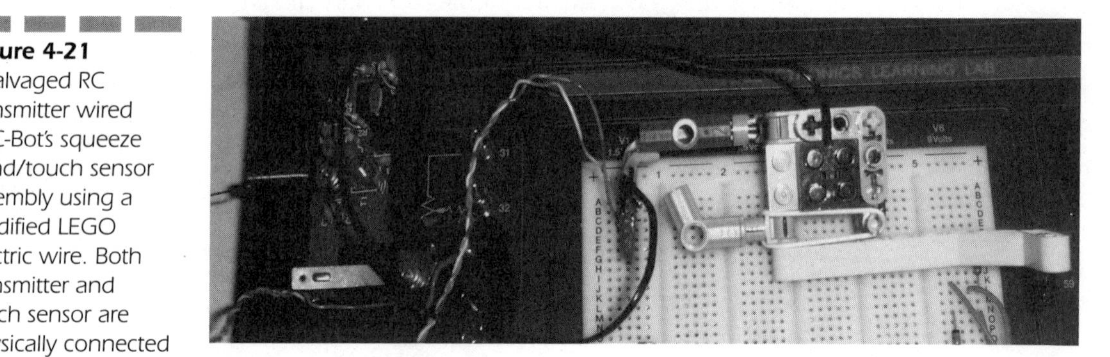

Figure 4-21 A salvaged RC transmitter wired to C-Bot's squeeze hand/touch sensor assembly using a modified LEGO electric wire. Both transmitter and touch sensor are physically connected using a solderless breadboard system.

4. Adjust the dual DC power supply for the +9- and +3-volt DC source using the DMM.
5. Use the +6-volt DC source from the Electronics Learning Lab breadboard to power the transistor relay circuit.
6. Connect the DMM to the base circuit of the 2N3904 or equivalent NPN transistor.
7. Turn on the dual DC power supply and the +6-volt DC source from the breadboard.
8. Squeeze and hold C-Bot's hand. The relay's contact should click. Record the DMM voltage reading (2.19 volts).
9. Release C-Bot's hand. Record the DMM voltage reading (−0.37 volts).

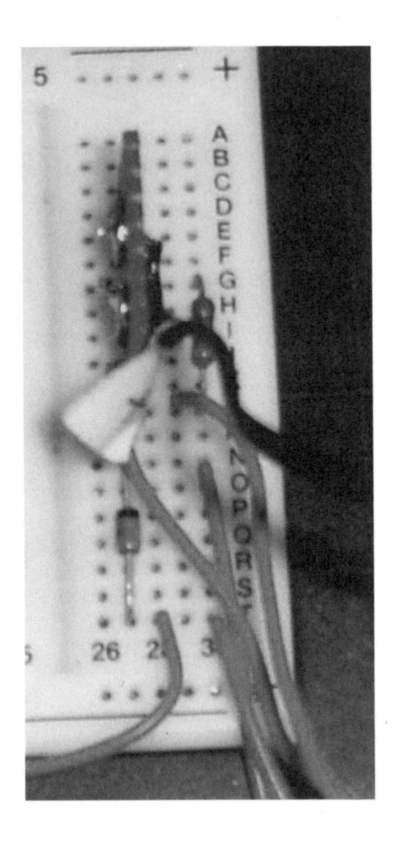

Figure 4-22
Input 1 to C-Bot is made using modified LEGO electric wire, one resistor, and a connection to electromechanical relay via 1N4818 silicon diode.

10. Turn on the RCX P-Brick. Run C-Bot's code located in program location 1.
11. Squeeze and release C-Bot's hand several times noting the conveyor's directional change and the DMM volts reading.

The next lab project outlines a procedure that illustrates the process of using a DMM with a PC interface and a salvaged RC transmitter-receiver pair for detecting the activation of a Mindstorms touch sensor.

Wireless Lab Project Procedure

This lab project demonstrates how a wireless sensor detection system can be built using a DMM with a PC interface, a salvaged RC transmitter-receiver pair, and C-Bot. The record and data view features of the Scope-View PC Interface software will be demonstrated as well within the lab project procedure.

BOM

- One DMM with PC interface software

The procedure for this lab is as follows:

1. Set the data-logging equipment using DMM with a PC interface. See Figure 4-23.
2. Set up the logging session controls as shown in Figure 4-24.
3. Click on the Record button. A dialog box will appear on the screen.
4. Click on the appropriate drive and move the cursor to the filename entry box. Type c_botwir.txt. Hit Enter. The data log control settings dialog box should appear.
5. Click the Scope button. The Scope Plotting window should appear.
6. Turn on the dual DC power supply, C-Bot, and the breadboard +6-volt DC supply.
7. Push Run on C-Bot.
8. Click Start on the Scope Plotting window.

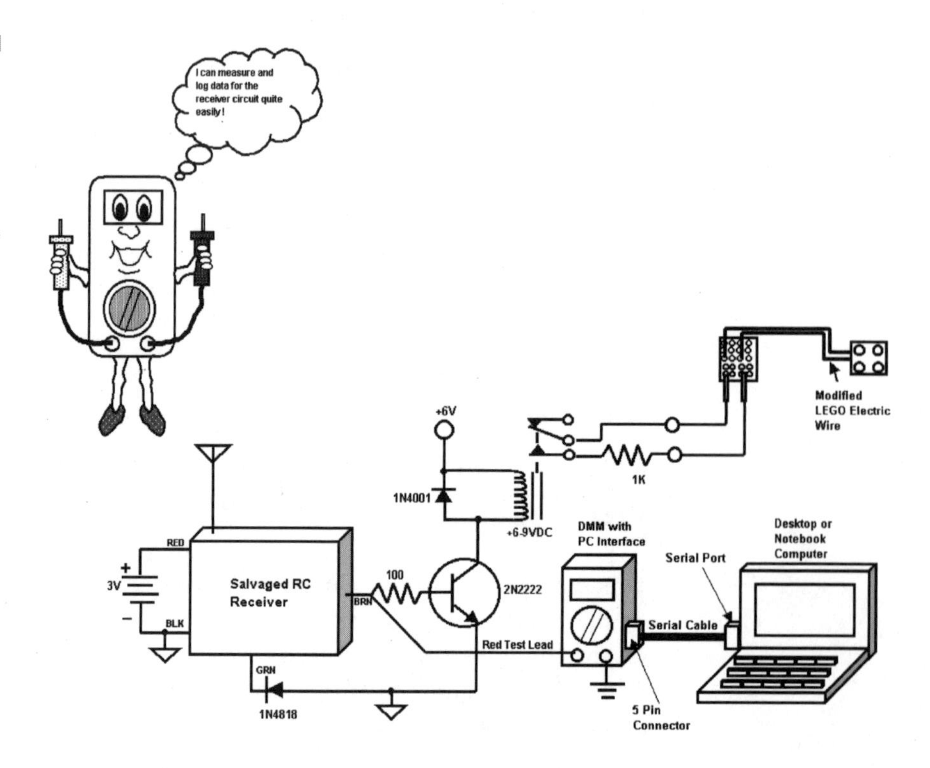

Figure 4-23 Using DMM with a PC interface to obtain voltage data from a Mindstorms RC receiver input circuit

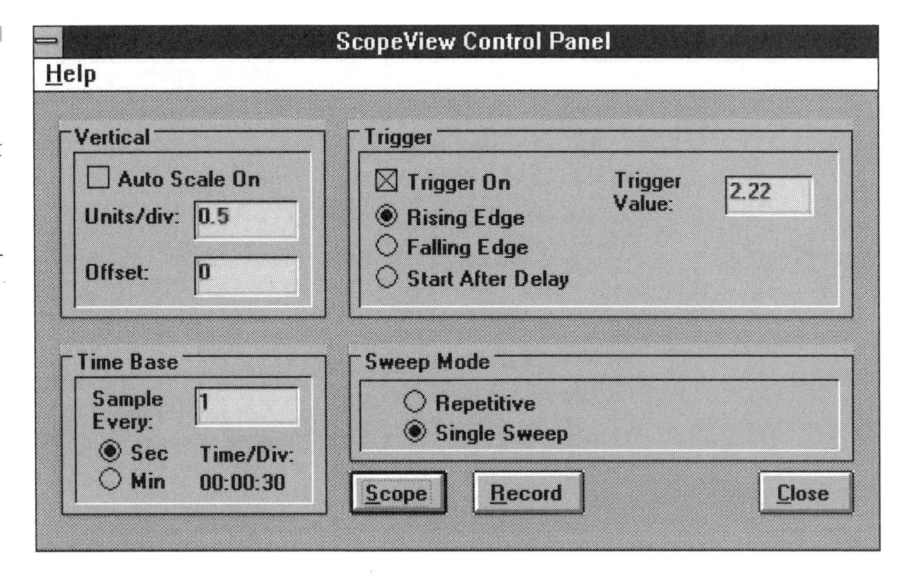

Figure 4-24 ScopeView Control Panel setup for the voltage measurement of the RC receiver input circuit interface to C-Bot

9. Squeeze and hold C-Bot's hand for about 10 seconds. The ScopeView *Software* (SW) should start creating a signal trace on the screen.
10. Release C-Bot's hand for about 10 seconds. Note that the signal falls during this wave cycle.
11. Repeat Steps 9 through 10 for several data-logging iterations. Stop the logging session of ScopeView.
12. Plot the data recorded from the file c_botwir.txt using Excel. Use the Area plot within the Chart Wizard to obtain a good representation of the data.

    Compare the results with the actual waveform displayed in ScopeView.
13. Turn off the dual DC power supply, the breadboard 6-volt DC supply, and C-Bot.

Wireless Lab Analysis

This lab project illustrates how sensor detection can be obtained wirelessly and graphically. The ScopeView software provides a method of detecting C-Bot's squeeze hand touch sensor based on monitoring the base voltage being applied to the RC receiver. The waveform displayed on the screen would be the signature of the sensor being activated. The rising and falling edges of the waveform represent one pulse width of the sensor being activated. Therefore, a simple monitoring system for robotic detection is created using salvaged RC transmitter-receiver electronics and a transistor relay input driver circuit. The POC here is the solution of remote obstacle detection using simple wireless technology. The wireless lab project also demonstrates how C-Bot's squeeze arm provided a control signal for the transmitter, illustrating its capability to be an RDA in the lab as well.

In the next section, C-Bot will be rewired so that the touch sensor connects directly to input 1. A two-transistor audio oscillator will be wired to output A and will sound after 20 seconds of conveyor rotation.

Electronic Output Switching

Up to this point, the lab projects have been based on electronic input switching. In Chapter 3, electromechanical controls were used to logically switch the inputs of RCX or Scout P-Brick. The previous sections of this chapter looked at how bipolar transistors and PMOSFETs enhance the electromechanical relay switching capability. As stated by the title of this book, output interface switching should be explained to round off the fundamental concept of electronic switching for Mindstorms P-Bricks. The following discussion illustrates the ease of output interfacing for electronic switching using a simple audio oscillator for tone generation.

Output Switching Control of an Audible Tone Generator

Figure 4-25 shows the system block diagram for controlling a tone generator using an audio oscillator. The 7805 Linear Regulator circuit decreases the voltage produced by the output of the Mindstorms P-Brick. The voltage regulator has enough current source capability to drive a small audio oscillator—in this case, a two-transistor circuit. This output switching of the P-Brick opens a wealth of unique audible indicator-warning projects for Mindstorms robots. The oscillator's output frequency can be changed based on the component selection of the RC pair. A 1-mega-ohm potentiometer provides frequency control via rotating counterclockwise or clockwise. The tone can be selected to provide a certain warning or alert status to the operator/designer of the robot. Therefore, the Mindstorms robot can provide an audible monitoring feedback system alerting the condition of the smart machine. An interesting feature about this tone generator is that sound can

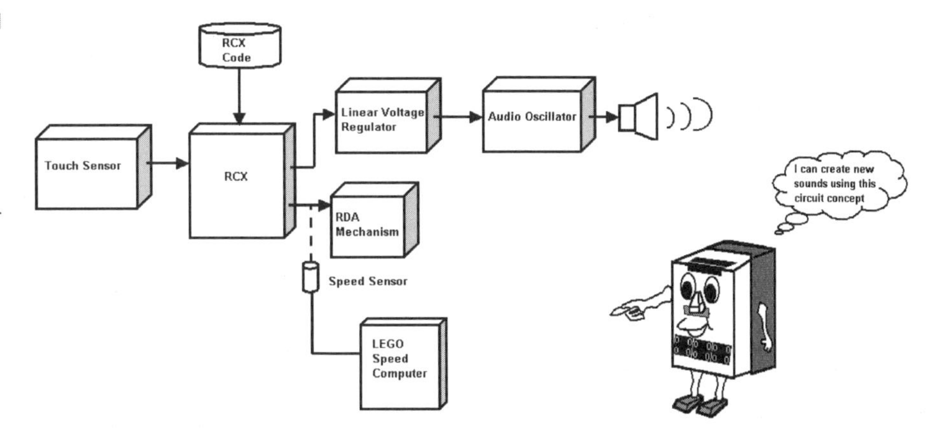

Figure 4-25 RCX discusses the system block diagram for the C-Bot-controlled audio tone generator.

be programmed for input detection from a touch or light sensor. Upon detection of an object or light, the tone generator can be programmed to sound off. The following lab project illustrates the basic concepts of output switching using an electronic interface and providing an audible tone to enhance the presence of an alarm state using C-Bot.

An Audible Tone Generator Lab Project Procedure

This lab project investigates the electronic output switching for C-Bot by using a 7805 linear voltage regulator to control a simple two-transistor audio oscillator. C-Bot's embedded RCX code will be modified to provide a conveyor stop alarm upon the machine stopping after a 20-second rotation cycle.

BOM

- One 7805 linear voltage regulator IC
- One 1-mega-ohm potentiometer
- One 8-ohm speaker or equivalent
- One 100-kilo-ohm resistor
- One 0.01 µF capacitor
- Two 0.1 µF capacitors
- One 2N3906 PNP transistor or equivalent

- One 2N3904 NPN transistor or equivalent
- C-Bot
- One touch sensor with the LEGO electric wire
- One modified LEGO electric wire

The procedure used for this lab project is as follows:

1. Wire the audio oscillator, as shown in Figure 4-26.
2. Attach a modified LEGO electric wire to output A of C-Bot. Make sure the wire is oriented as shown in Figure 4-27 to ensure proper operation.
3. Replace the modified electric wire on C-Bot's input 1 to the squeeze arm touch sensor with a standard wire.
4. Turn on C-Bot. Press and hold the forward A button on the remote.
5. Did the circuit produce a tone? If no, recheck the wiring and repeat Step 4.

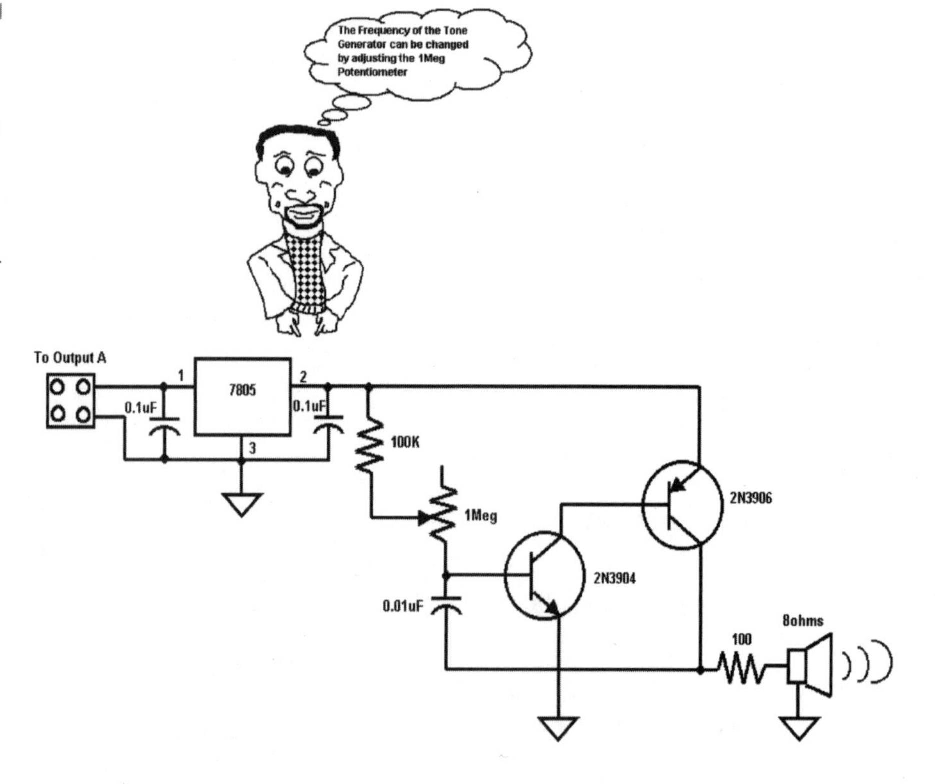

Figure 4-26 Mr. Don discusses how to control the frequency of a simple two-transistor oscillator using 1-mega-ohm potentiometer.

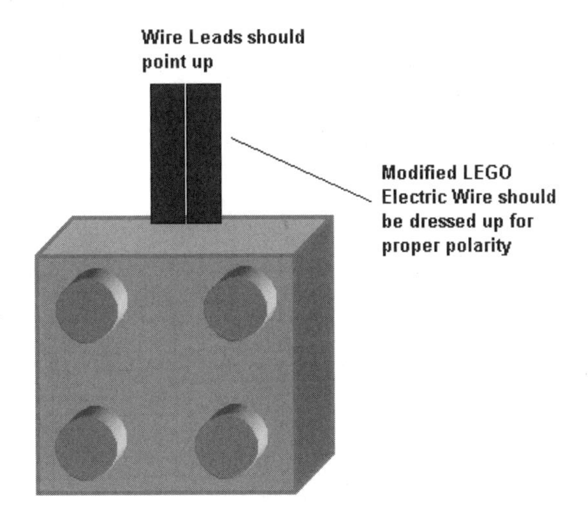

Figure 4-27 Modified LEGO electric wires need to be positioned upward in order for proper linear voltage regulation to occur with the two-transistor oscillator.

6. Within the RCX program environment, build and download the code shown in Figure 4-28 to location 1.
7. Press the Run button on C-Bot. The conveyor should start moving immediately.
8. Squeeze and release C-Bot's hand. The conveyor should move in the opposite direction. After 20 seconds, the conveyor should stop. The tone generator should sound for 10 seconds.

   Next, the code must be modified for two short tone pulses:

9. Modify the code using Figure 4-29 as a guide. Notice that the repeat watcher is integrated into the my command block named ToneBeep.
10. Download the new code to program location 1.
11. Repeat Steps 7 and 8.
12. The audio generator should produce two short tone pulses.
13. Try changing the frequencies of the oscillator by adjusting the 1-mega-ohm potentiometer. Run the code and notice the change in sound output.

Audio Tone Generator Analysis

This lab demonstrates the output interface switching capability of the Mindstorms P-Brick. The 7805 linear voltage regulator IC is not only a DC power supply for the two-transistor oscillator, but it also provides electronic

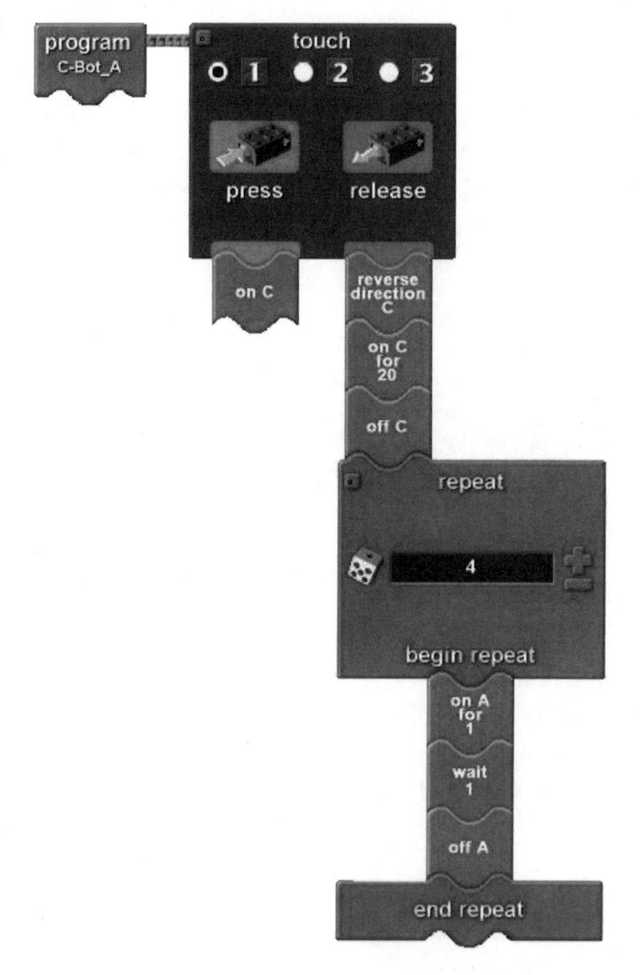

Figure 4-28 RCX code for a C-Bot-controlled audio tone generator

switching to the electrical circuit. The 7805 IC can provide a regulated +5-volt DC output capable of driving the tone generator circuit. Software gives the tone generator an enhanced sound output by switching the output A for two 2-second intervals, creating short audible pulses. This gives the conveyor-complete alarm a nice subtle way of getting the operator's/designer's attention. Also, the my command programming tool within the RCX code environment is used to manage the number of software blocks on the screen. Once the audio tone generator is wired to C-Bot's output, a variety of sounds can be programmed into the RCX P-Brick.

The last lab project on the subject of electronic switching looks into creating a simple diagnostics tool for the output actuation of C-Bot using a graphics calculator and a DCU interface. The tool will use input 2 of C-Bot

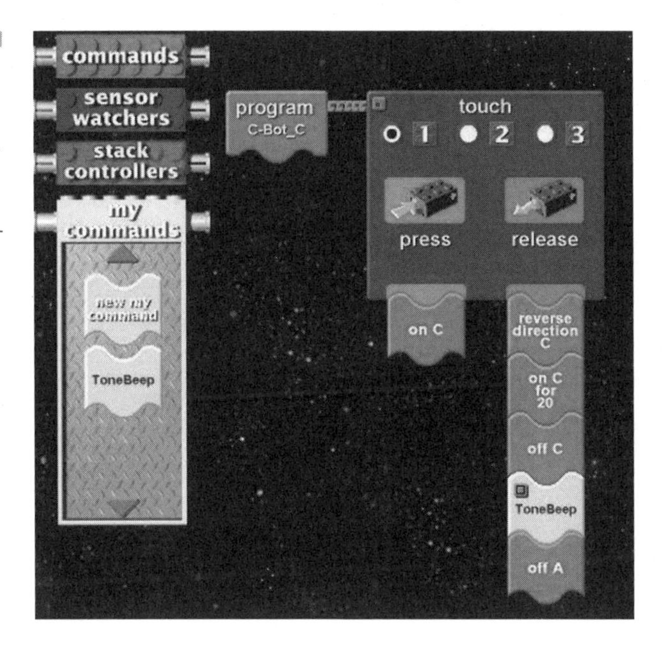

Figure 4-29 RCX code minimization for audio tone generator using the my command block

to turn on output A, which will drive a seven-segment display showing the diagnostics status of C-Bot. The CBL will be introduced to the hobbyist or experimentalist as a viable tool for the electronic switching interface control of Mindstorms robots.

CBL-DCU for Robot Diagnostics

By using a *Texas Instruments* (TI) graphics calculator, a small handheld laboratory can be built to monitor and manage data being received from the CBL. The CBL is an electronic interface made by TI that can obtain electrical and mechanical properties such as voltage, current, heat, light, sound, and force. The CBL can also be used to send out digital data for control applications. The control feature of the CBL provides a wealth of projects for Mindstorms interfacing because of its capability to switch a robot's features and functions on or off.

In order to use the physical switching feature for the CBL, a small company named Vernier Software and Technology created the DCU. The DCU is a small electronic switching interface box with six outputs. The output current from each electronic driver is 600 mA and is managed using an

Figure 4-30 Complete diagnostics and control setup using a (clockwise) TI graphics calculator on top of a CBL connected to the DCU for digital output control of the electromechanical relay input circuit. The +6-volt DC required to operate the DCU is provided by an alternating current (AC) adapter.

ULN2003A IC. The box communicates with the CBL using a cable that plugs into the digital-out connector. Figure 4-30 shows the complete setup of the CBL, DCU, and graphics calculator. In the lab project, the CBL and DCU are used to create a simple diagnostics tool for C-Bot. The tool provides a method of testing the conveyor and diagnostic status after completing the short testing event.

A DCU Diagnostics Tool Lab Procedure

The purpose of this lab project is to illustrate how a diagnostics tool can be built for a Mindstorms robot. This tool will test C-Bot's conveyor using one digital output of the DCU as an electronic switching input to C-Bot. Also, the lab project is intended to expose the hobbyist and experimentalist to alternative electronics using a graphics calculator, CBL, and DCU to test a Mindstorms robot.

BOM

- One 7805 linear voltage regulator IC
- Two 0.1 µF capacitors
- One 100-kilo-ohm resistor

- One seven-segment display, common cathode
- Four 470 resistors
- One IRF630 N-Channel PMOSFET
- Two 6- to 9-volt DC electromechanical relays
- One IN4004 silicon diode or equivalent
- One CBL
- One DCU
- One TI graphics calculator (Models 73, 82, 83, 83+, 86, 89, 89+, 92, and 92+)
- Radio Shack Electronics Learning Lab breadboard or equivalent
- Two modified LEGO electric wires
- 120-volt AC to 6-volt DC output CBL power supply (Vernier Software order code TI-9201)

The procedure used for this lab is as follows:

1. Wire the diagnostics status indicator circuit on a Radio Shack Electronics Lab breadboard or equivalent, using Figure 4-31 as a guide.

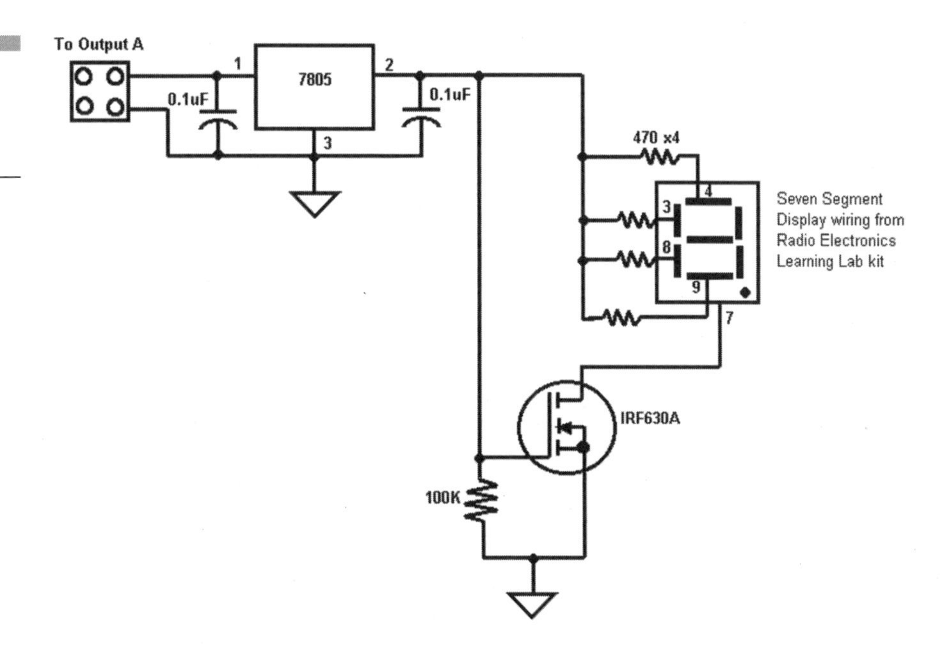

Figure 4-31 Diagnostics status indicator circuit for C-Bot

2. Wire the DCU output electromechanical switch on a Radio Shack Electronics Lab breadboard or equivalent, using Figure 4-32 as a guide.
3. Build and download the diagnostics code using Figures 4-33a and 4-33b.
4. Turn on the CBL and TI graphics calculator. The green *Enable* (EN) LED should be on.
5. Run both the DCUINIT and DCUTOGGL programs in the TI graphics calculator.
6. Run the diagnostics program for slot location 5.
7. Press 1 on the calculator. The status should be ON. The DCU LED should be on. The conveyor should move forward for 10 seconds and then stop. It will move in the opposite direction for 10 seconds. After 10 seconds, the seven-segment display will show the letter C. It will flash for five cycles and then stay lit.
8. Press 1 on the calculator. The status should be OFF. The seven-segment display will turn off. The LED on the DCU (1) should be off.
9. Repeat Steps 7 through 8 several times noting the time sequences between the conveyor turning on and the diagnostics status indicator.
10. Turn off C-Bot, DCU, and CBL.

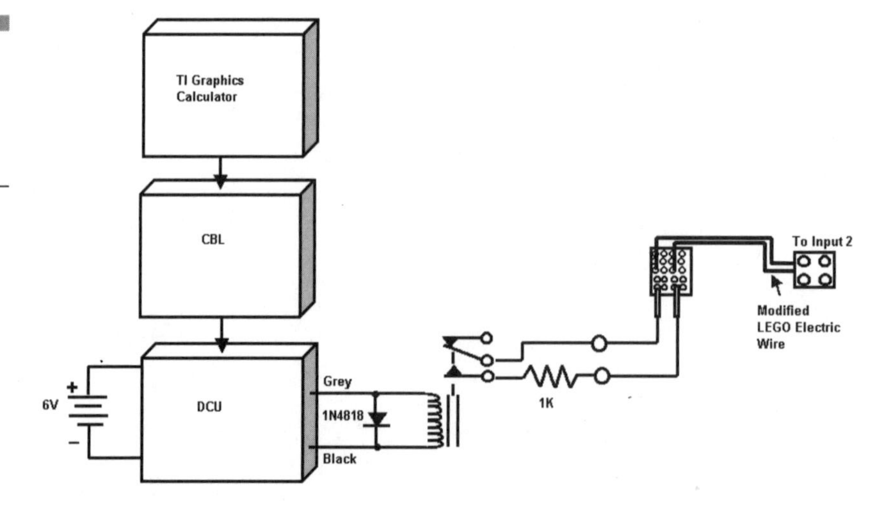

Figure 4-32 Functional block diagram and circuit for CBL-DCU robot diagnostics

DCU Diagnostics Lab Analysis

Alternative electronics to control switching for the Mindstorms robot interfacing is presented in this lab. The CBL-DCU controller is managed using the TI graphics calculator software programs DCUINIT and DCUTOGGL. The ability to use a math tool to control a Mindstorms robot provides a lot of flexibility in programming electronic switches for interfacing projects. The main intent behind this lab project is to expose the robotics hobbyist

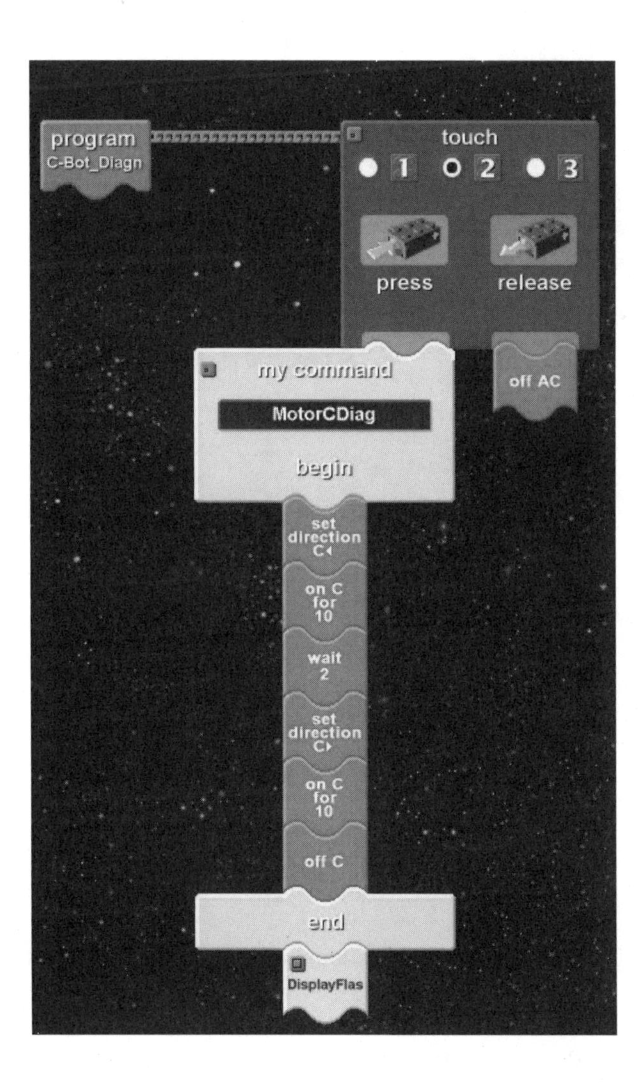

Figure 4-33a
RCX code for C-Bot diagnostics

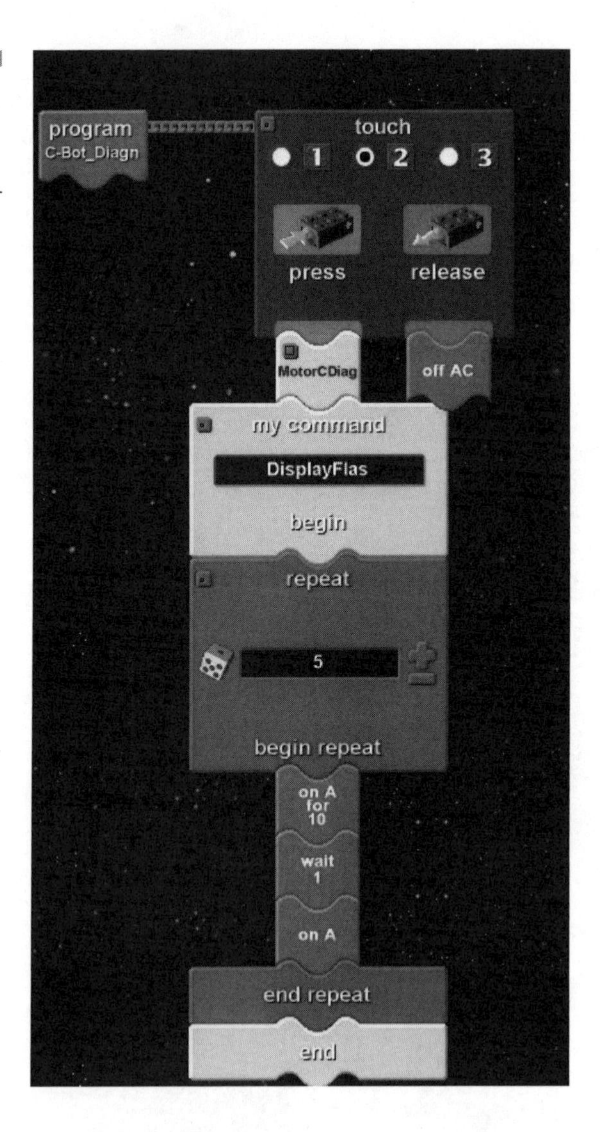

Figure 4-33b RCX code for C-Bot diagnostics continued

and experimentalist to a unique way of controlling Mindstorms robots using calculator-based technology.

In the next chapter, sensors are discussed and the usual hands-on projects are included for the hobbyist and experimentalist to build and experiment. A diode interface circuit used to connect analog signals to Mindstorms robot inputs is also discussed in the next chapter. More experimental techniques will be used to explore passive and active experimental sensors.

CHAPTER 5

Sensor-Interfacing Basics

Sensors are critical to the operation of a Mindstorms robot. They enable the robot to understand the environment that surrounds it. With sensors, a robot can obtain data that enable it to make logical decisions and take appropriate and responsive actions. The sensors that are packaged with the *Robotics Invention System* (RIS) or *Robotic Discovery Kit* (RDK) help the Mindstorms robot to perform responsive actions based on either touch (touch sensor) or light (light sensor). These actions occur because of the logical processes embedded in the software that resides in the microcontroller's memory structure. The small 9-volt motors accompanying the RIS and RDK enable the sensory data to become physical by providing locomotion for the robot. Therefore, an interactive interfacing environment is established because of the relationship among the *activate* (sensor signal), *interpret* (logic), *control* (energy conversion), and *driver* (actuator control) functions that exist between the robot's physical *input/output* (I/O) layer and internal processor structure.

This chapter explains sensor-interfacing techniques for Mindstorms robots. The objective of this chapter is to explain how the natural elements of light, heat, and sound can be detected and processed by a robot. The physical elements of force and pressure are also explored through the use of detection sensors and software. The following topics are discussed in this chapter:

- Sensor basics
- Building a sensory-activated *Robot Digital Assistant* (RDA)
- Characterizing sensors through electrical measurements
- The *Silicon Diode Interface Circuit* (SDIC)
- Coin-operated *Task Robot* (T-Bot)
- Rheostat-sensor-controlled T-Bot
- Vision Command LEGO Camera as an advanced sensor

Sensor Basics

A sensor is a device that samples the physical properties of nature (sound, light, temperature, motion, and touch) and delivers a proportionate current or voltage of the natural stimuli. The RDK and RIS have two varieties of sensors that are capable of detecting and responding to the presence of light and touch stimuli: the light and touch sensor, respectively. Other sensors called temperature and motion sensors can be purchased to detect temperature and motion stimuli.

Sensors can be classified in two categories: passive or active. Passive sensors do not require an external *Direct Current* (DC) source to detect physical stimuli. The signals generated influence the energy within the network they are wired to. Another name used for passive detection is *Go-No-Go sensors*. Go-No-Go sensors are simple mechanical or resistive devices that provide an on-off, yes-no, or binary event of the measured stimulus detected. The resistive switching discussed in Chapter 3, "Electromechanical Controls Interfacing," is an example of a Go-No-Go sensor. If the appropriate resistor element is selected, the *Robot Command Explorer* (RCX) or Scout *programmable brick* (P-Brick) can treat the input signal as a binary event of the sensing device. Other passive or Go-No-Go sensors include magnetic proximity, photoresistor, photodiode, thermistor, piezoelectric, and vibration sensors. Figure 5-1 shows some of these sensing components.

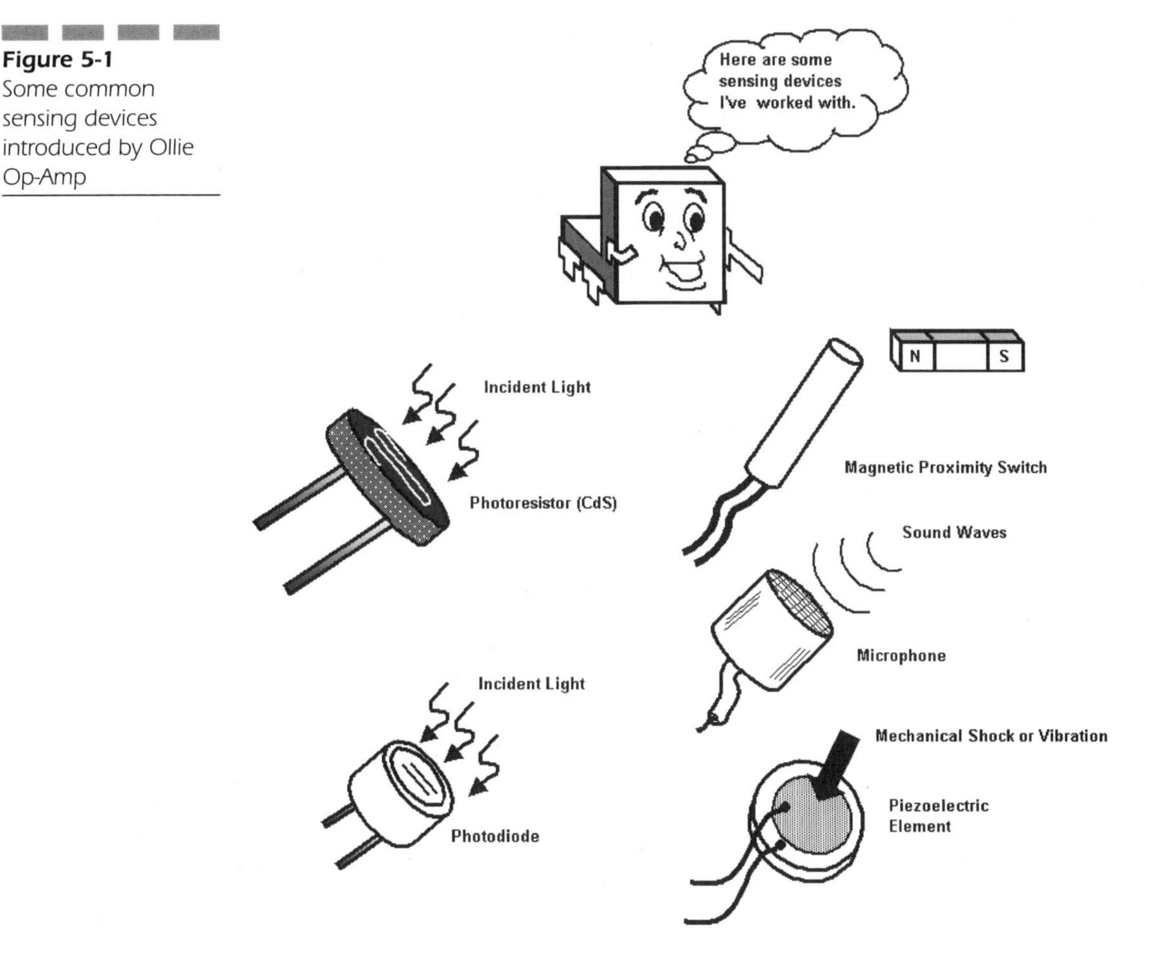

Figure 5-1 Some common sensing devices introduced by Ollie Op-Amp

Active sensors require an external DC source to generate a signal and usually need some electronics to perform a proper detection operation. Active sensors can be used as Go-No-Go detection devices when they are wired to an amplifier and switching circuit. The amplifier is required to provide signal amplification and conversion. The switching circuit is activated if the amplified signal exceeds or falls below a threshold. Usually, signal conversion is necessary when interfacing to a microcontroller or a *single computer board* (SCB). The Scout and RCX P-Bricks use the Hitachi H8 microcontroller and therefore require an input signal conversion in order to provide the appropriate output response. The amplified analog signal is converted into a digital equivalent so the microcontroller or SCB can process the data naturally. Another word associated with active sensors is *analog*.

The following lab project uses an RDA that responds to the appropriate physical stimulus by rotational tracking through its base. A generic sensory robot is used in order to provide an intelligent output device that can be programmed to respond to various sensing devices wired to a P-Brick's inputs. Therefore, a reconfigurable sensory input detection system will be created so we can experiment with standard and homebrew sensors. Before the first lab project is discussed, a brief discussion on sensory-activated RDAs is in order.

Building a Sensory-Activated RDA

As mentioned briefly in the last section, the intelligent machine used in this chapter is a robot that has the capability to track physical stimuli detected by its input sensor. The robot's movement is translated using a rotational base of one *degree of freedom* (DOF). T-Bot is the name of the RDA used to assist in the forthcoming lab projects. Figures 5-2 through 5-11 show the mechanical parts of T-Bot. The squeeze hand provides a simple diagnostics switch for quickly checking the robot's rotational tracking base drive assembly. T-Bot is a combination of C-Bot's upper-body design and V-Bot's rotational base drive assembly. (See Chapter 2, "Developing GUIs: Software Control Basics," for base drive assembly instructions.)

Although hobbyists and experimentalists have the freedom to create their own tracking robotic system, T-Bot will be used as a sample RDA throughout the chapter to give the discussion of sensor-interfacing basics a physical identity. Also, the Scout P-Brick can be substituted for the RCX unit quite easily. Therefore, this exercise is left as a design challenge to the hobbyist or experimentalist.

Figure 5-2
Top or plane view of the T-Bot's squeeze hand using a touch sensor

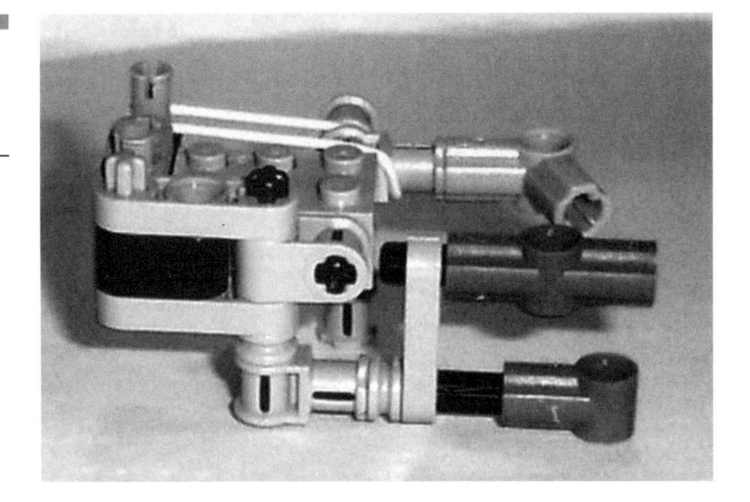

Figure 5-3
Side view of T-Bot's squeeze hand using a touch sensor

Recycling of Mechanical Assemblies (ROMA) Concept

The *Recycling of Mechanical Assemblies* (ROMA) concept reduces the amount of development time when building Mindstorms robots. This concept is very similar to building an electronic bookshelf where core circuits

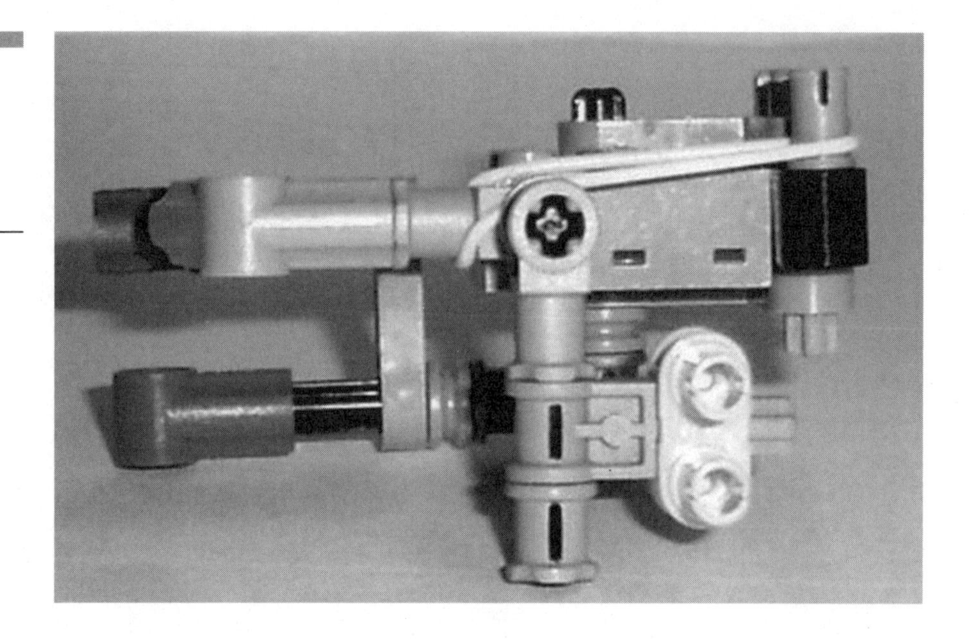

Figure 5-4 Opposite side showing the dual yellow connector pegs of T-Bot's squeeze hand

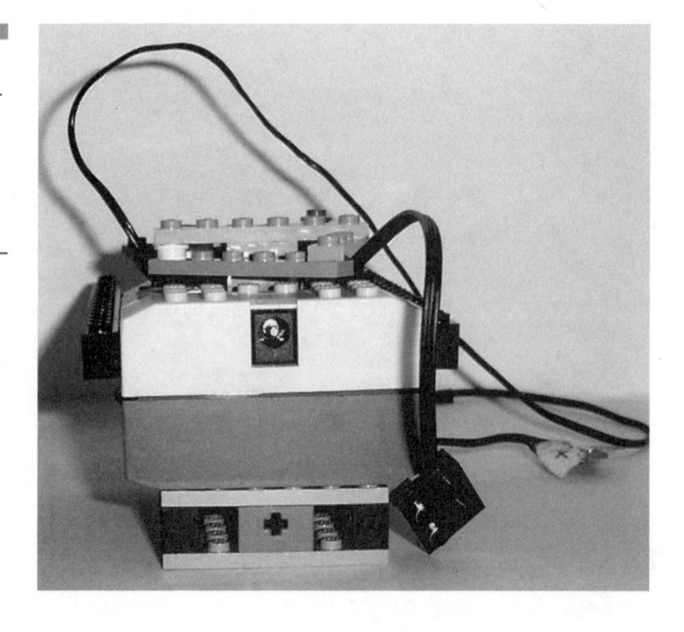

Figure 5-5 Bottom view of T-Bot. Note the Technic brick with a crossbar hole for mounting the motorized drive assembly.

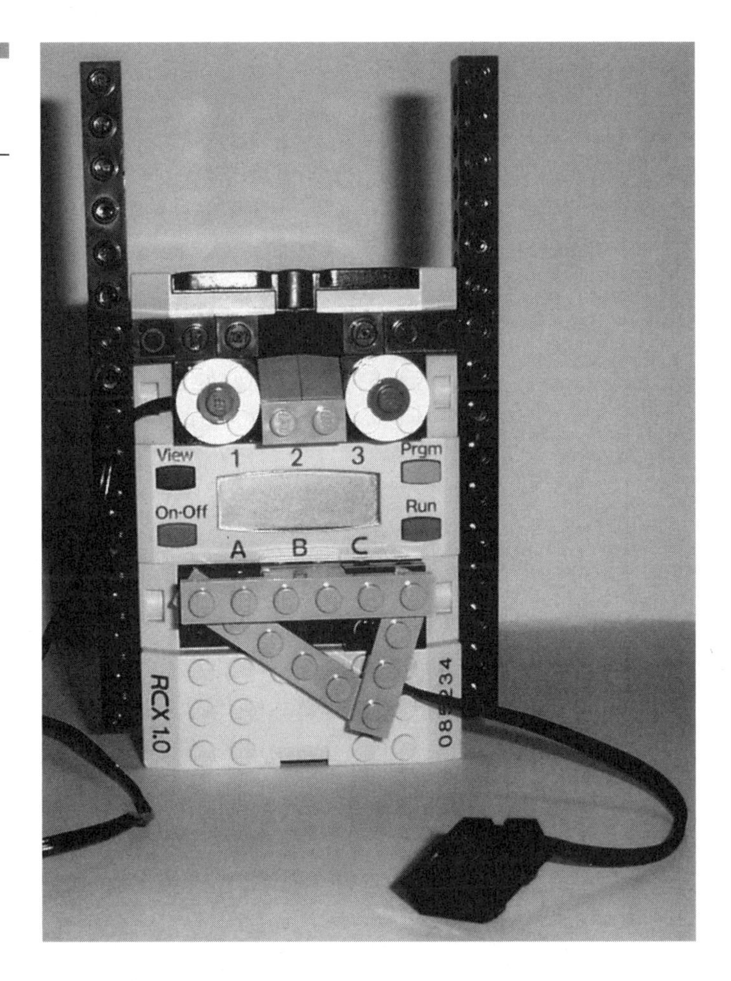

Figure 5-6
T-Bot with vertical Technic bricks

are stored for future applications. Mechanical building blocks can be used to explore new designs by taking existing proven subsystems. Figure 5-12 shows the ROMA concept diagram for T-Bot. To create a recycling core diagram, begin by using the input sensing mechanical assembly to the robot's body. The motorized drive assemblies should be an output from the body. Finally, any other mechanical supports and mounts should be shown in the

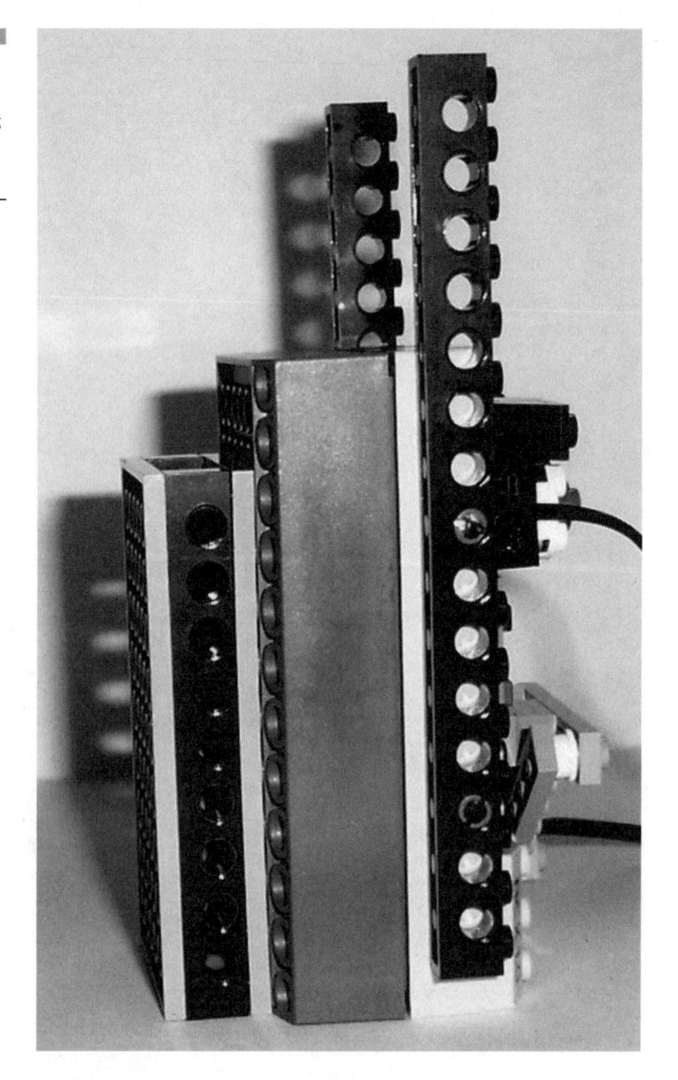

Figure 5-7
Side view showing vertical Technic bricks mounted to the RCX P-Brick

vicinity of the robot's body. The diagram should help hobbyists or experimentalists build subcomponents that will eventually be attached to each mechanical block, completing the target robot. T-Bot was built using this diagram; therefore, T-Bot is the first real application of the ROMA concept for Mindstorms robotics.

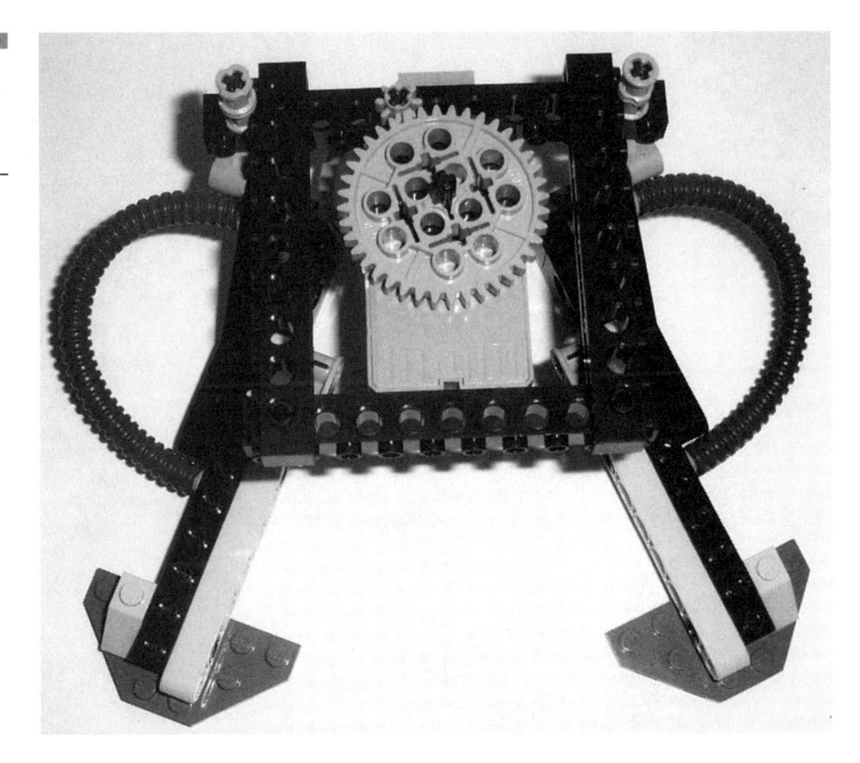

Figure 5-8
Top or plane view of the motorized drive assembly for T-Bot

Characterizing Sensors Through Electrical Measurements

To understand a sensor's operation, its electrical signals must be measured. The measurement activity provides insight into the behavior of the sensor and its reaction to physical stimuli. The information produced by the sensor is usually the voltage or current over time. With this data, a characteristic curve or plot can be created. The curve or plot is a graphical signature of the sensor's behavior as it reacts to physical stimuli. The characteristic curve also provides the data necessary to design the sensor for a given product application.

The following lab project outlines a procedure for characterizing several standard sensors. The RCX software is used to set up the sensor and enable

Figure 5-9
Bottom view of the motorized drive assembly for T-Bot

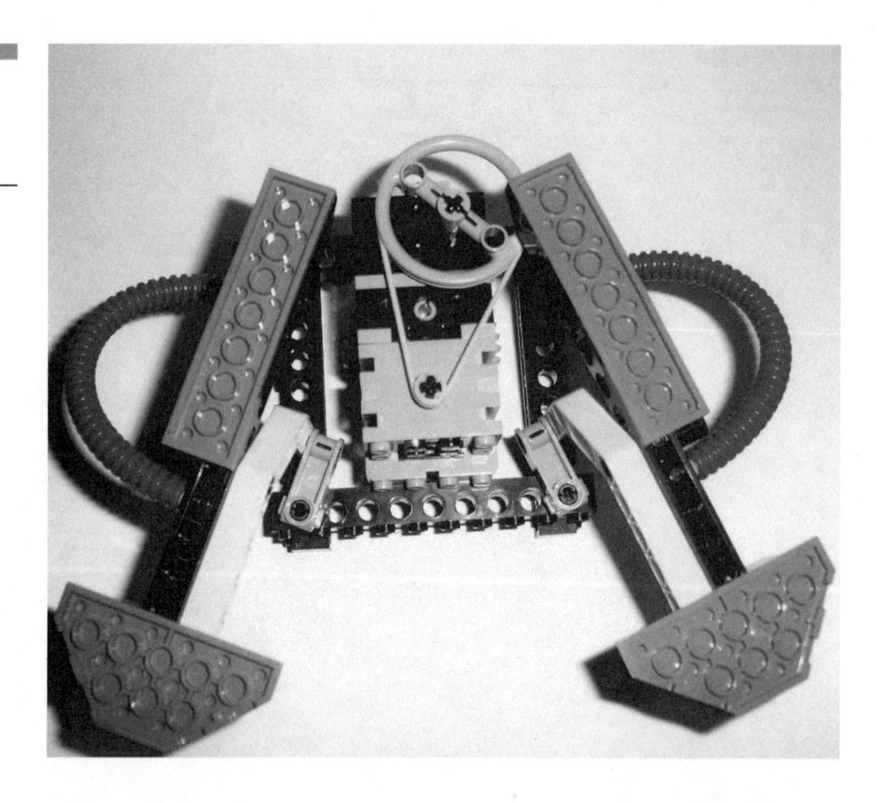

Figure 5-10
Side view of the motorized drive assembly for T-Bot

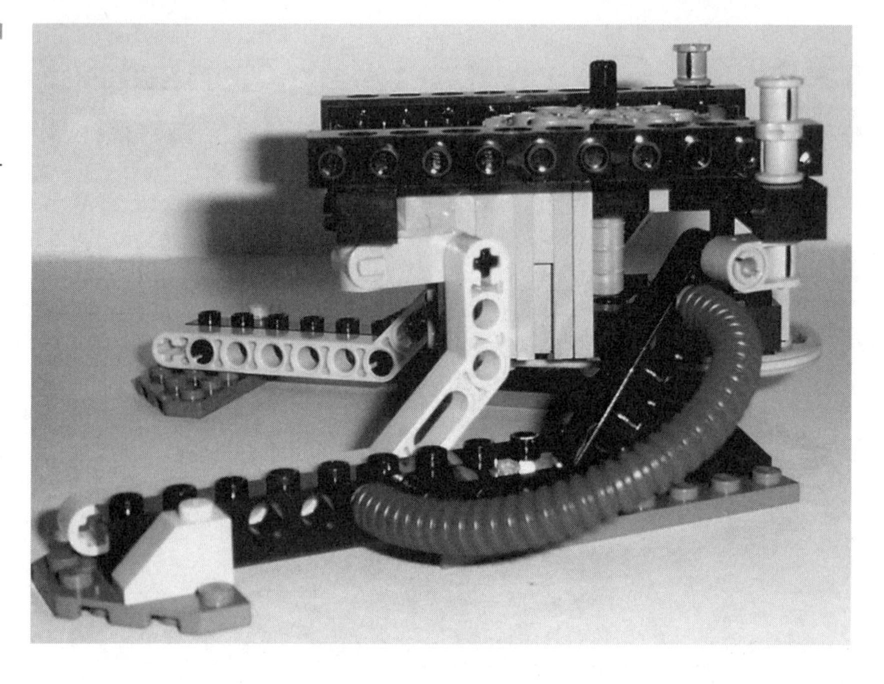

Figure 5-11
The completed T-Bot

T-Bot to perform specific tasks based on the physical stimulus present. Although the RCX P-Brick is used for this lab project, the Scout P-Brick can easily be substituted as well.

Characterization Lab

In this lab, the following standard sensors are characterized: touch, light, temperature, and rotation devices. A standard measurements setup is used to capture the sensor's signal signature. This logged data is saved as an *American Standard Code for Information Exchange* (ASCII) file and reviewed in Excel.

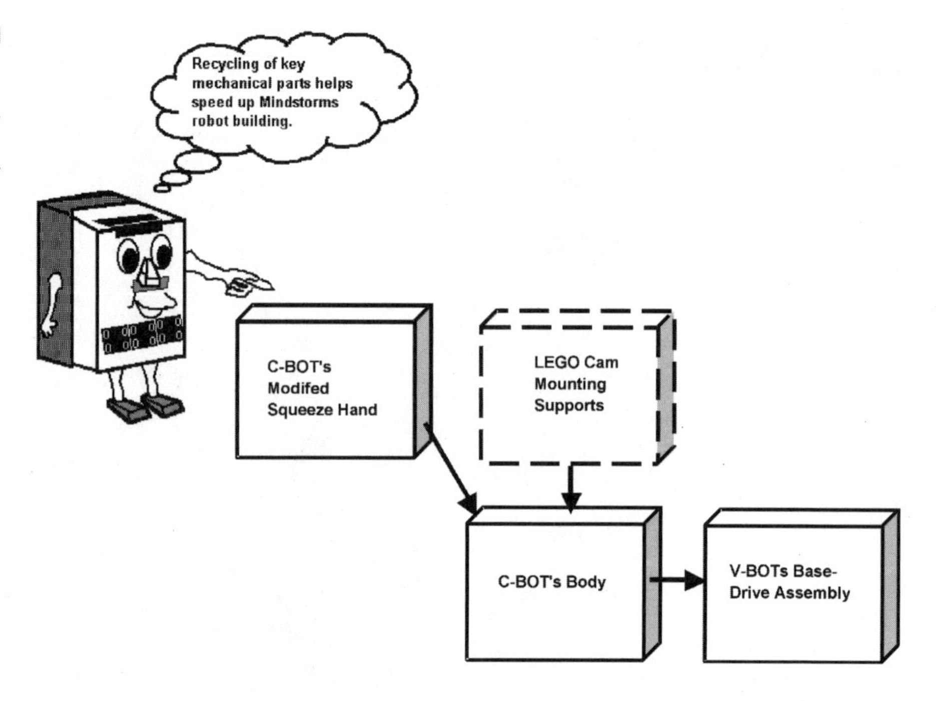

Figure 5-12 The ROMA block diagram explained by RCX

Bill of Materials (BOM)

- One touch sensor
- One temperature sensor
- One light sensor
- One rotation sensor
- One modified LEGO electric wire
- One standard LEGO electric wire
- One LEGO infrared remote control
- One flashlight
- One Constructopedia Manual versions 1.5/2.0
- One *Digital Multimeter* (DMM) with PC interface
- One ScopeView software
- One Radio Shack Electronics Learning Lab breadboard or equivalent

Touch Sensor Procedure

1. Build T-Bot using Figures 5-2 through 5-11 as an assembly guide. Also use pages 64 through 68 of the Constructopedia for building the motorized drive assembly of T-Bot.
2. Download the RCX code into program location 1, using Figure 5-13 as a guide.
3. Attach a standard electric wire from the touch sensor to input 1 of the RCX P-Brick. Attach the touch sensor of T-Bot's squeeze hand to the modified electric wire by placing it on top of the standard electric wire.
4. Insert the modified electric wire into the Electronics Learning Lab breadboard. Attach the test leads of the DMM to the modified electric wire using the Electronics Learning Lab breadboard as measuring points.

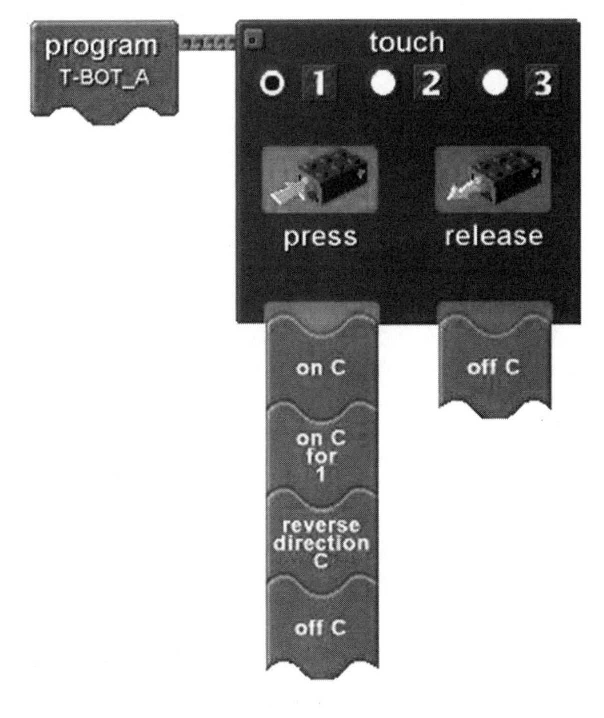

Figure 5-13 The RCX code for the touch sensor characterizing lab

5. Turn on the DMM and run the ScopeView software. Figure 5-14 shows the measurement setup for characterizing sensors.
6. Turn on T-Bot.
7. Calibrate the motorized drive assembly using the infrared remote control by pressing and holding the down-arrow C button. The direction traveled should be right. T-Bot should be positioned, as shown in Figure 5-15.
8. Start a logging session within ScopeView using Figure 5-16 as a guide for setting up the controls. Change the trigger value to 4.98 and the units/div to 1.
9. Click the Record button on the control panel. The Save File dialog box should appear on the screen.
10. Select the appropriate drive for saving the data. I used the a: drive to save data onto a 3.5-inch floppy.
11. Click on the Filename: box and type t_botsen.txt. Click OK.
12. Click the Scope button.
13. Turn on T-Bot and run the RCX code stored at program location 3. Push the Run button. T-Bot should have turned for a quick jog cycle to the left.
14. To quickly check out T-Bot's program, squeeze and release the hand. T-Bot should have turned to the calibrated position in Step 7.

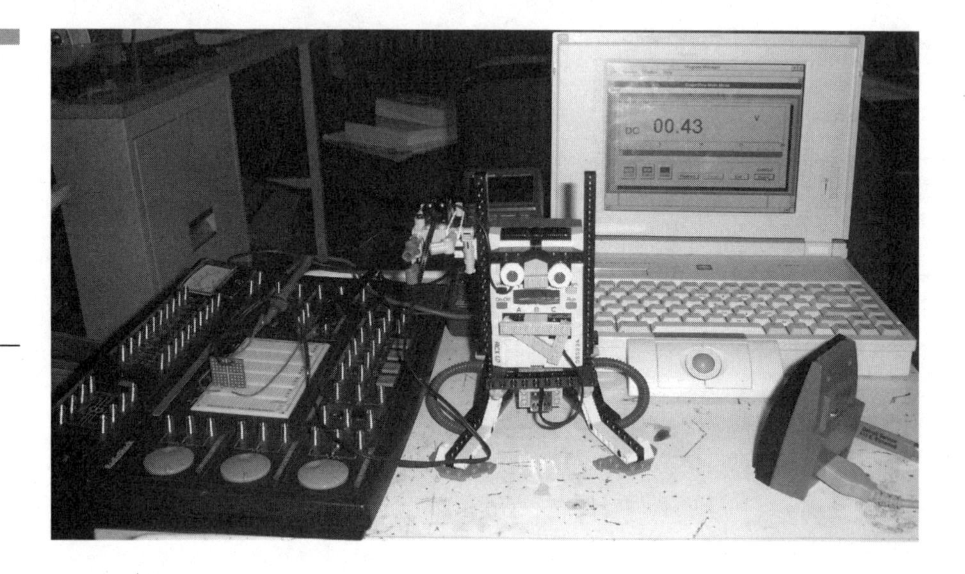

Figure 5-14 The measurement setup for characterizing sensors. Note the resurrection of a 486 notebook computer running the ScopeView software.

Figure 5-15
The position of T-Bot for proper swing motion. The LEGO infrared remote control was used to locate T-Bot to this position.

15. Click Run on the ScopeView Output window to start a data-logging session.
16. Squeeze and hold T-Bot's hand for 20 seconds. The software should start plotting a trace on the screen. The rising edge should occur at the trigger value of 4.98 volts.
17. Release the hand for about 20 seconds. Repeat Steps 16 through 17 until one sweep is completed.
18. Click the Close button on the control panel to stop the data-logging session.

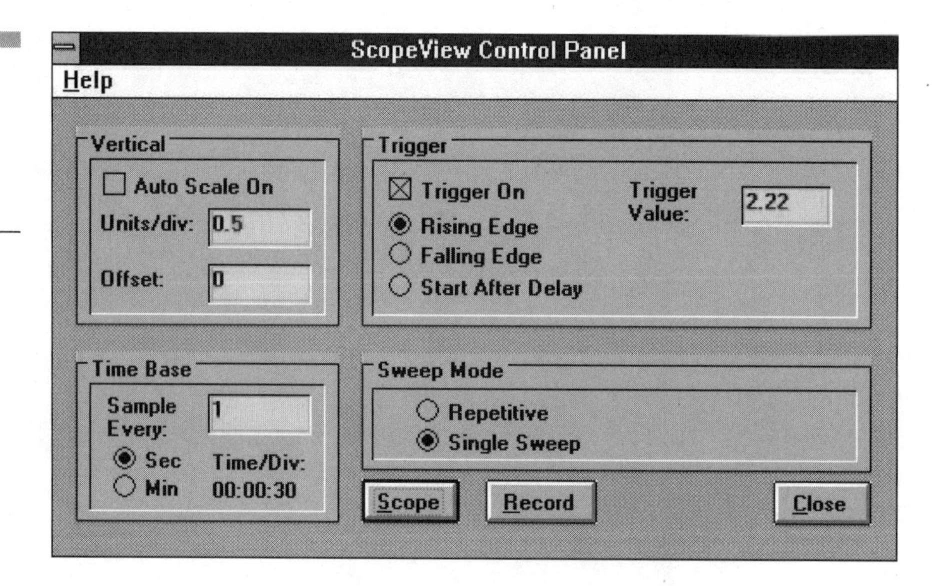

Figure 5-16 The control panel settings for the ScopeView data-logging session of the touch sensor

Reviewing and Playing Back the Logged Data

The logged data can be reviewed or played back within ScopeView as follows:

1. Click Close on the control panel *Graphical User Interface* (GUI). The Main Menu should be displayed on the screen.
2. Click Playback on the Main Menu panel. The Data File Reviewer GUI should be displayed.
3. Click the Open File button. The Read File dialog box should be displayed.
4. Click the t_botsen.txt file. Click OK. The Data File Reviewer should start plotting the data on the screen. Figure 5-17 shows the results of the T-Bot's squeeze hand signal measurement.
5. Plot the ASCII data from the t_botsen.txt file in Excel or an equivalent spreadsheet application. (See Figure 5-18.) Compare the results with Figure 5-17.
6. Squeeze and release T-Bot's hand and compare the movement and timing with the RCX code shown in Figure 5-13.
7. Stop T-Bot.

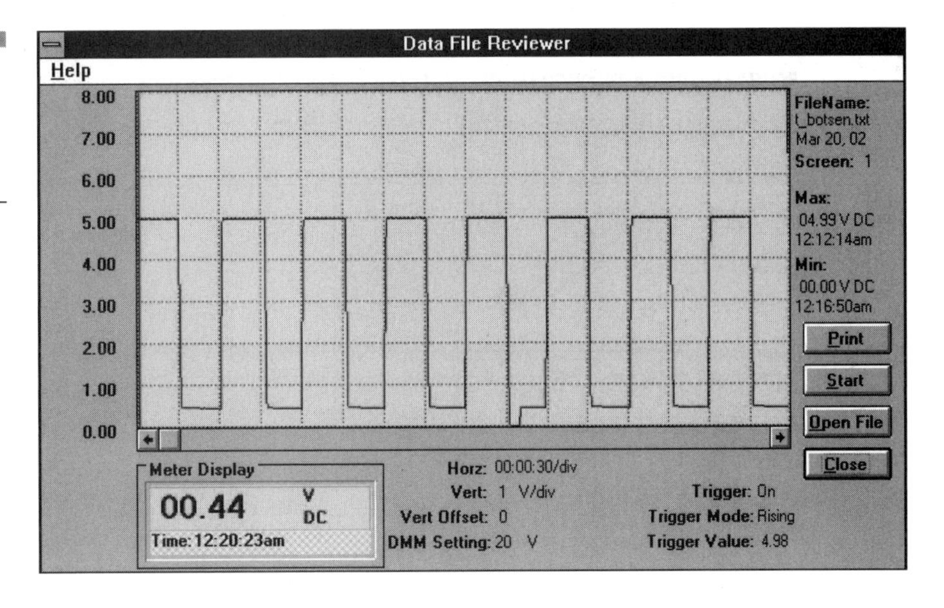

Figure 5-17 Output plot of the touch sensor data logged within ScopeView software

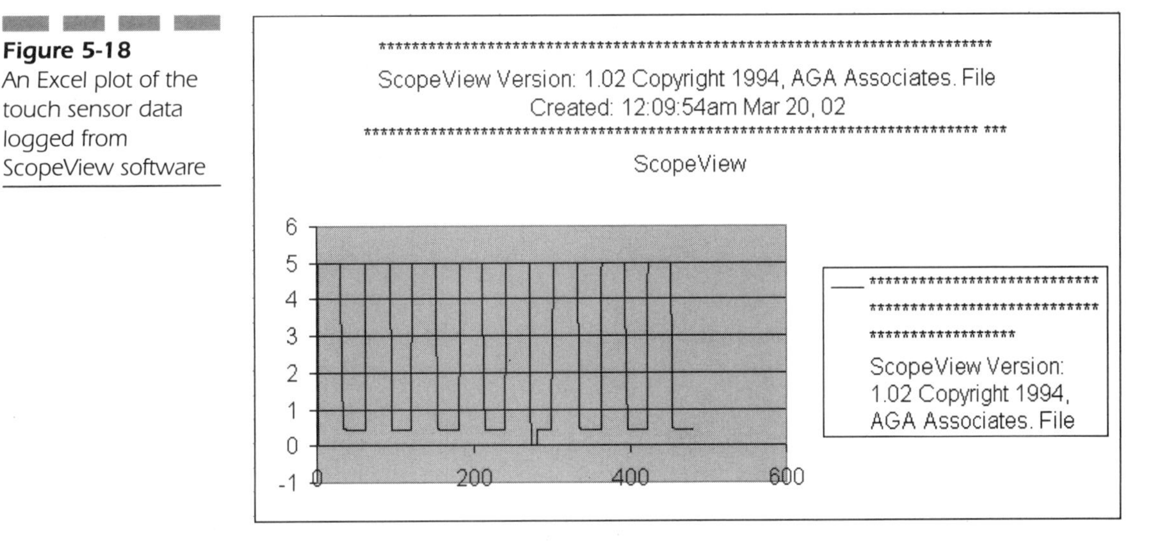

Figure 5-18 An Excel plot of the touch sensor data logged from ScopeView software

Light Sensor Procedure

The following steps characterize the light sensor:

1. Attach the light sensor (as shown in Figure 5-19) on top of the touch sensor connected to the squeeze hand.

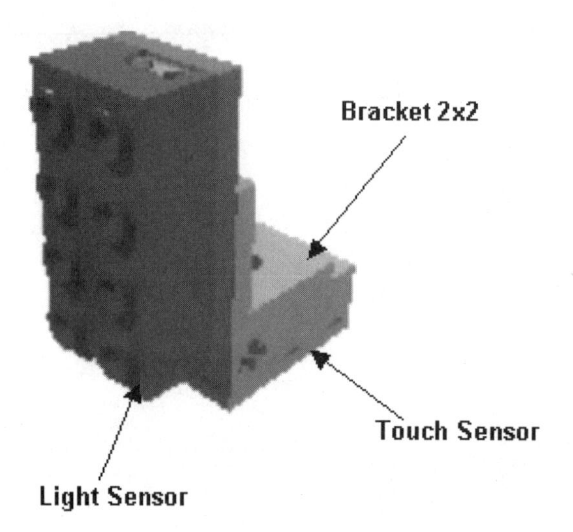

Figure 5-19 Mounting the light sensor on top of the touch sensor for the light sensor characterizing lab. The light and touch sensor model was created with the Michael Lachmann Computer Aided Design (MLCAD) software.

2. Insert the modified electric wire into the Electronics Learning Lab breadboard. Attach the test leads of the DMM to the modified electric wire using the Electronics Learning Lab breadboard as measuring points.
3. Build the RCX code shown in Figure 5-20 and download it to T-Bot.
4. Turn on T-Bot.
5. Turn on the DMM and run the ScopeView software.
6. Start a logging session within ScopeView. Change the units/div to 1.
7. Click the Record button on the control panel. The Save File dialog box should appear on the screen.
8. Click on the Filename: box and type l_botsen.txt. Click OK.
9. Click the Scope button.
10. Run the RCX code stored at program location 5.
11. Click the Run button on the ScopeView Output window to start a data-logging session.
12. Swing the flashlight back and forth slowly above the light sensor. T-Bot should track the movement of each pass of the flashlight.
13. Allow the software to log data for one complete sweep.
14. Click Close on the control panel to end the logging session.
15. Follow Steps 1 through 5 of the section "Reviewing and Playing Back the Logged Data" to view and plot the logged light sensor data. Figure 5-21 shows the data plotted for the light sensor.
16. Stop T-Bot and the ScopeView software.

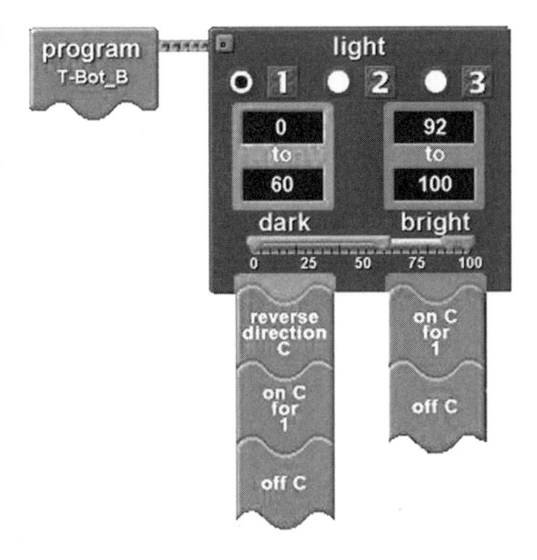

Figure 5-20 RCX code for the light sensor characterizing lab

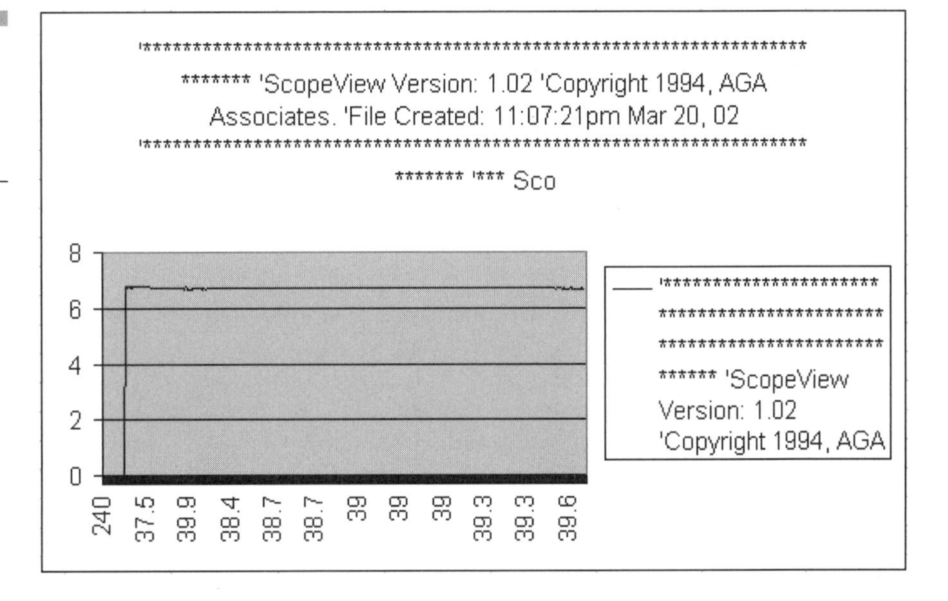

Figure 5-21 An Excel plot of the light sensor data obtained from ScopeView software

Activating the Temperature Sensor

The temperature sensor must be activated before it can be used in Mindstorms Robotics applications:

1. Go to the Main Menu of the RIS RCX software.
2. Click on the Getting Started button.

3. Click on Set Up Options.
4. Click on Advanced at the bottom of the screen.
5. Click on the gray button with an X across from the Unlock Temperature Sensor button. The button will turn red with a yellow checkmark.
6. Save the changes by clicking the green accept box.

Temperature units can be changed to Fahrenheit or Celsius:

1. Go to Set Up Options.
2. Click Advanced at the bottom of the screen.
3. Click on the Degrees Fahrenheit button across from Temperature Measure In. This enables degrees Fahrenheit to be displayed in the *Liquid Crystal Display* (LCD) of the RCX P-Brick.
4. The acceptable temperature range of the sensor is from −4 to +140 degrees Fahrenheit and −20 to +50 degrees Celsius.

Temperature Sensor Procedure

The following steps characterize the temperature sensor:

1. Attach the temperature sensor (as shown in Figure 5-22) on top of the touch sensor connected to the squeeze hand.

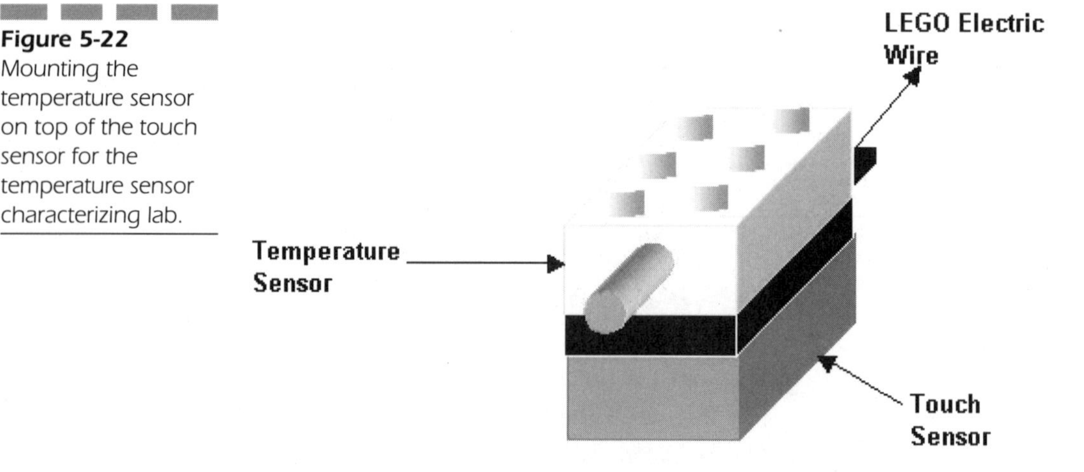

Figure 5-22 Mounting the temperature sensor on top of the touch sensor for the temperature sensor characterizing lab.

2. Insert the modified electric wire into the Electronics Learning Lab breadboard. Attach the test leads of the DMM to the modified electric wire using the Electronics Learning Lab breadboard as measuring points.
3. Build the RCX code shown in Figure 5-23 and download it to T-Bot.
4. Turn on T-Bot.
5. Turn on the DMM and run the ScopeView software.
6. Start a logging session within ScopeView. Change the units/div to 1.
7. Click Record on the control panel. The Save File dialog box should appear on the screen.
8. Click on the Filename: box and type h_botsen.txt. Click OK.
9. Click the Scope button.
10. Run the RCX code stored at program location 4.
11. Click Run on the ScopeView Output window to start a data-logging session.
12. With the infrared remote control, forward and reverse T-Bot. This forward and backwards sweeping motion stirs up the temperature surrounding the sensor.
13. Repeat Step 12 for one complete sweep of data logging.
14. Click Close on the control panel to end the logging session.
15. Follow Steps 1 through 5 of the section "Reviewing and Playing Back the Logged Data" to view and plot the logged light sensor data. Figure 5-24 shows the data plotted for the temperature sensor.
16. Stop T-Bot and the ScopeView software.

Figure 5-23
RCX test code for checking the temperature sensor

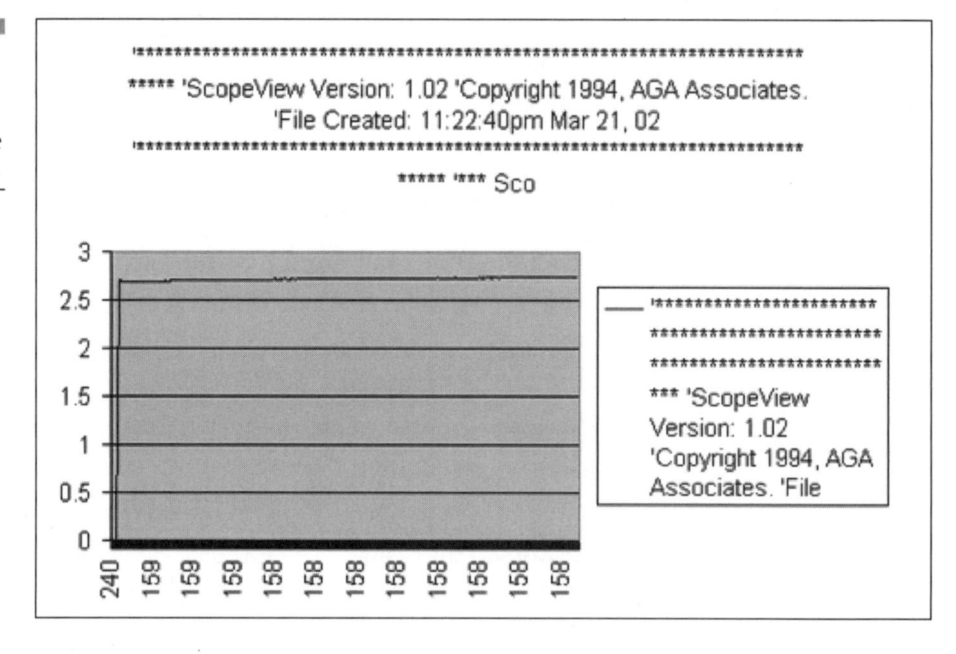

Figure 5-24 An Excel plot of the temperature sensor data logged from the ScopeView software

What Is a Rotation Sensor?

A rotation sensor keeps track of how many times an axle inserted in the electromechanical device rotates. The sensor measures rotation in increments of 16. In other words, for one complete rotation of the axle, the sensor counts up to 16. For half of a rotation, the sensor count is 8, and for two rotations, the sensor value is 32. Figure 5-25 shows the mathematics behind calculating the number of counts for a given rotation. The rotation sensor can count both clockwise and counterclockwise directions. If the axle rotates in the opposite direction, the sensor counts backwards and the LCD displays a negative number.

Activating the Rotation Sensor

The rotation sensor must be activated before it can be used in Mindstorms robotics applications:

1. Go to the Main Menu of the RIS RCX software.
2. Click on the Getting Started button.

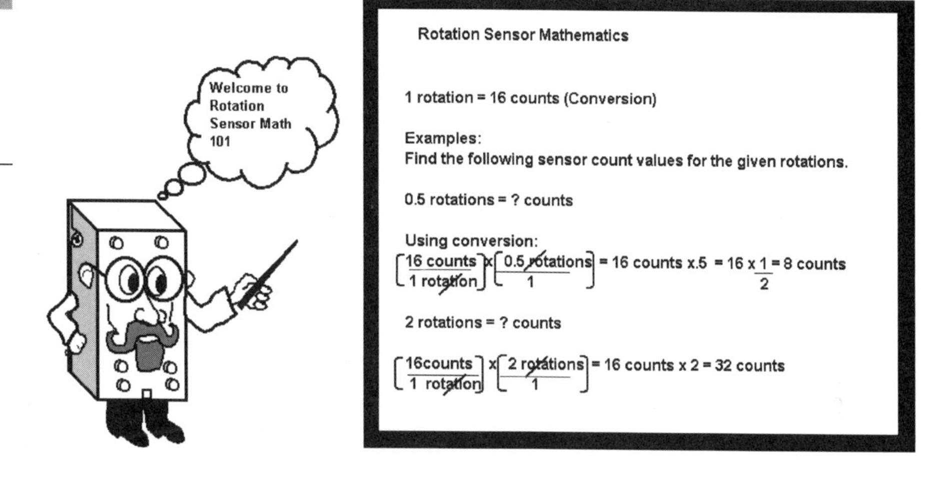

Figure 5-25 Professor Roto explains rotation sensor count mathematics.

3. Click on Set Up Options.
4. Click on Advanced at the bottom of the screen.
5. Click on the gray button with an X across from the Unlock Rotation Sensor button. The button will turn red with a yellow checkmark.
6. Save the changes by clicking the green accept box.

Rotation Sensor Procedure

The following steps characterize the rotation sensor:

1. Attach the temperature sensor (as shown in Figures 5-26 and 5-27) on the bottom of T-Bot's motorized drive assembly.
2. Insert the modified electric wire into the Electronics Learning Lab breadboard. Attach the test leads of the DMM to the modified electric wire using the Electronics Learning Lab breadboard as measuring points.
3. Build the RCX code shown in Figure 5-28 and download to T-Bot using program location 4.
4. Turn on T-Bot.
5. Turn on the DMM and run the ScopeView software.
6. Start a logging session within ScopeView. Change the units/div to 1.

Figure 5-26
T-Bot slightly leans because of the rotation sensor mounted on the bottom of its motorized drive assembly.

7. Click the Record button on the control panel. The Save File dialog box should appear on the screen.
8. Click on the Filename: box and type r_botsen.txt. Click OK.
9. Click the Scope button.
10. Run the RCX code stored at program location 4.
11. Click Run on the ScopeView Output window to start a data-logging session.
12. With the infrared remote control, forward and reverse T-Bot. This forward and backwards sweeping motion enabled the motorized drive assembly axle inserted in the sensor to rotate.
13. Click the Close button on the control panel to end the logging session.

Figure 5-27
Bottom view of the motorized drive assembly showing the location of the rotation sensor

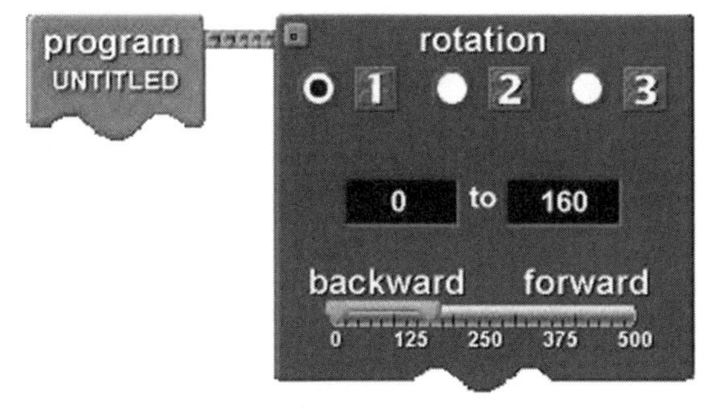

Figure 5-28
RCX test code for checking the rotation sensor

14. Follow Steps 1 through 5 under the section "Reviewing and Playing Back the Logged Data" to view and plot the logged light sensor data. Figure 5-29 shows the data plotted for the temperature sensor.

15. Stop T-Bot and the ScopeView software.

Characterization Lab Analysis

The objective of this lab is to obtain data on each of the standard LEGO sensors that can be used for developing experimental sensing devices. The plots generated provide hobbyists and experimentalists with signal

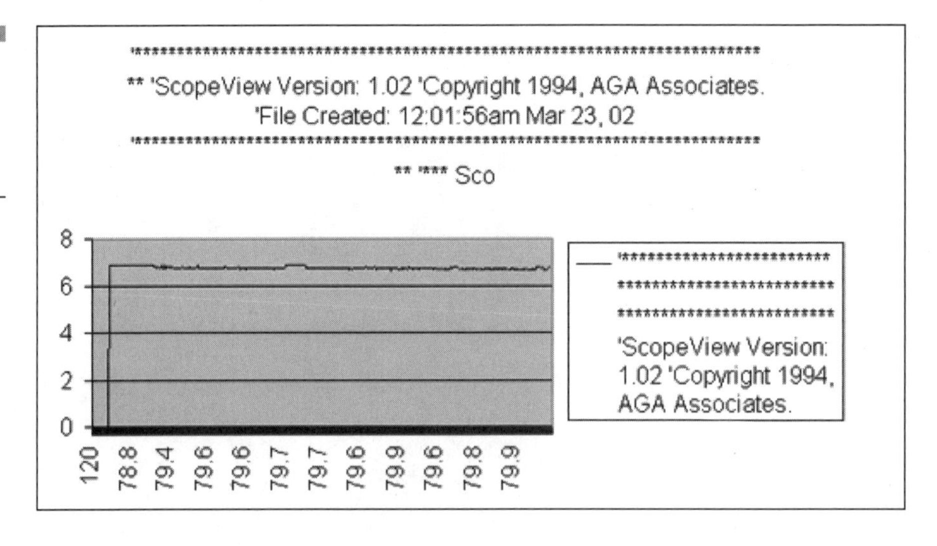

Figure 5-29 An Excel plot of the rotation sensor data logged from the ScopeView software

characteristics of each sensor that can be simulated using passive and active electronic circuits. Looking at the plots, the light and rotation sensors produced a steady DC voltage that was greater than 6 volts. The reason for this high voltage signal is based on how the LEGO RCX P-Brick reads outside physical stimuli and powers the active sensors using an input. Basically, the input of the RCX P-Brick provides a *pulse-width modulation* (PWM) to maintain a constant operating voltage for the sensor and establish a signal path for the sensor's data readings using a switching scheme controlled by the H8's embedded firmware. The RCX firmware is the RCX's internal software that defines how it behaves as a smart LEGO brick. The firmware orchestrates how the electrical I/O physical layer works with the software components of the H8 microcontroller. Basically, it's the operating system for the P-Brick. To accomplish this switching scheme between reading sensor data and providing a constant DC voltage source, a diode interface circuit is used for signal and power management. The next section discusses how this circuit can be used to implement homebrew sensors for the RCX P-Brick.

The Silicon Diode Interface Circuit (SDIC)

The SDIC is capable of receiving power from the RCX-powered interface and applying it to the sensor. It can also read its value using a single pair of wires. The SDIC is able to perform this task by switching between the functions via the RCX-powered interface. The RCX P-Brick applies an internal 8 volts to the sensor port. The duration for this voltage at the port is 3 milliseconds. The SDIC passes the measured sensor data to the powered interface, as it would for the touch sensor, with a duration of 100 microseconds. The data-measuring and port-powering event repeats as long as the active sensor is present.

As shown in Figure 5-30, the SDIC is a bridge rectifier circuit. Normally, bridge rectifiers are used to convert *Alternating Current* (AC) to DC. The diodes labeled D1 through D4 form the power circuit for the sensor rectifier interface. Current from the internal +5-volt battery cannot flow through diode D1 because of reverse bias. Diode D2 enables current to flow out (forward bias) to the external circuit powering it to 5 volts. To complete the current flow path back to the negative potential of the internal battery, diode D2 assists in this electrical return path. Reversing the LEGO electric wire onto the input of the RCX P-Brick would not allow the current to flow through diode D2 because of reverse bias. Diode D4 would be the new current path for powering up the external circuit instead of diode D3, and diode D1 would provide the return path to the internal +5-volt DC source.

The circuit enables the sensor to work regardless of how it's connected to the RCX P-Brick's input port. The circuit shown in Figure 5-30 can be built on a *Printed Circuit Board* (PCB), but I recommend building it on an experimenter's board to make it easier for homebrew sensor experimentation. Figure 5-31 shows a permanent board for the SDIC. By building a small protoboard of this circuit, the development time for experimental circuit

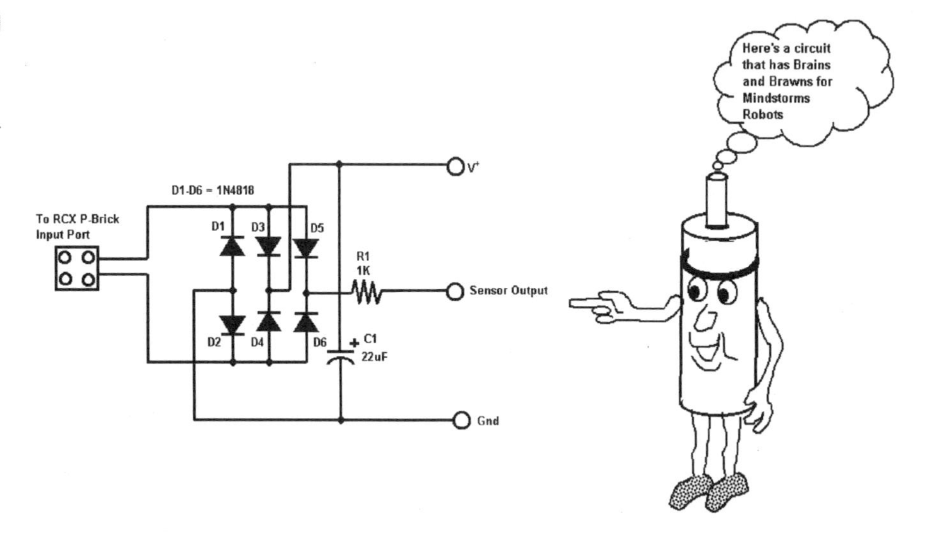

Figure 5-30 Danny Diode and the SDIC

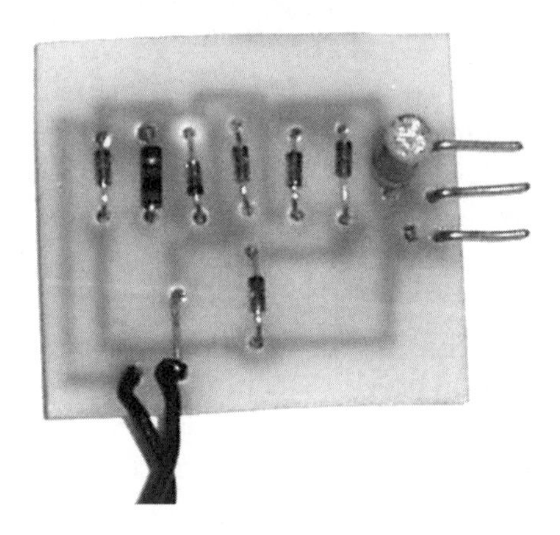

Figure 5-31 Experimenter's board (top) and PCB (bottom) of the SDIC

building is reduced significantly. The following lab projects illustrate how homebrew sensing circuits can be wired to the SDIC to control T-Bot.

Coin-Operated T-Bot

The passive sensor used to make a coin-operated robot is a homebrew electric switch made from LEGO bricks and aluminum foil. The concept for this circuit and lab project came from the section "Coin Detector" in the book *Extreme Mindstorms*[1], which is written by Dave Baum, Michael Gasperi, Ralph Hempel, and Luis Villa. On page 222, Michael Gasperi explains how a simple coin detection sensor can be built using LEGO bricks and aluminum foil. The detector has been modified to keep the coin confined to the sensor by adding bricks in front of the aluminum foil/brick switch contacts. A 1-kilo-ohm resistor is included for the proper wetting current of the alu-

[1]Extreme Mindstorms was published by Apress Publishing, copyright 2000.

minum foil/brick switch contacts. Figure 5-32 shows the circuit schematic for the coin detection sensor. The homebrew sensor illustrates the *Proof Of Concept* (POC) prototyping development quite proficiently. In the POC development of products, prototypes are usually crude but very functional. When developing Mindstorms robots, input/output sensors and actuators should be breadboarded initially so the functionality of circuit interfaces and software can be verified. Therefore, this sensor demonstrates how additional LEGO bricks can assist in the development of specialized robots.

Coin-Operated T-Bot Lab Project

The following lab project tests out the sensor for detecting coins using simple RCX code. In this lab project, a coin-operated robot is built using a

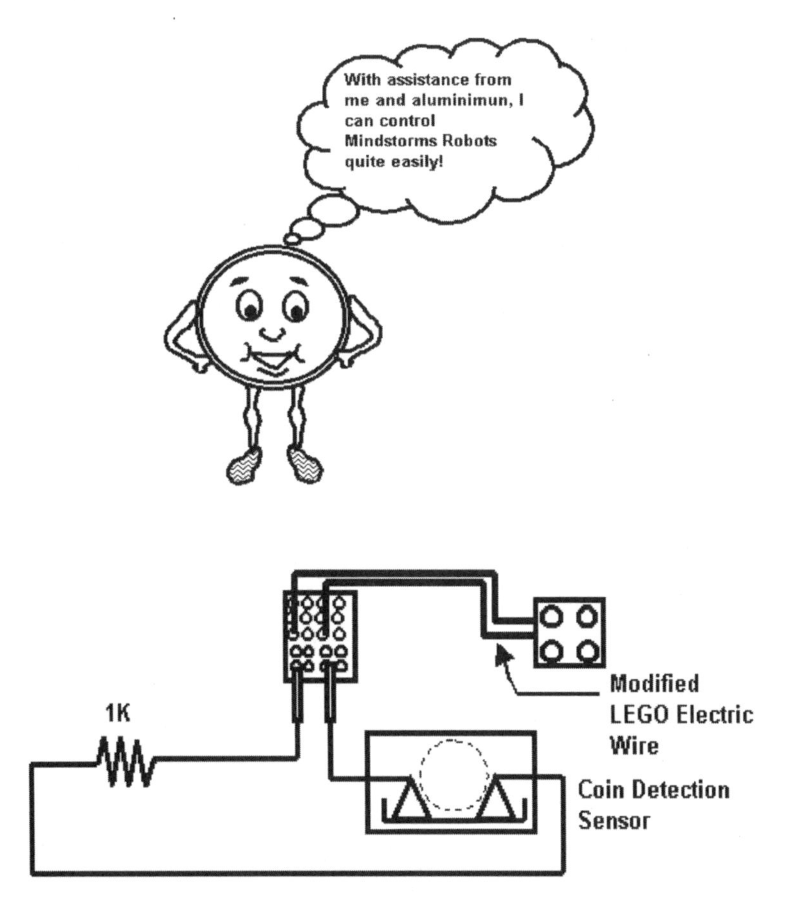

Figure 5-32 The circuit schematic of the coin detection sensor

simple sensor built with a few LEGO bricks and aluminum foil. The coin detection sensor is a homebrew variation of an electromechanical switch (see Chapter 3) that uses LEGO bricks for the physical packaging of the aluminum contacts. The coin completes the circuit by enabling electrons to flow into the RCX P-Brick's internal interface circuit.

BOM

- Aluminum foil
- Alligator test leads
- Two 2×6-stud bricks
- Two 1×4-stud bricks
- Two 1×2-stud bricks
- Two 1-stud sloped roof bricks
- 4×8 plate
- Modified LEGO electric wire
- Infrared remote control
- Radio Shack Electronics Learning Lab breadboard system or equivalent
- One 1-kilo-ohm resistor

Coin-Operated T-Bot Procedure

1. Build the coin detection sensor using the additional bricks described in the BOM. Use Figures 5-33 through 5-35 as assembly guides.
2. Build the circuit in Figure 5-32 using the Radio Shack Electronics Learning Lab breadboard system or equivalent with the modified LEGO electric wire.
3. Connect the circuit to input 1 of T-Bot using the modified LEGO electric wire.
4. Build the code shown in Figure 5-36 and download it to program location 3.
5. Using the infrared remote control, position T-Bot where it's looking straight ahead. This is the starting point from which T-Bot will swing to the right and left of its rotational axis.

Figure 5-33
Inside of the coin detection sensor. Note the aluminum foil that covers the two LEGO bricks.

Figure 5-34
Coin detection sensor with a quarter resting across the two aluminum foil LEGO bricks

Figure 5-35
The completed coin detection sensor

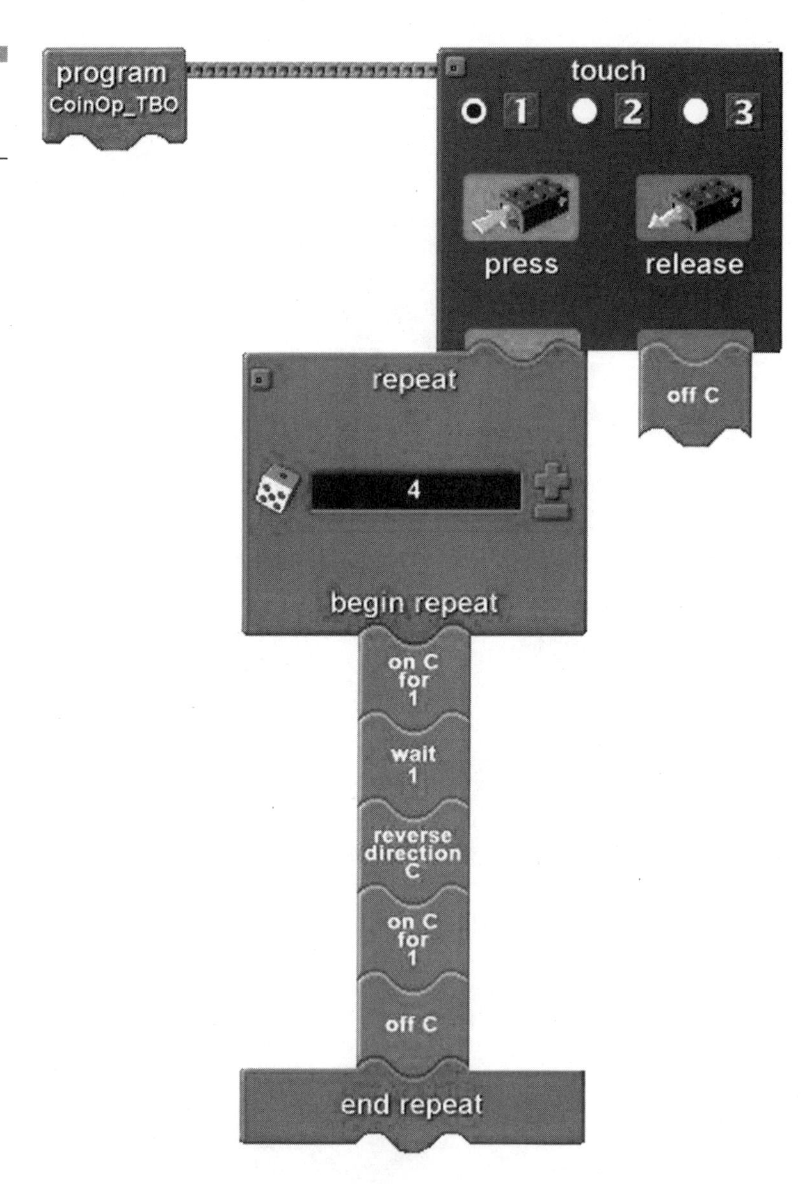

Figure 5-36
RCX code for the coin-operated T-Bot

6. Push Run and drop a coin into the sensor. T-Bot should begin to rotate to the left and right of its rotational axis four times before coming to a complete stop.
7. Try various coins to test the sensor detection function of T-Bot.
8. Stop the code and turn off T-Bot.

Coin-Operated T-Bot Lab Project Analysis

The sensor in this lab project is a simple variation of the electromechanical switch discussed in Chapter 3. Various coins enable T-Bot to execute the code quite well with no hesitation. If the coin touches the two 1-stud sloped aluminum foil covered roof bricks, the 1-kilo-ohm resistor becomes physically attached to the internal interface circuit of the RCX P-Brick. The code receives the control signal and enables T-Bot to swing on its rotational axis four times.

This sensor opens up a wealth of opportunities in object-detection schemes. For example, the sensor could be used to discern between metal and nonmetal objects and materials, creating a homebrew-simplistic metal detector system. An Excel *Visual Basic of Applications* (VBA) monitor/GUI could be built to display and record metal versus nonmetal objects. The data could then be printed on a worksheet for further statistical analysis.

The RCX P-Brick could be replaced with the Scout P-Brick quite easily. *Stand-Alone Code* (SAC) would be used to allow mobility for the robot. The prepackaged motion programs could be configured to give the Scout P-Brick some interesting mobility patterns. This design suggestion is left as a lab project for hobbyists or experimentalists to perform on their own.

Rheostat-Sensor-Controlled T-Bot

The rheostat is a variable resistor that restricts the flow of electrical current. The rheostat is a converted potentiometer used to change the current flowing in a DC or AC circuit. The traditional potentiometer has three terminals (top, bottom, and center) coming out of a cylindrical body. A silver shaft appears on the top of the body. By rotating the shaft of the potentiometer, the resistance varies based on the movement of the center terminal wiper arm. When one of its terminals and the center are used, a rheostat is configured for current control. When connected to the input of the RCX P-Brick, the device behaves like a simulated manually controlled temperature sensor. Using the RCX code temperature sensor watcher, thresholds or boundaries can be established for turning the outputs of the RCX P-Brick on or off. An interesting thing to note about this interface is that once the position of the robot is found, the equivalent temperature value can be programmed into the temperature sensor watcher. After the robot receives the equivalent temperature value, it can carry out the specific task. Therefore, rheostat can also assist in developing temperature sensor control robots by acting as a physical simulator.

Rheostat-Sensor-Controlled T-Bot Lab Project

The following lab project demonstrates how the rheostat can be used to simulate a temperature sensor for controlling T-Bot. In the procedure, the circuit schematic is left as an exercise for the hobbyist or experimentalist to draw. Figure 5-37 is used as a guide for drawing the circuit schematic with RCX code for testing the electrical interface to T-Bot.

BOM

- One 1-mega-ohm potentiometer
- One 1-kilo-ohm resistor
- Modified LEGO electric wire
- Radio Shack Electronics Learning Lab breadboard or equivalent

Rheostat-Controlled T-Bot Procedure

1. Draw the circuit schematic for the rheostat sensor control interface using Figure 5-37 as a reference.

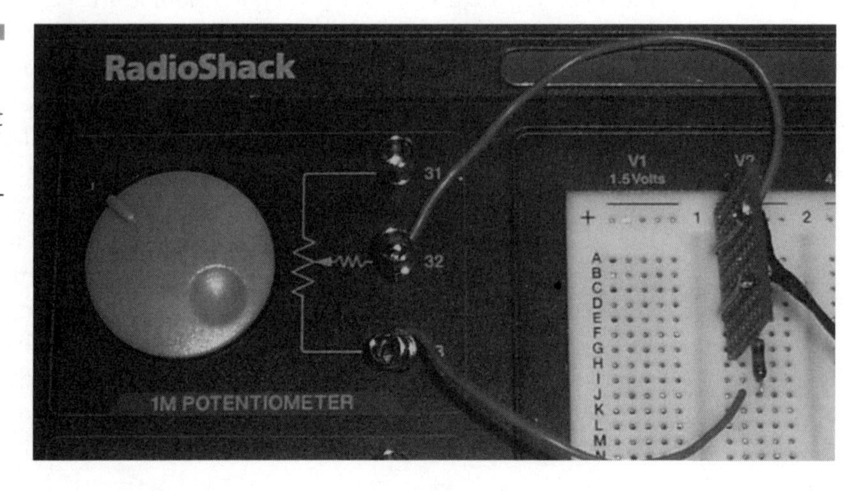

Figure 5-37 Wiring of the 1-mega-ohm rheostat to the modified LEGO electric wire

2. Build the sensor interface circuit using the Radio Shack Electronics Learning Lab breadboard or equivalent.
3. Connect the circuit to input 1 of T-Bot using the modified LEGO electric wire as shown in Figure 5-37. Figure 5-38 is a preview circuit using the SDIC for the Analog-Digital Control Lab project on page 193.
4. Build the code shown in Figures 5-39 and 5-40. Change the temperature values to 50 (low) and 55 (high) and download it to program location 2.
5. Run the code and push View to read the temperature value (Fahrenheit) on the LCD.
6. Slowly rotate the 1-meg rheostat and note the readings on the LCD. Set the rheostat for 50 degrees. T-Bot should execute the embedded RCX code. The swing motion should repeat four times.
7. Adjust the rheostat above 55 degrees. Did T-Bot move?
8. Change the boundaries of the temperature sensor watcher and download the new code into T-Bot.
9. Slowly rotate the rheostat to approach the new setting and notice T-Bot. The repeat stack controller can be changed for a new X number for the code to loop. Download the new change into the robot and notice its movement patterns.
10. Stop the code and turn off T-Bot.

Figure 5-38
The analog-digital sensor (ADS) breadboard circuit with an SDIC board attached to input 1 of T-Bot

Figure 5-39 RCX my command code for swinging T-Bot to the right and left of its motorized drive assembly

Rheostat-Sensor-Controlled T-Bot Lab Analysis

This lab project enables hobbyists or experimentalists to draw a rheostat sensor interface circuit using a photograph of a wired prototype. The temperature sensor watcher enables T-Bot to perform the embedded task based on detecting a threshold value range of 50 to 55 degrees Fahrenheit. Other values can be programmed into the sensor watcher and tested as well using the 1-meg rheostat as a simulated temperature sensor. The pro-

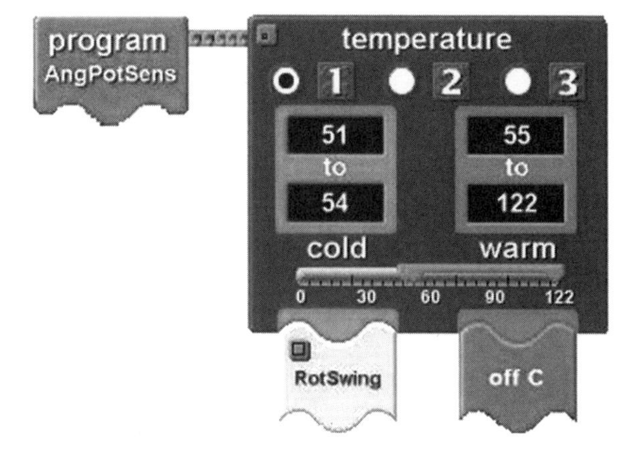

Figure 5-40 RCX code for the rheostat-sensor-controlled T-Bot lab project

cedure outlined could be used to calibrate the robot for the development of temperature-sensing applications and controls. The last lab project in this chapter explores the use of an active sensor to detect an analog voltage level and generate a digital signal for robotic control. The Vision Command Camera will be used as a visual eye for T-Bot.

Analog-Digital Sensor (ADS) Control for T-Bot

The ADS is a hybrid circuit that consists of analog components capable of generating a digital control signal. The heart of the sensor is a voltage comparator circuit used to monitor the analog DC voltage being applied to its input pin. Figure 5-41 shows a functional block diagram for T-Bot control using an ADS. The analog DC signal is created by the potentiometer operating as a voltage divider circuit. This signal is compared with a known reference voltage (V_{REF}). A comparator is an electronic circuit that produces either a high or low output based on a comparison between an input voltage and a reference voltage. This signal is compatible with digital logic interfacing, making it a simple analog-digital switch. With either a high or low signal present at the RCX P-Brick, the embedded code will process this input event and produce the appropriate output control signal for driving the RDA mechanism. Figure 5-42 shows the complete circuit schematic for the ADS interface.

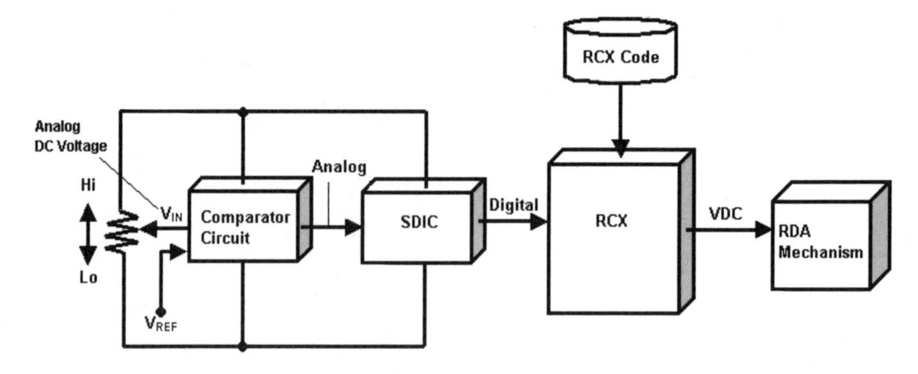

Figure 5-41 System block diagram for the ADS control lab project

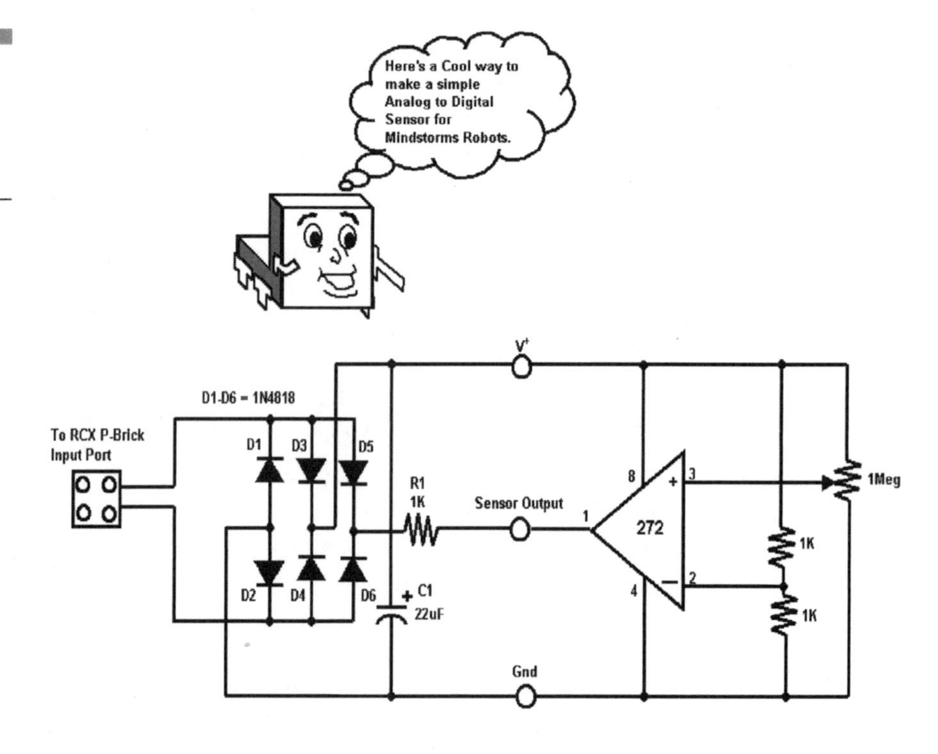

Figure 5-42 The complete circuit schematic for the ADS presented by Ollie Op-Amp

The circuit is a modified version of Forrest Mims' comparator shown on page 80 of the "Basic Electronics Workbook 1" of the Radio Shack Electronics Learning Lab. The circuit works when the 1-mega-ohm potentiometer provides a variable input voltage range between 0 − +5.26 volts DC. The reference voltage (V_{REF}) is set to 2.62 volts. When the input voltage is greater than (V_{REF}), the output swings from 0 volts to approximately 3.71

volts at pin 1 of the TL272 operational amplifier. This analog voltage will be converted into a digital equivalent value that the RCX P-Brick will use to drive the motorized base assembly.

The RCX code provides a software interface for the P-Brick, enabling great flexibility in setting threshold values for Mindstorms robotic control. If the different V_{REF} is needed, changing the resistor values of the divider network accomplishes the task. The following lab provides a procedure for building and testing the ADS and gives information on how to add a Vision Command LEGO Camera that serves as a visual eye to T-Bot.

ADS Control Lab Project

This lab project demonstrates the *Analog-to-Digital* (A/D) conversion using the ADS to control T-Bot. This procedure outlines how to build and test the sensor and provides information on how to add the Vision Command LEGO Camera to T-Bot.

BOM

- One 1-mega-ohm potentiometer
- Two 1-kilo-ohm resistors
- One TLC272 operational amplifier
- One SDIC
- Vision Command LEGO Camera
- LEGO Camera software
- DMM
- LEGO infrared remote control

ADS Control Lab Procedure

1. Build the sensor interface circuit shown in Figure 5-42 using the Radio Shack Electronics Learning Lab breadboard or equivalent. Note: Before powering the circuit, check your wiring carefully. Do not power the circuit interface with a secondary DC power supply because the RCX P-Brick's input can be damaged.
2. Build and download the code into program location 1 using Figure 5-43. Note: The comparator circuit will not be powered until this program is loaded into the RCX P-Brick. The my command Swing is stored on the

RCX programming library; therefore, the subprogram can be used with the light sensor watcher, as shown in Figure 5-43.

3. Turn on T-Bot and position the robot using the LEGO infrared remote control so it's facing straight ahead.
4. Run the code at program location 1. Depending on the position of the 1-mega-ohm potentiometer, T-Bot might begin to swing from right to left.
5. Press View on the P-Brick. LCD should display 84 through 85.
6. Rotate the potentiometer until you get an LCD reading of 0.
7. Rotate the potentiometer slowly until the T-Bot starts to swing.
8. Using a DMM, measure the input voltage at pin 3 of the TLC272 op-amp. Record the voltage for $V_{IN\text{-}PIN\ 3}$.
9. Using a DMM, measure the voltage at pin 2 of the TLC272 op-amp. Record the voltage for $V_{REF\text{-}PIN\ 2}$.
10. Measure the output voltage at pin 1 of the TLC272 op-amp. Record the voltage for $V_{OUT\text{-}PIN\ 1}$.
11. Adjust the 1-mega-ohm potentiometer for an LCD reading of 0. Record the voltages for the following: $V_{IN\text{-}PIN\ 3}$, $V_{REF\text{-}PIN\ 2}$, and $V_{OUT\text{-}PIN\ 1}$.
12. Play with the potentiometer and notice T-Bot executing the embedded code via the swing motion of its body.
13. Change the repeat value for different loop cycles for T-Bot and validate the correctness of this data value.
14. Turn off T-Bot.

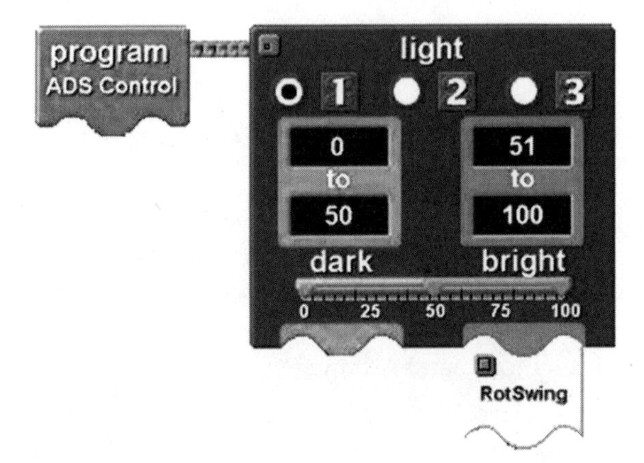

Figure 5-43 RCX code for the ADS lab project

Vision Command Camera Lab Procedure

The following steps outline the procedure for adding a Vision Command LEGO Camera to T-Bot:

1. Add the four appropriate connector pegs to the Vision Command LEGO Camera, as shown in Figure 5-44.
2. Attach the LEGO Camera to T-Bot using Figure 5-45 as an assembly guide.
3. Attach the camera to the *Universal Serial Bus* (USB) port of the notebook or desktop computer.
4. Run the LEGO Camera software through the Vision Command GUI.
5. Turn on T-Bot and run the ADS sensor control program.
6. Adjust the 1-mega-ohm potentiometer to start T-Bot's swing motion. The display window within LEGO Camera software should show different views of the target room.
7. Readjust the potentiometer several times to view the target room with the camera.
8. Experiment with other features of the software using the ADS control interface circuit for T-Bot.
9. Stop the program and turn off T-Bot.

Figure 5-44
The four connector pegs mounted to the Vision Command LEGO Camera

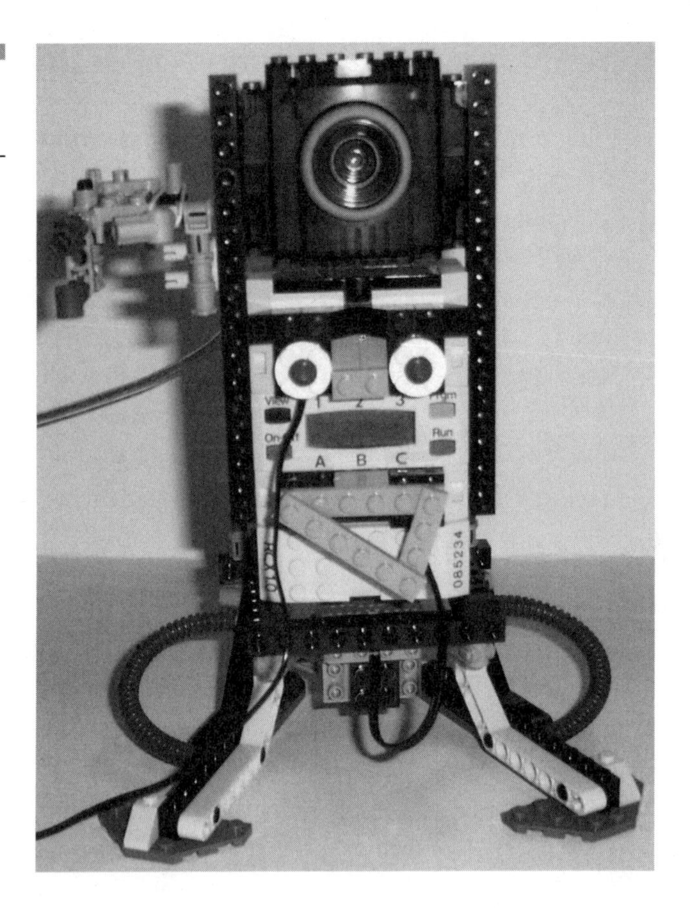

Figure 5-45 The LEGO Camera attached to T-Bot

ADS Control Lab Project Analysis

The data measured using the DMM and recorded in Steps 8 through 11 describe a noninverting comparator with the output behavior of $V_{OUT-HIGH} = V_{IN\text{-}PIN\,3} > V_{REF\text{-}PIN\,2}$. The output voltage of 3.71 volts is at a signal level that is appropriate for digital control interfacing. T-Bot would execute the code because of the logic High input signal received. The LEGO Camera provides a visual eye for T-Bot, enabling the robot to see the environment it lives in. As stated in the procedure, other visual activities can be explored with T-Bot as an RDA. The ADS is a versatile sensing device that is only limited by the hobbyist and experimentalist's imagination. The following are some suggestions for further experimentation using T-Bot and sensors:

- Try logging the output transition or swing voltage of the comparator circuit using the DMM with a PC interface measurement method.
- Try some of the design challenges discussed in the lab project procedures and record the results in an engineering notebook.
- Try using the Vision Command software along with the sensors discussed to trigger the LEGO Camera upon receiving the right input stimulus.

And by all means have fun with sensor interfacing with homebrew devices for controlling Mindstorms robots.

The lab projects discussed in this section enable hobbyists and experimentalists to explore the basic operation of traditional sensors as well as develop and test homebrew devices. The circuits used in the book were modified using existing designs out of the "Basic Electronics Workbook 1," written by Forrest Mims and packaged with the Radio Shack Electronics Learning Lab. The main focus of the chapter was to illustrate how passive and active sensors give the RXC P-Brick greater flexibility in mobility and functionality. Using RCX code enables hobbyists and experimentalists to test homebrew sensors quickly by characterizing their performance via logging signal data. The data can then be imported into a spreadsheet and a plot can be created for further review and analysis of the sensing device. The POC methodology of sensor development is based on creating a prototyping test system that can quickly acquire data and provide a characteristic curve related to the sensor's sensing performance. A Mindstorms robot's autonomous feature is based on its capability to maneuver within the target environment using electronic sensors. The homebrew sensors discussed in this chapter can be packaged into a standard Technic brick for advanced robotic sensing-control applications.

In Chapter 6, "Using Procedural Languages for Mindstorms Robot Control," alternative programming languages for the Scout and P-Brick are explored. The RCX code is a very good programming language for developing test code and basic tasks for the Mindstorms robot. Higher programming languages enable hobbyists and experimentalists to develop behaviors that are more flexible and allow somewhat of an artificial intelligence to be embedded inside the smart machine. Therefore, the Mindstorms robot is a research tool for exploring computer science and software development techniques. The state machines discussed in the Chapter 3 and Chapter 4, "Electronic Switching Circuits," provide a gateway into software development tools. Chapter 6 also further investigates how to use state diagrams to assist in writing high-level language programs.

Also, the format changes in that details on circuit interfacing are left as exercises for the hobbyist and experimentalist to develop. Block diagrams are shown to illustrate the circuit-interfacing concept to assist the experimentalist and hobbyist in building a successful Mindstorms robot. Basic circuit-interfacing techniques will be referenced to the appropriate previous chapters in the book. Only specialized circuits are provided with a full schematic in the Chapter 6.

CHAPTER 6

Using Procedural Languages for Mindstorms Robot Control

The Merriam Webster's definition of a procedure is "to lay down a rule and a way of doing something." When building a robot, the task it is given describes activities that it will carry out following a set of logical patterns or behaviors. A rule is an accepted method of behavior or procedure. Robots are primarily procedural machines because their programming is based on a set of rules on how to do something. The behavior is embedded within the rules, giving the robot a defined personality.

The RCX code packaged with the *Robotics Invention System* (RIS) kit is a graphical procedure because it establishes basic rules that enable the Mindstorms robot to do a specific task using colorful blocks. Each block contains specific instructions or rules that the robot must follow to carry out the overall function established by the programmer. The *Robot Command Explorer* (RCX) code is a great programming tool for testing or calibrating sensors (see Chapter 5, "Sensor-Interfacing Basics") or validating simple behaviors and functions for robots. However, as the task becomes large or complex, a higher-level programming language is needed. The language must permit events to be conditionally wired so that the decision-making activity touches on the environmental reality in which the robot must maneuver. Therefore, the language should enable the programmer to program a well-defined set of procedures that the robot can use to meet these environmental challenges.

This chapter examines several procedural languages. An emphasis is placed on writing small bytes of code that can serve as building blocks for more sophisticated functions. The *Proof Of Concept* (POC) of software development follows the same architectural concept as systems engineering, which was discussed in the previous chapters. Therefore, a similar approach for writing and building code for Mindstorms robots is adopted in this chapter. This chapter examines the following procedural languages:

- *Not Quite C* (NQC)
- *LEGO Assembly* (LASM)
- *Interactive C version 4* (IC4)

The RCX code previously used serves as a springboard for writing and building robot control applications using the procedural languages.

Experimental Design Format for This Chapter

The last three chapters presented the basics of hardware circuit interfacing to the RCX and Scout *programmable bricks* (P-Bricks). The philosophy behind building hardware circuits was to apply the knowledge to create advanced robots and controls. The circuits and their electrical characteristics were captured using simple alternative data acquisition measuring instruments. The knowledge obtained from the previous lab projects should be easily applied to this chapter. The topics of electromechanical and electronic interfacing described in this chapter draw from the circuits discussed in Chapter 3, "Electromechanical Controls Interfacing," Chapter 4, "Electronic Switching Circuits," and Chapter 5; therefore, detailed information about operation and performance are kept to a minimum.

The lab project procedures instruct hobbyists and experimentalists on how to design an interface circuit that provides a signal with certain voltage characteristics. Functional block diagrams assist in the design of the electromechanical and electronic interface for robot control. This format enables hobbyists and experimentalists to apply what was learned in the last three chapters to this chapter's discussion on using procedural languages for Mindstorms robot control applications. Let's kick off this chapter with a discussion on NQC language and its application to Mindstorms robot controls.

NQC

NQC was developed by Dave Baum[1] (www.enteract.com/~dbaum/lego/nqc) to assist the LEGO Mindstorms Online Community *LEGO Users Group Network* (LUGNET) (www.luqnet.com/robotics/) in writing code with no limitation in structure and function. NQC can be thought of as an application C language for the RCX P-Brick. The same programming format found in C language can be applied to writing code for Mindstorms robotic applications with NQC. An interesting point to make about NQC is that it gives you the ability to write code on the microcontroller level. The ability to

[1]The Definitive Guide to LEGO Mindstorms was published by Apress Publishing, copyright 2000.

manipulate bits related to *liquid crystal display* (LCD) segment control, conditional and relationship programming of sensors, as well as the speed control of motors and specialized actuators gives NQC a greater advantage over the prepackaged RCX language.

NQC Overview

Dave Baum, the creator of NQC, wrote a book called *The Definitive Guide to LEGO Mindstorms*. Chapter 2 of his book, "Introduction to NQC," explains the story behind NQC in detail. This section is a short summary of the programming concepts of NQC. The following lab project gives hobbyists and experimentalists an opportunity to explore some of the basic concepts of NQC by taking a simple RCX code used in a previous lab project and converting it to an NQC program. The overview discussion explains the basic programming style and syntax used to write NQC code for Mindstorms robots.

Basic Tools for NQC Code Development

Programming in NQC requires two tools: a compiler and text editor. A text editor such as Notepad, WordPad, or MS Word enables the program to be typed using plain English syntax. The use of English text helps the programmer to write code that can be understood by another developer and simplifies the debug effort during error correction. The compiler takes the English format of a program known as a *source file* and converts it to byte code that the microcontroller understands. The code can be downloaded to the RCX P-Brick using the Infrared Tower. Once loaded into the RCX P-Brick, the code can be executed by running the specified program location. To compile the code, an MS-DOS command prompt must be used to enter the basic instructions that convert the source file into byte code. A Windows tool is also available that combines the compiler and text editor into one programming environment known as the *Brick Command Center* (BricxCC) developed by John Hansen (http://members.aol.com/johnbinder/bricxcc.htm/). The BricxCC Windows tool is used in this book to build code for Mindstorms robotic applications. Let's take a look at the basic programming style for writing the NQC code used in Mindstorms robots.

Programming Style for NQC

Basically, the programming style for NQC consists of the following main sections: the program title, variable declaration, task, the main body, and functions. The program title is usually a short one-line description of the code. The syntax usually looks like the following:

```
// name of program.nqc
```

where name of program is the title of the code given by the programmer and .nqc is the source file extension written in NQC language. The variable declaration is the name used to store data within an NQC program. The RCX P-Brick uses 32 locations to store data. They can be declared or made known to the rest of the program either locally (within a specific task) or globally (accessible to the entire program). The key word used in declaring a variable is int. The syntax for int looks like the following:

```
int load;    // global variable load
```

Therefore, this block of the code is used for variable declaration. Once a variable is declared, it may be assigned a value using the following assignment statement:

```
load=12; // set the value of load to 12
```

The program takes the integer value 12 and stores it into a memory location with the variable name load. Other variable names can be created that have a definitive meaning related to the operation or process NQC executes. The #define command makes the program easier to read, debug, and modify. The syntax for #define is as follows:

```
#define Volt_Out      OUT_A
```

Using On(OUT_A): is correct, but using #define enables this command to take on a meaningful name that relates to the function of output A of the RCX P-Brick.

In C language, the main() function is where the execution of the program begins. Statements inside of the open curly braces are executed when the main() starts the code listing. The program stops when the closed curly braces are reached with the main().

Dave Baum explains what tasks are in his book *The Definitive Guide to LEGO Mindstorms*[2] using several robot programs as an example on pages 50 and 51. My definition of a task is a small procedure used with "main" to describe the name of the function to be executed within the curly braces." Basically, a task can be thought of as the name of a user-defined function or behavior embedded within an NQC program to be called by the main(). The task is usually written before the main(), creating a structure-programming format. This programming style makes code development and debugging easy to manage because each task can be tested in sequential or random order, enabling validation and modification to be seamless development activities for Mindstorms robots. The syntax for task and main() is as follows:

```
Task main()
{
     instructions go here
}
```

Functions are used to break up large programs into smaller manageable code blocks. When used in this manner, functions make NQC code easier to read, maintain, and reuse (see Chapter 2, "Developing GUIs: Software Control Basics," on code reusability). The syntax for a function looks similar to that of a task, but the keyword void is used instead of task:

```
              void Convert( )
}
       function statements go here
}
```

The function can then be called from within a task (or another function):

```
task main()
```

Figure 6-1 shows the programming style graphically for NQC.

The best way to learn a programming language is to play with it. The following lab project enables hobbyists and experimentalists to write a small NQC program using an existing RCX code as a basic template for NQC code conversion.

[2]The Definitive Guide to LEGO Mindstorms was published by Apress Publishing, copyright 2000.

RCX to NQC Code Conversion Lab Project

RCX code is a good way for a first-time Mindstorms user to become acclimated to software development. RCX code can be used as test code that checks a specific robot control task or investigates a special behavior for a specialized robotic application. Some Mindstorms robot developers believe RCX code has nothing to offer in programming style and accessibility to the inner workings of the microcontroller. If used as an NQC code development aid, RCX language can serve as a tool for writing functions for Mindstorms robots. The sensor watchers are good programming tools for developing specialized functions for traditional and homebrew sensing devices for Mindstorms robotic interfacing. After reading the text version of an RCX program, it becomes apparent that a wealth of code information can be converted into NQC. The following lab project outlines a reverse-engineering procedure for reviewing RCX code and then converting it to an equivalent NQC program.

Bill of Materials (BOM)

- BricxCC software
- Notepad or a simple text editor
- Smart Switch.rcx program
- RCX P-Brick

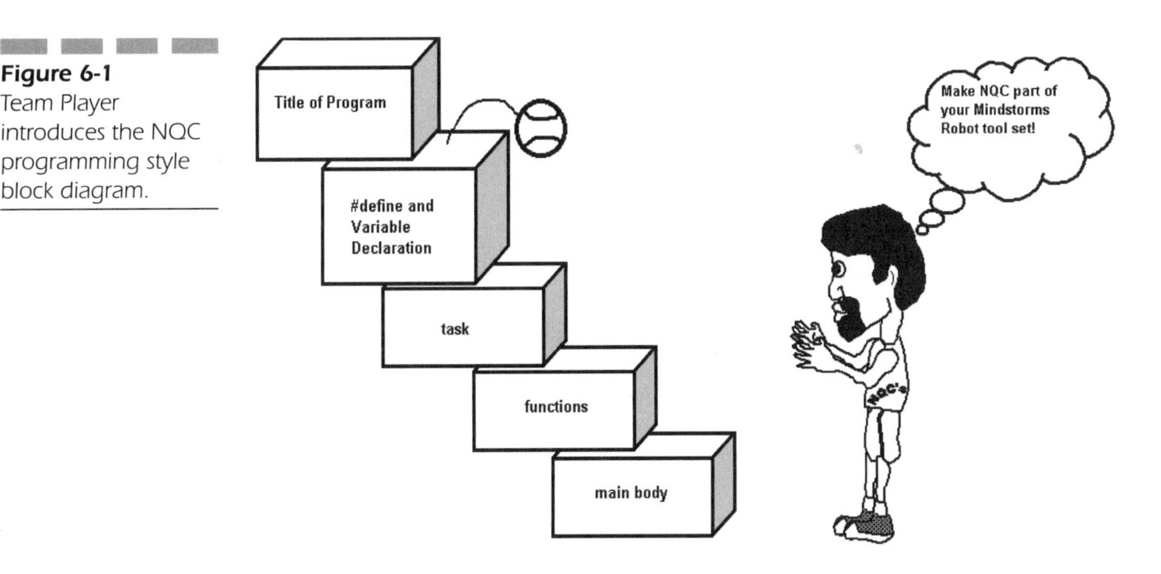

Figure 6-1 Team Player introduces the NQC programming style block diagram.

- Two LEGO electric wires
- One touch sensor
- One LEGO small 9-volt motor

RCX to NQC Code Conversion Lab Procedure

1. Open the Smart Switch.rcx stored on your hard drive or build the code using Figure 3-18 as a guide (see the section "Smart Switch Lab Procedure" in Chapter 3).
2. Using Notepad, view the text version of the Smart Switch.rcx program. See Figure 6-2. Note the number of lines for the program.
3. Open the BricxCC software. Within the editor window, enter the Smart Switch.nqc version of code shown in Figure 6-3. Notice the number of lines used in the program and the instructions used to read the sensors and control the motors. Compare this number to what was observed in Step 2.

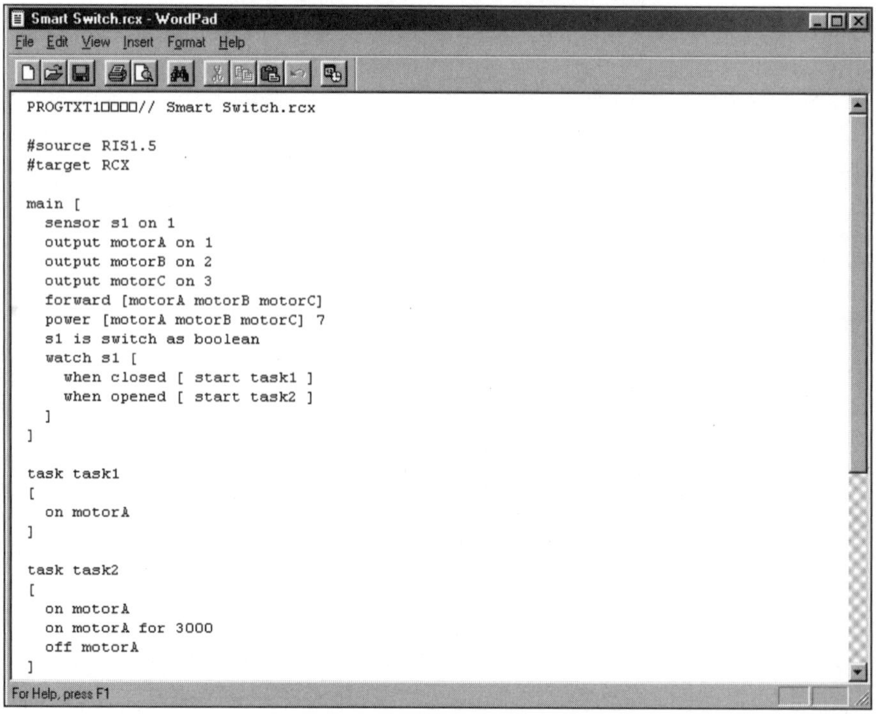

Smart Switch.rcx - WordPad

File Edit View Insert Format Help

```
PROGTXT1□□□□// Smart Switch.rcx

#source RIS1.5
#target RCX

main [
  sensor s1 on 1
  output motorA on 1
  output motorB on 2
  output motorC on 3
  forward [motorA motorB motorC]
  power [motorA motorB motorC] 7
  s1 is switch as boolean
  watch s1 [
    when closed [ start task1 ]
    when opened [ start task2 ]
  ]
]

task task1
[
  on motorA
]

task task2
[
  on motorA
  on motorA for 3000
  off motorA
]
```

For Help, press F1

Figure 6-2 The source-text code from the Smart Switch.rcx program

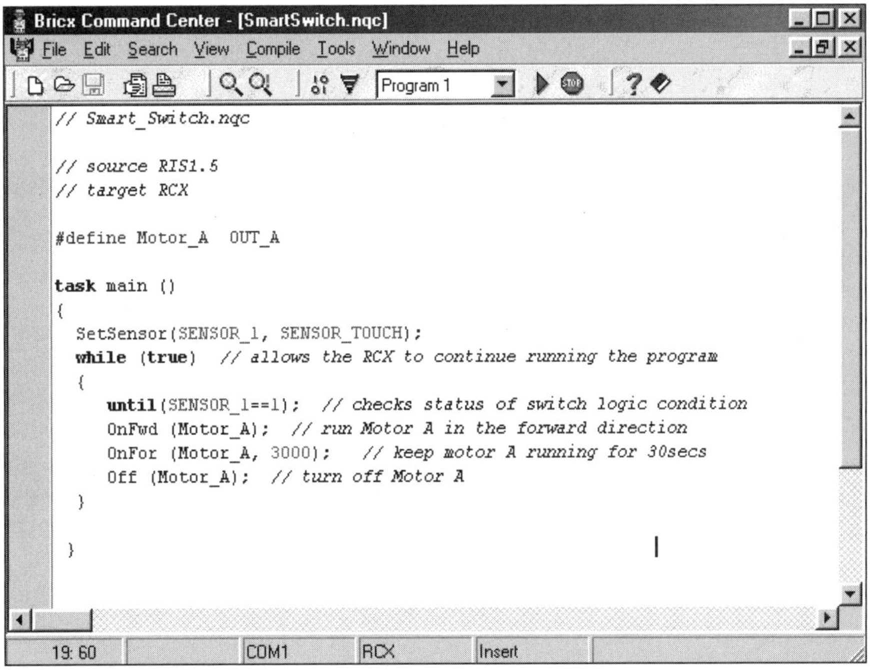
Bricx Command Center - [SmartSwitch.nqc]
File Edit Search View Compile Tools Window Help
Program 1
// Smart_Switch.nqc
// source RIS1.5
// target RCX
#define Motor_A OUT_A
task main ()
{
SetSensor(SENSOR_1, SENSOR_TOUCH);
while (true) // allows the RCX to continue running the program
{
until(SENSOR_1==1); // checks status of switch logic condition
OnFwd (Motor_A); // run Motor A in the forward direction
OnFor (Motor_A, 3000); // keep motor A running for 30secs
Off (Motor_A); // turn off Motor A
}
}
19: 60
COM1
RCX
Insert

Figure 6-3 The NQC code version of the Smart Switch.rcx program

4. Download the program by highlighting Compile on the BricxCC menu bar and click Download. Store the code in program location 1.
5. Add the touch sensor to input 1 of the RCX P-Brick and the LEGO small 9-volt motor to output A using the two LEGO electric wires.
6. Push Run on the P-Brick, and press and release the touch sensor. The RCX P-Brick executes the code stored at program location 1. The LEGO small 9-volt motor should turn on after the sensor is pressed and released. After 30 seconds, the motor should turn off. The little man should still be running as displayed on the LCD, signifying that the code is operating.
7. Repeat Step 6 and notice the little man on the LCD.
8. Edit the while(true) statement by adding // in front of it. Compile the code and download the program into the RCX P-Brick.
9. Press Run on the RCX P-Brick to execute the code. The outcome should be the same as in Step 6 with the exception of the code not running after the task is completed.
10. Repeat Step 9 and observe the LCD. Is the code still running?

11. Remove the // from the while(true) statement, and compile and download the code into the P-Brick.
12. Repeat Step 6 and notice the LCD. Is the program running after the task is completed?
13. Stop the program by pressing Run on the P-Brick.

RCX to NQC Code Conversion Lab Project Analysis

This lab project is a basic introduction to writing NQC code for the RCX P-Brick. The RCX code viewed in text shows the procedural language used to read the touch sensor, process the digital input data, and turn on output A for 30 seconds. The RCX program uses 16 lines of instruction to carry out a basic control function. The NQC code uses seven lines of instruction to perform the same control function. Therefore, NQC can be used to write programs using small amounts of code to perform simple or complex Mindstorms robot controls.

The while() statement is used to tell all of the RCX P-Bricks to continue executing their program if the argument within the braces is true. By setting the argument inside () to true, the program continues to run even though the control function or task has ended. In Step 6 of the lab project, the little man displayed in the LCD runs because of the while(true) statement. The program stops running after the control function is executed because of the //. The // is used to add comments in the code for short descriptive notes about how the program works.

Commenting can also be used to remove the code instruction without deleting it from the program. In Step 8 of the lab project, this activity enables you to see the code execute and the program stop once after the execution of the control function.

Although the lab project was simple to build and execute, the concept is a basic building block for developing more sophisticated software controls for Mindstorms robots.

The seven lines of code can be easily expanded by adding a second touch sensor and enabling a logic control function to be created using NQC code.

Figure 6-4 shows the code for creating an AND logic smart switch. The && is used to initiate a logical AND function. The program, when executed, does not turn on the motor at output A until both touch sensors are pressed. The logical OR function can easily be implemented by replacing && with ||. These symbols are used in C language and have the same syntax struc-

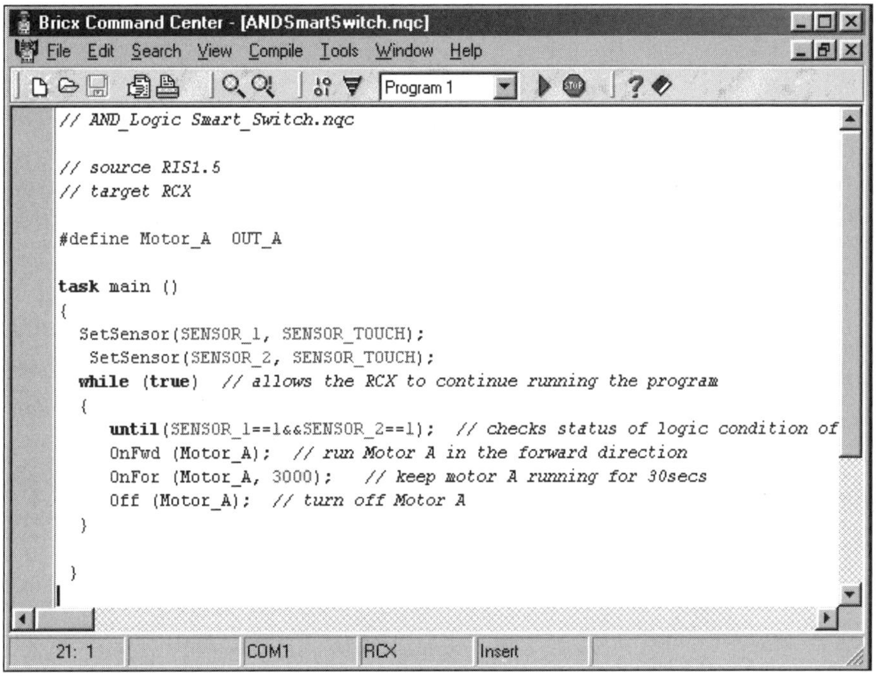

Figure 6-4
The AND logic smart switch NQC code

```
Bricx Command Center - [ANDSmartSwitch.nqc]
File Edit Search View Compile Tools Window Help
Program 1

// AND_Logic Smart_Switch.nqc

// source RIS1.5
// target RCX

#define Motor_A  OUT_A

task main ()
{
  SetSensor(SENSOR_1, SENSOR_TOUCH);
   SetSensor(SENSOR_2, SENSOR_TOUCH);
  while (true)  // allows the RCX to continue running the program
  {
     until(SENSOR_1==1&&SENSOR_2==1);  // checks status of logic condition of
     OnFwd (Motor_A);  // run Motor A in the forward direction
     OnFor (Motor_A, 3000);   // keep motor A running for 30secs
     Off (Motor_A);  // turn off Motor A
  }

 }

21: 1    COM1    RCX    Insert
```

ture of NQC. Therefore, NQC is a user-friendly way to learn C language. As an additional lab project, try the logical OR function on your own and see if the program enables each touch sensor to turn on the motor at output A. Implementing this function using RCX code requires quite a bit of graphical blocks to perform the logical task. This feature of small code development makes NQC an appealing software tool for Mindstorms robotic applications.

LASM

Back in the old days of mainframe computers, assembly language was used to communicate with the machine's microprocessor internal functions such as its *arithmetic logic unit* (ALU). The internal shift registers and counters used to perform the mathematics could be manipulated by moving logic bits of 0s or 1s with assembly language. In assembly, opcode/nmenonic codes are used to develop the software. Opcode/nmenonic codes have a textual interface that uses three- to four-letter words representing specific commands for performing math, controlling program flow, and manipulating logic bits.

The LEGO Company has developed a version of assembly language for the Scout P-Brick called LASM. In Chapter 1, "Wireless Basics," some of the LASM opcode/nmemonics were explored. Direct user commands such as out and playt are used to control the outputs of the Scout P-Brick and its internal tone generator's frequency. In the Scout *Software Development Kit* (SDK) *User Guide and Reference*[3], the opcode/nmenonics used in LASM are discussed on pages 33 through 44. LASM is a little tricky to use, which makes it challenging to learn. Several programs are scattered throughout the pages of the Scout SDK. The best way to learn software programming is to find a similar program application and modify it to meet the project requirements. Listing 1 is a modified version of the sample program on pages 29 through 30 of the Scout SDK:

```
Listing 1:
#include  "ScoutDef.h"
#define TASK_MAIN  0
#define TASK_TOUCH 1

dels              //Deletes all subroutines from memory
delt            //Deletes all tasks from memory

task TASK_MAIN
  ; initialization
  out      OUT_OFF,  OUTLIST_AB               ;  turn motors off
  setfb  SRC_CON,  FBMASK_NO_FB               ; shut the system up
  tmrs     0, SRC_CON,  CR_SEC_1              ; wait a second
  ; start sensor watchers
  start TASK_TOUCH
endt

task TASK_TOUCH
starttask_label:

mone   SRC_CON,  EVENT_TPR,  watchercode_label    ;  sets the event
"EVENT_TRP" to interrupt program execution  and jump to address label

forever_label:
   jmp     forever_label    ; wait here until event "EVENT_TRP" happens

watchercode_label:

   plays    SND_BEEP           ; plays sound beep
   out       2, 1              ; turns on Motor A
   wait      2, 3000           ; Motor A runs for 30 secs
   out       1, 1              ; turns off Motor A

   jmp       starttask_label   ; goto beginning of task TASK_TOUCH
endt
```

[3]Published November 1999.

The code is equivalent to the smart switch software programmed in RCX code and NQC language. The only difference in this code application is the SND_BEEP argument for the plays opcode/nmenonic. If the touch sensor is pressed, the internal sound generator produces a beep tone before turning on the motor at output A. LASM has similar syntax traits to C=language. Both indicate comments by using the // and #define statement.

The programming style for LASM is comparable to that of NQC. The program starts off with a header file called ScoutDef.h. This C-language file contains all of the constant declarations for the Scout P-Brick resources. It's basically the library of the constant opcode declarations used to program the P-Brick. Pages 1 through 14 located after Sample06.asm of the Scout SDK shows all of the constant declarations used with LEGO LASM. To add this library to the LEGO LASM user application program, use the #include statement at the beginning of the Scout P-Brick code:

```
#include   "ScoutDef.h"
```

The #define statement is used to create names for the individual tasks that the Scout P-Brick uses to perform various robotic controls. To ensure that all memory and previous tasks are erased from the internal registers of the Scout P-Brick, the dels and delt opcodes are used. This ensures that no garbage from an old program interferes with the new code for the Mindstorms robot. The task TASK_MAIN and additional tasks help to manage the program flow and behavior of the robot. The endt opcode stops the task after completing the defined function or behavior. Figure 6-5 shows the programming style for LEGO LASM.

This lab project enables hobbyists and experimentalists to experiment with LEGO LASM using the Scout P-Brick as a software development board. By following the instructions of the lab project, a smart switch application can be built using software and implemented on the Scout P-Brick. The Scout *Integrated Development Environment* (IDE) tool enables text to be entered by using the embedded editor screen, as shown in Figure 6-6. Once the code is successfully entered, translated, and downloaded to the Scout P-Brick, the operation should be equivalent to the RCX-code and NQC-language program applications of the RCX P-Brick version of the smart switch.

LEGO LASM Smart Switch Lab Project

This lab project illustrates how LEGO LASM can be used for Mindstorms robotic applications with the Scout P-Brick. After the code is entered into

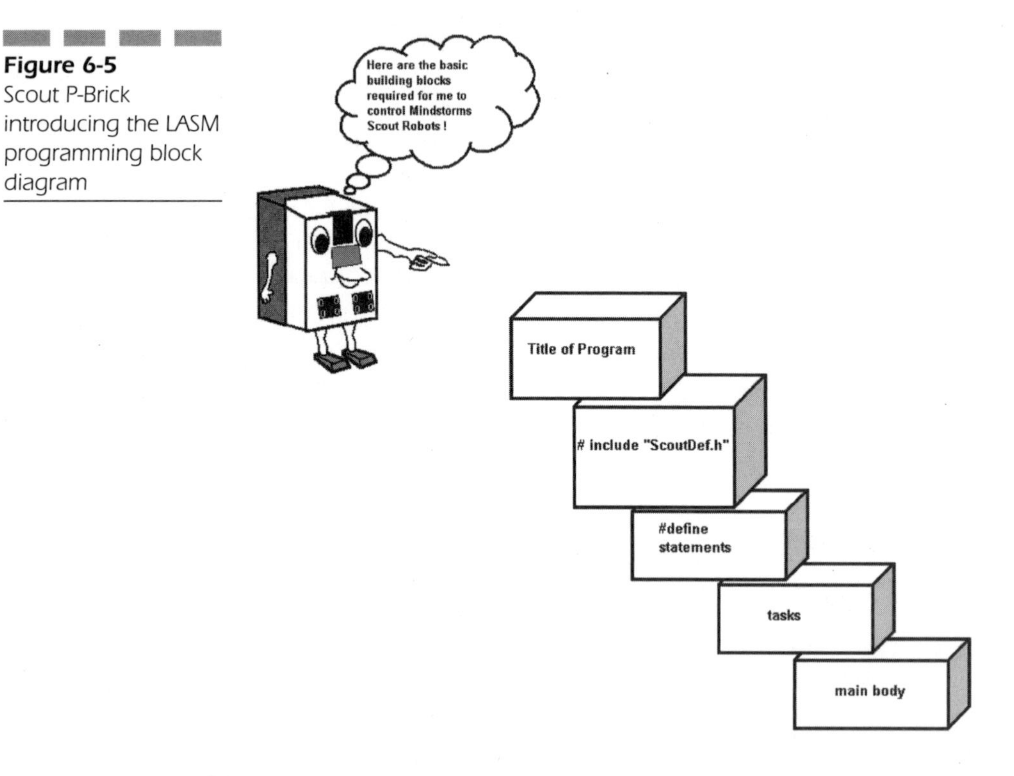

Figure 6-5 Scout P-Brick introducing the LASM programming block diagram

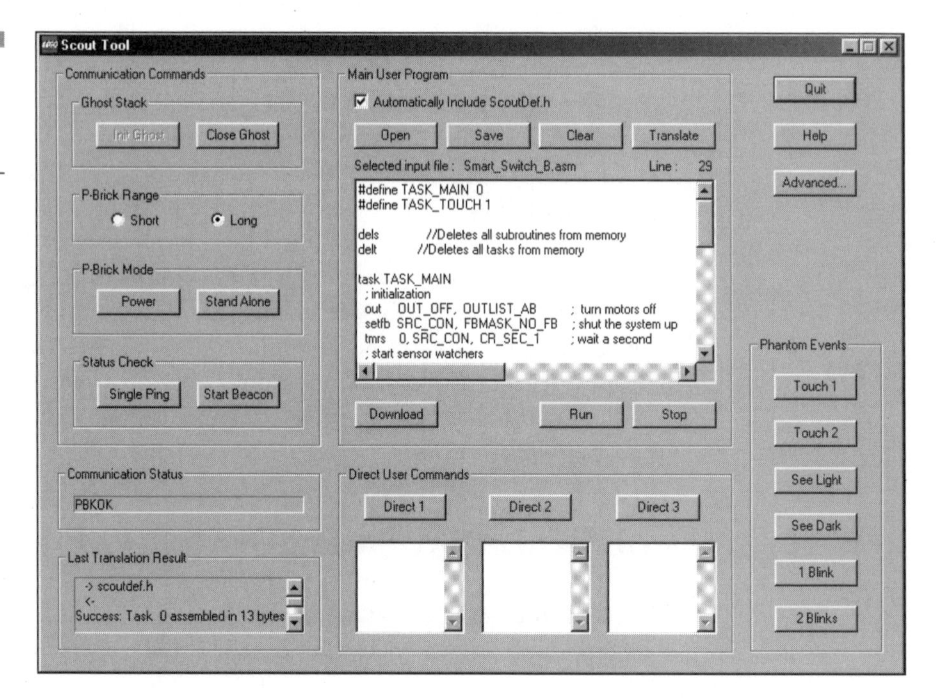

Figure 6-6 Using the Scout IDE tool's embedded text editor to enter code

the Scout IDE tool editor, and translated and downloaded to the Scout P-Brick, the smart switch control function can be implemented within the P-Brick.

BOM

- The Scout IDE tool
- Scout P-Brick
- Two touch sensors
- One LEGO small 9-volt motor
- Three LEGO electric wires
- The Scout SDK

The LEGO LASM Smart Switch Lab Procedure

1. Open the Scout IDE tool by clicking on the icon located on the desktop or the listing of Windows program files.
2. Add the touch sensor to inputs 1 and 2 of the RCX P-Brick and the LEGO small 9-volt motor to output A using the three LEGO electric wires.
3. Turn on the Scout P-Brick.
4. Click the Power Mode button on the IDE tool.
5. Enter Listing 1 into the Scout IDE tool editor, as shown in Figure 6-6.
6. Click on the Translate button located on the Scout IDE tool. If no errors are present, download the code to the Scout P-Brick. After the code is downloaded, a small folder should be displayed on the Scout P-Brick's LCD.
7. Run the code, and press and release Touch Sensor 1. Did the motor at output A turn on?
8. Press and release the each touch sensor. Did the motor at output A stop?
9. Wait 30 seconds and press each touch sensor. Did the motor at output A start?
10. Repeat Steps 7 through 9 several times to observe the program operation of the LEGO LASM code.

11. Use the Run and Stop buttons on the IDE tool to execute and terminate the smart switch code embedded within the Scout P-Brick.
12. Review the original code shown on pages 29 through 30 of the Scout SDK. Review the #include "ScoutDef.h" listing shown on pages 1 through 14 after Sample06.asm of the Scout SDK.
13. Stop the smart switch LEGO LASM code and turn off the Scout P-Brick.

LEGO LASM Smart Switch Lab Analysis

LEGO LASM can be used to create a smart switch with the Scout P-Brick. The Scout IDE tool enters the code for the LASM application. The Scout P-Brick device works like the RCX smart switch in this lab. The lab also gives hobbyists and experimentalists the opportunity to develop software using assembly language as the core programming language for controlling the P-Brick. LEGO LASM also illustrates how more control function capability is possible with the Scout P-Brick. The ability to communicate with a brick's internal microcontroller using assembly language makes the Scout P-Brick quite versatile in advanced Mindstorms robotic applications.

The next section of this chapter shows how an interactive language can be used to control a Mindstorms RCX P-Brick.

IC4

IC is a derivative language of C that was created to run on small, self-contained robot controllers such as the RCX P-Brick. IC enables hobbyists and experimentalists to interact with the language using a command prompt text box. The interactive command prompt text box enables the user to enter functions to gain immediate control of the RCX P-Brick. IC has a compiler-interactive command prompt and runtime machine language module. IC has a subset of C control structures such as for, while, if, and else as well as local, global variables, arrays, timers, and pointers. IC enables hobbyists or experimentalists to test code by typing instructions into the command prompt text box and seeing the results immediately on the PC or notebook screen. The ability to test code while writing it helps to reduce debug time and enables any experimentation in optimization to be done right on the spot during the development of the application. Therefore, IC has more development benefits than the NQC programming language. IC4 can serve as a prototyping language because of the sections of code that can be tested immediately before the final code is compiled. This feature falls in line with the POC philosophy of prototyping circuits and testing the software to check out the basic function of the device under test. IC4 is a free software language that can be downloaded from the Web.

Getting IC4 from the Web

IC4 can be obtained at the *Uniform Resource Location* (URL) address www.kipr.org/ic/download/. The web site is similar to a *File Transfer Protocol* (FTP) location where files can be downloaded from the Internet. When the URL address is reached, the next step is to open the windows/ folder. This leads you to the windows-ic-4 010-ins..> file. The file is about 5.5MB, and depending on the speed of the computer and the type of download connection, the file download could take a few minutes to be stored on your hard drive. Figures 6-7 and Figure 6-8 show the main directory and files that enable you to obtain IC4 from the Web. The file is a self-extracting/installing program that provides a Windows icon for running the application once completed. After opening the file, an IDE is displayed on the screen. Figure 6-9 shows the IC4 IDE tool. Like the BricxCC tool, the IC4 development tool enables the code to be entered using an embedded text editor. Prior to using the tool for IC4 code development, the firmware needs to be installed to the RCX P-Brick. Once installed, the LCD on the RCX P-Brick displays an IC40 1 message. The RCX P-Brick is now ready to send and receive messages, instructions, and programs for Mindstorms robotic control applications. Let's take a look at some basic functions used by IC4 for controlling the RCX P-Brick.

Basic IC4 Functions for Immediate Control of the RCX P-Brick

A function not only manages small blocks of code, but it also provides a special purpose to the software application. In IC4, functions can be used to provide an immediate control response to the RCX P-Brick. Immediate control is highlighted because it is an advantage over NQC language. The functions created for IC4 can control the robot once they are entered in the command prompt text box. The RCX P-Brick then performs the function received from the Infrared Tower immediately. The first function that will

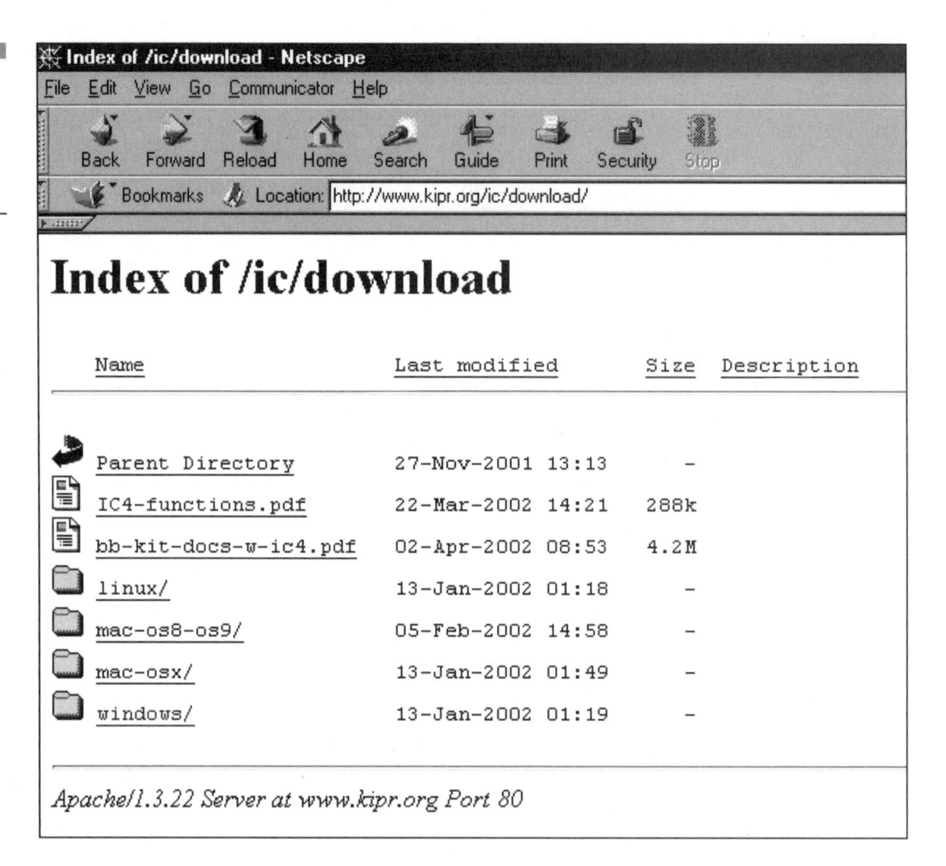

Figure 6-7
The FTP site for obtaining IC4 language. Select the windows/ folder.

be discussed is the printf(). By typing this function in the command prompt, the IDE tool sends a message to the RCX P-Brick requesting it to display the message between the braces on the LCD screen. The world-famous "Hello world" message seen in C-language programming books can be partially displayed on the RCX P-Brick's LCD screen. The first word in the message is displayed due to the size of the LCD displaying area. Try it! Have the RCX P-Brick display Hello in the LCD screen by typing printf("Hello") in the command prompt text box. After hitting the Enter key, the function displays Hello on the LCD screen. See Figure 6-10 to try out this function example using the RCX P-Brick. The word *HELL o* will be displayed on the RCX P-Brick's LCD screen.

The printf() function opens a wealth of LCD display opportunities for the visual status of Mindstorms robots. The LCD display can provide information about the input data that it has received. Analog and digital input signals can be viewed on the LCD using the printf() function. Diagnostics

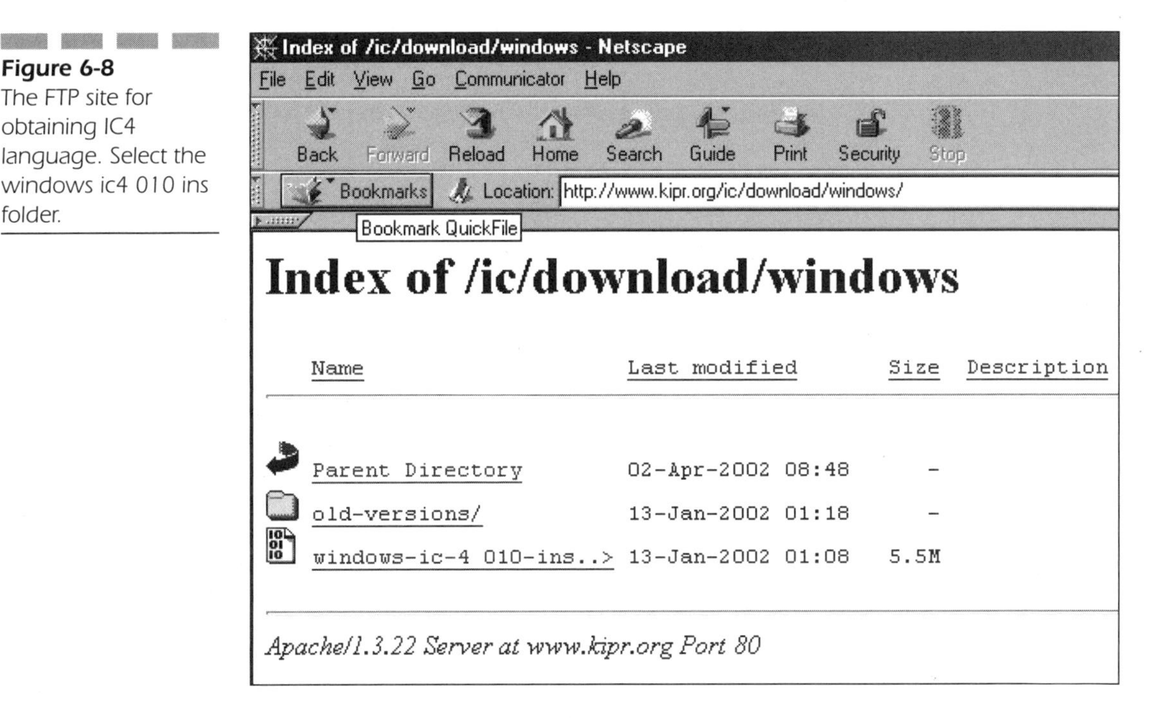

Figure 6-8
The FTP site for obtaining IC4 language. Select the windows ic4 010 ins folder.

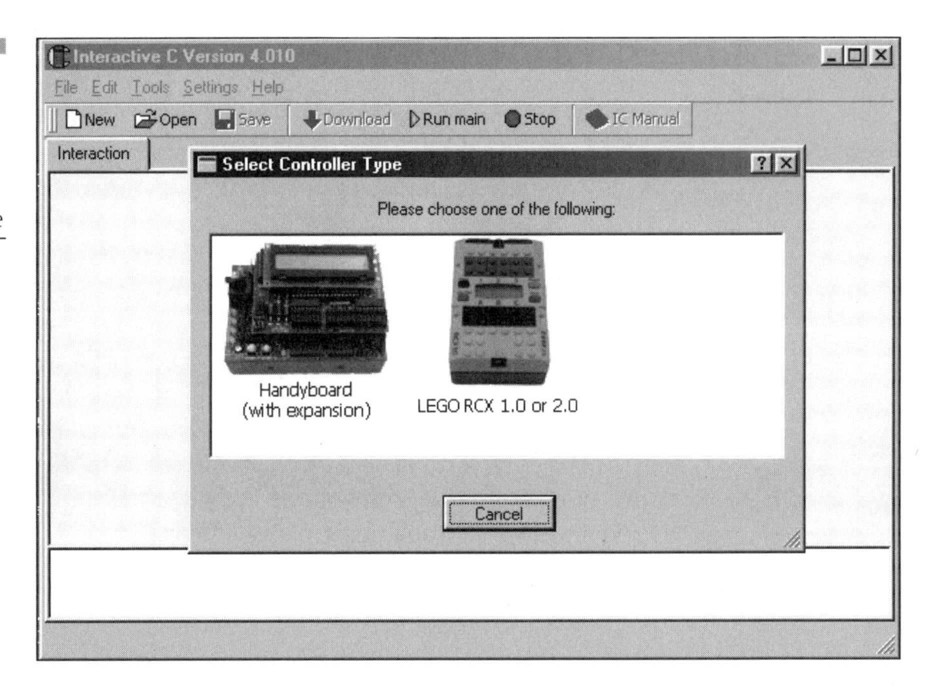

Figure 6-9
The IC4 IDE tool running on a notebook computer after installation of windows ic4 010.exe

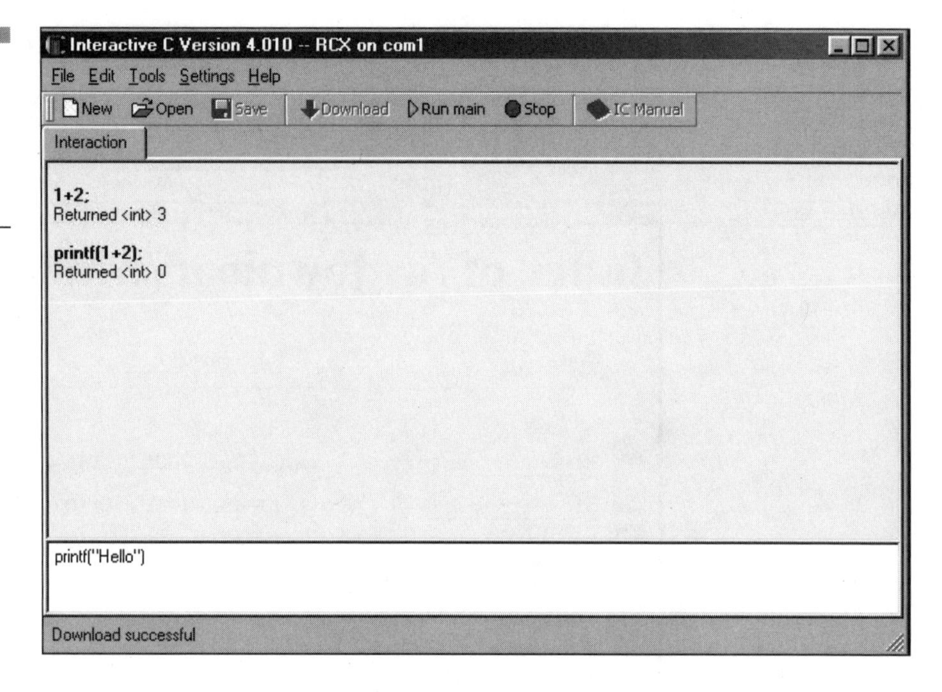

Figure 6-10 An example of using the command prompt text box for testing the printf() library function

messages can be created based on using the *digital control unit* (DCU) and supporting circuits to initiate Mindstorms robot control testing (see Chapter 3). Like the seven-segment display, creating some characters on the screen requires experimentation and an eye for creativity. The RCX LCD is small and can only display a maximum of five characters plus a few special symbols. Alpha characters look strange on the RCX LCD because the screen is only designed to display numbers. For example, the letter *m* looks like *n*. Therefore, character generation is more an art than a science when it comes to using the printf() function to display messages on the LCD. The only limiting factor of using the function is the imagination of the hobbyist or experimentalist.

Output control for LEGO *Direct Current* (DC) motors is quite easy to achieve with the IC4 language. The DC motors are labeled 1 through 3. Motors can be programmed to run in a forward and backward direction. The fd(int) function and bk(int) turn motor int on with the specified direction. To turn the motor attached to output A forward of the RCX P-Brick, use fd(1). Proper syntax for the function is as follows:

```
fd(1);
```

After entering this function in the command prompt text box and hitting the Enter key, the motor attached to output A spins in the forward direction. To make the motor spin backwards, use the following syntax:

```
bk(1);
```

The motor at output A begins to spin in the reverse direction. If you want to stop the motor by floating the output, then use off(int). The off(int) stops all the motors attached to the RCX P-Brick's outputs without floating them. Proper syntax for the function is as follows:

```
off(1);
```

To stop the motor without a gradual reduction in speed, use the brake(int) function. Proper syntax for using the function is as follows:

```
brake(1);
```

Another important function for Mindstorms robot control is sleep(int). This function causes the program to stop running for a given period of time. When this function is reached within the code, nothing happens for the specified time inside the parenthesis. Proper syntax for using the function is as follows:

```
sleep(1.0);
```

The decimal point is required. Without it, the compiler generates an error.

Input monitoring is also easy to achieve using the IC4 language. To read a touch sensor or an electromechanical control circuit (see Chapter 3), the digital(int) function is used. The int refers to integers 1 through 3 that are representative of inputs 1 through 3 on the RCX P-Brick. This function is used with the if statement to create conditional program control flow for Mindstorms robots. The programming style for using the digital(int) with an if statement is as follows:

```
if  (digital(1) ==1) {
       fd(1);
       sleep(5.0);
       off(1);
}
else {
       off(1);
}
```

When the switch at input 1 is pressed and held, the motor at output A spins in the forward direction for 5 seconds and then stops. When the switch is released, the motor at output A turns off. Analog inputs use the light(int) function to read sensor values. Like its digital brother, int refers to the input that the sensor is physically attached to. Therefore, int has the same numerical information as the digital(int) function for reading logic switching devices. To utilize the function within an IC4 program, use the following programming style:

```
if  (light(1) >100) {
          fd(1);
          sleep(5.0);
          off(1);
}
else {
          tone( 440.0, 1.0 );
}
```

When the analog value from the sensor is greater than 100, the motor spins in the forward direction and then stops. If the analog value from the sensor is less than 100, a 440 Hz tone beeps once.

More basic library functions can be found within the IC4 IDE tool's menu bar. There is an excellent online reference guide for programming the RCX P-Brick using IC4 at www.kipr.org/curriculum/programming.html. *The Interactive C Manual for the Handy Board* written by Fred G. Martin provides additional information on library functions, variables, operators, and multitasking that can be used for robots based on RCX P-Brick. IC4 and NQC languages have a similar programming style based on the use of parenthesis, semicolons, and indentations. The conditional statement of if-else uses similar indentation and placement of parenthesis in both procedural languages. Therefore, each language serves as a good reference to the other. In software POC studies, the need for code conversion is a critical item in the development cycle with regards to interfacing *input/output* (I/O) controls. Occasionally, the need to manually port code into another embedded development environment is important for developing a prototype. Therefore, porting code conversion requires languages and embedded development environments to share a common programming style.

The lab project demonstrates some of the basic IC4 functions discussed in this section. A basic RCX P-Brick along with a touch sensor, LEGO DC motor, and two electric wires are used throughout the demonstration. Although this chapter does not focus on the construction of a *Robot Digital Assistant* (RDA), hobbyists and experimentalists can build an advanced Mindstorms robot that can be used with the lab projects.

IC4 Basics Part 1 Lab Project

This lab shows some of the basic library functions working together in a simple Mindstorms robot controls program. The objective of the lab project is to illustrate the ease in which a software controls application can be created using IC4.

BOM

- RCX P-Brick
- Two LEGO electric wires
- One touch sensor
- One LEGO small 9-volt motor
- IC4 IDE tool

IC4 Basics Part 1 Lab Procedure

1. Open the IC4 IDE tool.
2. Click on the New tab and a clean text editor window will appear on the screen.
3. Using Figure 6-11, type in the code within the IC4 text editor. Make sure the firmware is downloaded to the RCX P-Brick prior to doing Step 4.
4. Save the code to your local hard drive. Download the code to the RCX P-Brick using the Download button on the menu bar.
5. Add the touch sensor to input 1 of the RCX P-Brick and the LEGO small 9-volt motor to output A using the two LEGO electric wires.
6. Turn on the RCX P-Brick. The IC40 1 logo should appear on the LCD screen.
7. Push Run on the RCX or click the Run Main button on the menu bar.
8. Press and hold the touch sensor and notice the motor at output A. Did the motor turn?
9. Press and hold the touch sensor while pressing the Run button on the RCX P-Brick. Did the motor turn?
10. Release the touch sensor and notice the motor. Did the motor float to stop after 5 seconds?

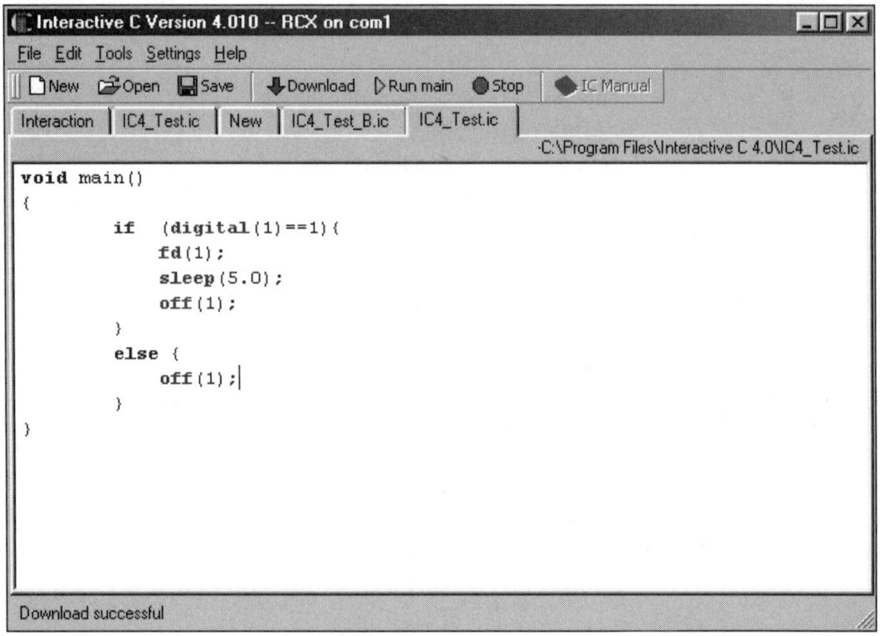

Figure 6-11
IC4 code for the IC4 basic functions part 1 lab project

Interactive C Version 4.010 -- RCX on com1
File Edit Tools Settings Help
New Open Save Download Run main Stop IC Manual
Interaction | IC4_Test.ic | New | IC4_Test_B.ic | IC4_Test.ic
C:\Program Files\Interactive C 4.0\IC4_Test.ic

```
void main()
{
        if  (digital(1)==1){
            fd(1);
            sleep(5.0);
            off(1);
        }
        else {
            off(1);
        }
}
```

Download successful

11. Repeat Step 9. Did the motor turn?
12. Stop the program by clicking Stop on the menu bar or pressing Run on the RCX P-Brick.

IC4 Basics Part 1 Lab Project Analysis

This lab project shows a simple controls program that was written in IC4 language. After downloading the code of Figure 6-11, the program can be executed by using the IDE tool's Run Main feature or pressing the Run button on the RCX P-Brick. The program did not start because of the control flow used in the code and the required operating sequence dictated by IC4. When the touch sensor is pressed and held before the program is run, the motor spins in the forward direction. The motor spins for 5 seconds before floating to stop. Next, the program is modified where the touch sensor can turn on the motor after the code has been executed using hard or soft run methods.

IC4 Basics Part 2 Lab Project

This lab illustrates how the while() statement can create an infinite loop, enabling the program to execute by pressing the touch sensor.

IC4 Basics Part 2 Lab Procedure

1. Click on the New tab and a clean text editor window will appear on the screen.
2. Using Figure 6-12, type in the code within the IC4 text editor.
3. Save the code to your local hard drive. Download the code to the RCX P-Brick using the Download button on the menu bar.
4. Turn on the RCX P-Brick. The IC40 1 logo on the LCD screen should appear.
5. Push Run on the RCX or click the Run Main button on the menu bar.
6. Press and hold the touch sensor and notice the motor at output A. Did the motor turn?

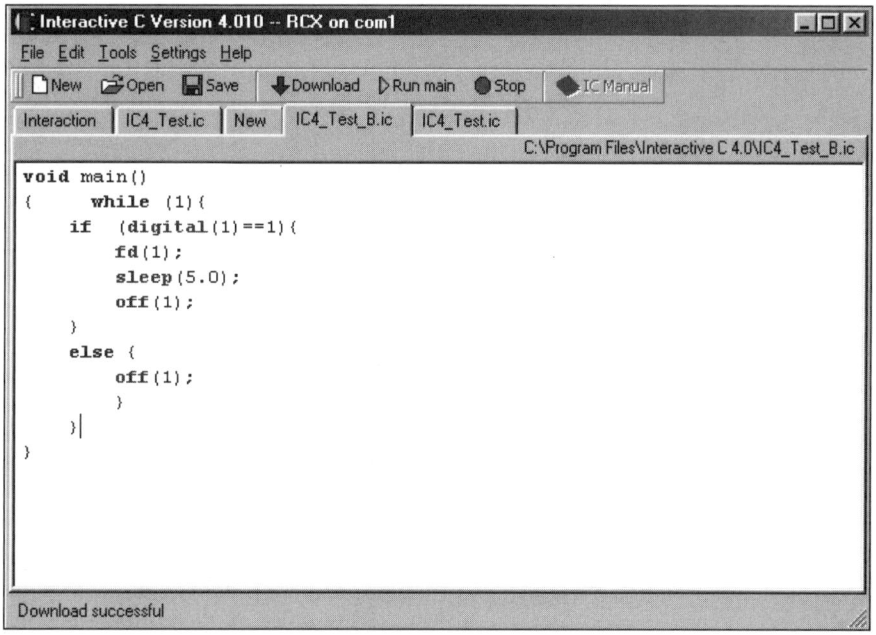

Figure 6-12 Modifying the code to include the while() loop statement

Interactive C Version 4.010 -- RCX on com1

File Edit Tools Settings Help

New Open Save Download Run main Stop IC Manual

Interaction | IC4_Test.ic | New | IC4_Test_B.ic | IC4_Test.ic

C:\Program Files\Interactive C 4.0\IC4_Test_B.ic

```
void main()
{      while (1){
    if  (digital(1)==1){
        fd(1);
        sleep(5.0);
        off(1);
    }
    else {
        off(1);
        }
    }|
}
```

Download successful

7. Release the touch sensor and notice the motor. Did the motor float to stop after 5 seconds?
8. Repeat Step 6. Did the motor turn?
9. Stop the program by clicking Stop on the menu bar or pressing Run on the RCX P-Brick.

IC4 Basics Part 2 Lab Analysis

This lab shows the modification of a program to include the while() statement. With the statement being true [(digital(1)==1)], the program can run continuously. This looping condition enables the RCX's output A to start and stop the motor depending on the condition of the digital(1) function. Thus, the motor can spin whenever the touch sensor is pressed as opposed to requiring the Run button to be pressed first. The while(1) is analogous to while(true) in NQC language. Therefore, programming ideas and code for one language can be translated to the other quite easily. The final lab project investigates the light(int) function for reading active sensing devices such as the light sensor.

IC4 Basics Part 3 Lab Project

This lab project shows how the light(int) function can be used to read an experimental active sensor and control output A. Also, the value read from the sensor is displayed on the LCD screen using the printf() function.

BOM

- One 1-mega-ohm potentiometer
- Two 1-kilo-ohm resistors
- One TLC272 operational amplifier
- One *Silicon Diode Interface Circuit* (SDIC)

IC4 Basics Part 3 Lab Procedure

1. Build the sensor interface circuit shown in Figure 6-13 using the Radio Shack Electronics Learning Lab breadboard or equivalent. Before powering the circuit, check your wiring carefully. Do not power the

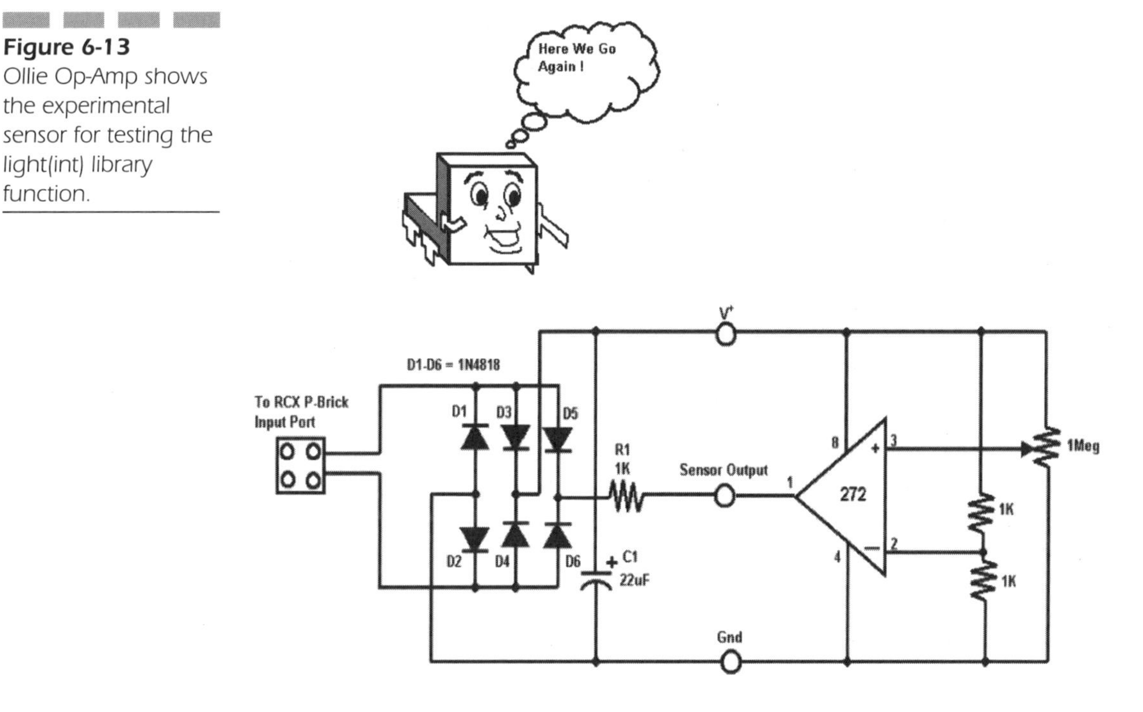

Figure 6-13 Ollie Op-Amp shows the experimental sensor for testing the light(int) library function.

circuit interface with a secondary DC power supply because the RCX P-Brick's input could be damaged.

2. Remove the touch sensor from input 1 of the RCX P-Brick and replace it with the experimental sensor interface circuit using the modified LEGO electric wire.
3. Click on the New tab and a clean text editor window will appear on the screen.
4. Using Figure 6-14, type in the code within the IC4 text editor.
5. Save the code to your local hard drive. Download the code to the RCX P-Brick using the Download button on the menu bar.
6. Turn on the RCX P-Brick. The IC40 1 logo on the LCD screen should appear.
7. Push Run on the RCX or click the Run Main button on the menu bar.
8. Turn the 1-mega-ohm potentiometer slowly until the motor starts spinning. The value on the LCD screen should be 491.

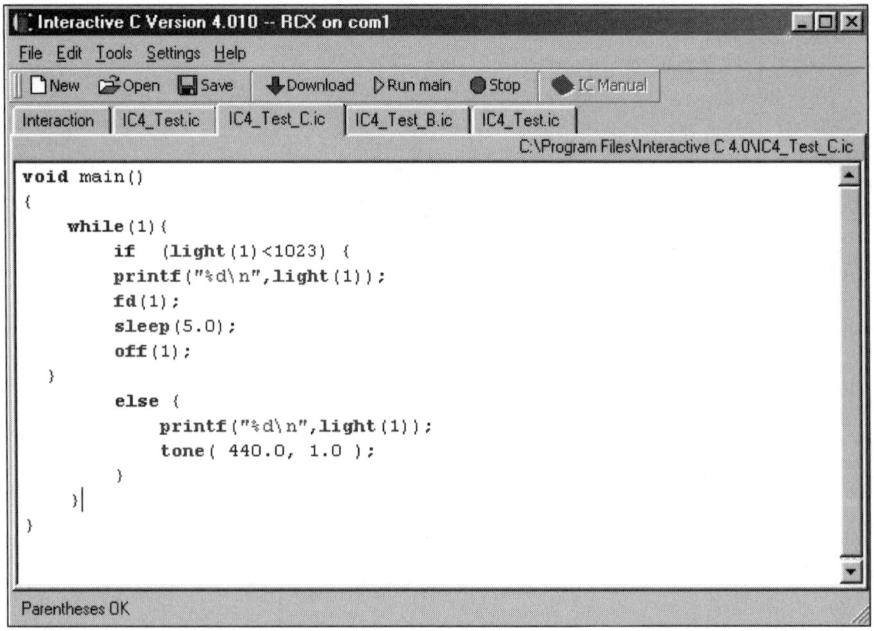

Figure 6-14 An improved analog sensing code with the printf() statement for reading the data of the sensor and displaying it on the LCD screen of the RCX P-Brick

Interactive C Version 4.010 -- RCX on com1

File Edit Tools Settings Help

New Open Save Download Run main Stop IC Manual

Interaction | IC4_Test.ic | IC4_Test_C.ic | IC4_Test_B.ic | IC4_Test.ic

C:\Program Files\Interactive C 4.0\IC4_Test_C.ic

```
void main()
{
     while(1){
        if  (light(1)<1023) {
        printf("%d\n",light(1));
        fd(1);
        sleep(5.0);
        off(1);
   }
        else {
            printf("%d\n",light(1));
            tone( 440.0, 1.0 );
        }
     }
}
```

Parentheses OK

9. Turn the 1-mega-ohm potentiometer slowly in the opposite direction until a 440 Hz tone is heard from the RCX P-Brick. The value on the LCD screen should be 1023.
10. Repeat Steps 8 through 9 and follow the logic of the code shown in Figure 6-14.
11. Stop the program by clicking Run on the RCX P-Brick or clicking Stop on the menu bar of the IC4 IDE tool.

IC4 Basics Part 3 Lab Analysis

This lab investigates the light(int) library function. The experimental sensor provides the necessary signal interface for detecting code and controlling the motor and internal tone generator of the RCX P-Brick. An interesting item to note about the light(int) function is that this statement is for active sensors (see Chapter 5). Another library function that is used to read passive sensors (see Chapter 5) is light_passive(int). Therefore, a simple rheostat can be used as a passive sensor and its data values can be read using the light_passive(int) library function. The printf() function enables the data from the experimental sensor to be displayed on the LCD screen. Another way to read sensor data using light(int), light_passive(int), or digital(int) library functions is to store the data in a variable. Variable expressions consist of the name of the variable that is equal to the digital or analog library function. The proper syntax for making variable expressions is as follows:

```
data= light(1);
```

The sensor value received from the light(1) library function is stored in the variable data. The printf() function must be used to read this value. The proper syntax for reading the variable data is as follows:

```
printf("%d\n", data);
```

As an additional lab project, modify the code in Figure 6-14 to use a variable expression to store the data and display it on the LCD screen using the printf("%d\n", variable name) library function. Also, replace the experimental sensor with a rheostat and use the light_passive(int) library function to read the values produced by the passive sensing device. These POC lab projects open the door for the development of smart robo-instruments using the RDA philosophy discussed in Chapter 2.

Chapter 7, "Client-Server Controller for Mindstorms Robots," outlines assembly procedures for building a tabletop, self-contained client-server controller for Mindstorms robots. The project is based on using Python, which is an open-source *Object-Oriented Programming* (OOP) language that enables the software developer to create a soft bridge to other languages and Windows applications as a small robot commands database. This virtual server is used to pass data to an Excel spreadsheet (client). Basically, a small *Robot Command Language* (RCL) is created using this software-interfacing technique to control a Mindstorms robot.

CHAPTER 7

Client-Server Controller for Mindstorms Robots

Client-server controllers are mostly used for web and database applications. The host system is remotely located and tasked with managing files and data that enable users to access information from satellite locations. Users can also store their information on the host system using a hardwire data transmission controller, which serves as a gateway to other host machines. Another name for host is *server* and another name for satellite is *client*. A *proof of concept* (POC) study that can help provide an understanding of data transmission topologies is to build a tabletop version using two software languages that emulate a client and server. A Mindstorms robot controller can also be built using this data transmission scheme.

In this chapter, the POC study involves building a client-server controller using an open-source language as the host and a standard Windows application language as the satellite. Data is passed from the server and received by the client. Information produced by the client can be viewed by the server, creating a bidirectional network. This viewed data can be used as a diagnostic tool that can check the status of the robot automatically. The objective of the lab project in this chapter is to build a client-server controller that enables you to manipulate the behavior of a Mindstorms robot.

The following topics are included in this chapter's discussion on building a client-server controller:

- Electrical/electronics and mechanical design
- What is Python language?
- Building a mini server in Python language
- Building a mini client in Excel *Visual Basic for Applications* (VBA)
- Putting the controller together

Let's begin the discussion of building a mini client-server controller by building a Mindstorms robot.

Electrical/Electronics and Mechanical Design

The mechanical design of the robot is key to experimenting with client-server controls. Therefore, we begin the discussion with a design process for building a Mindstorms robot that will be controlled by a mini client-server controller.

The robot that will be controlled using the client-server controller is a primitive derivation of Stegoclubber from the Extreme Creatures Expansion kit. The Extreme Creatures Expansion Kit, Catalog #9732, has additional LEGO bricks for building exotic robot insects and animals that react to light or touch. The Extreme Creatures Constructopedia provides an experimental robot to build for controlling using the Client-Server Controller discussed in this chapter. The robot's Creature Mover 1 mechanism is the focus of the mechanical design portion of this lab project. Again, the methods of control and design are used as examples; any Mindstorms robot can be built using this technique.

The first step in the mechanical design of the Mindstorms robot is to define the motion of the robot. This step is important because the mechanical interface and how the components work together provide a foundation for developing the electronics and software controls of the mechanical machine. Also, once a mechanical assembly has been tested, it becomes part of the *Recycling of Mechanical Assemblies* (ROMA). For Mindstorms robots, motions can be as simple as moving back and forth or be as complex as spinning in elaborate geometric patterns. The Creature Mover 1 mechanism shown on page 4 of the Extreme Creatures Constructopedia has back and forth motion. References such as those in Mindstorms Constructopedia provide a wealth of mechanical motions and mechanisms that can achieve the movements of a specific task.

Once the movement has been identified, the next activity is to build the mechanism. The Creature Mover 1 was built from pages 6 through 8 of the Extreme Creatures Constructopedia. The convenient feature of this Creature Mover mechanical drive is that you have the ability to use either the Scout or *Robot Command Explorer* (RCX) *programmable brick* (P-Brick) as the robot's brain. The ability to switch P-Bricks enables you to explore various programming languages, as discussed in Chapter 6, "Using Procedural Languages for Mindstorms Robot Control." The robot then becomes an educational trainer in software design and development. The personality of the robot can be built using eyes, antennas, fiber-optic whiskers, or a combination of the three.

The novice Mindstorms builder is encouraged to use references such as the Constructopedia as a source of mechanical design inspiration. Within the pages of the Constructopedia, a novice can find examples of design tricks, and photos showing mechanical and natural pictures that can be used to spark new Mindstorms robot mechanics and smart machines. The advanced Mindstorms builder may have another source for building robots. Whatever resource you choose to use for building your robots, the key concept is to have a design that provides a robotic motion that meets the application and that can be used as the foundation for the mechanical interface

Figure 7-1 Side view of the wheels and reciprocating leg of the Creature Mover robot

to the Mindstorms P-Brick. The Creature Mover 1 mechanical drive is shown in Figures 7-1 through 7-4.

Once the robot is built, the mechanical motion must be evaluated. The LEGO infrared remote control is a good testing tool for evaluating the mechanical motion of Mindstorms robots. As discussed in earlier chapters, the infrared remote control can be used to assist in designing a particular mechanical movement by selecting the appropriate outputs and pressing the forward and backward buttons on the unit. The infrared remote control is a mechanical motion design tool. The robot's movement is created by pressing a combination of the buttons on the unit and observing the action of the mechanical system.

The development phase enables hobbyists and experimentalists to design some outrageous mechanical movements as well as simple ones. Once the correct motion has been identified, the movement can be programmed in the RCX, *LEGO Assembly* (LASM), *Interactive C version 4.0* (IC4), or *Not Quite C* (NQC) software languages. Once the motion is programmed and running on the P-Brick, the robot's mechanical drive system can be tested. The debug involved at this stage consumes as little as a few minutes to an hour or two. Patience is very important when correcting software; therefore, add comments to sections of code that might be complicated to remember from memory. Once the robot performs as designed by the software, tasks can be assigned to sections of code related to the behavior and function of the robot.

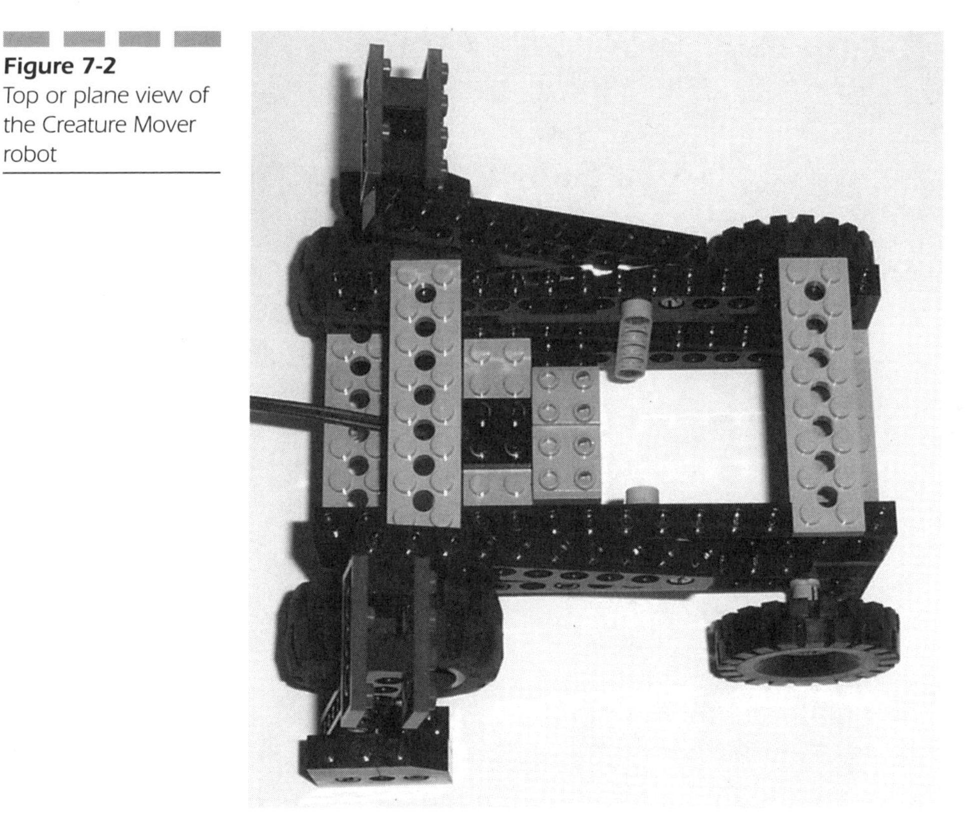

Figure 7-2
Top or plane view of the Creature Mover robot

The client controls the robot using specified tasks. The RCX *Software Development Kit* (SDK) has a StartTask (*number*) program execution command (see page 48 of the SDK document). This command is embedded within the client's software structure, enabling it to execute the behavior or function that is initiated by the server. Also, the tasks embedded within the control code can provide a software entry point for testing the client. This technique is discussed later on page 48 under the section, "Building a Mini Client in Excel VBA."

After the mechanical design has been completed, the electrical/electronics design can be built for the robot. As discussed in Chapter 4, "Electronic Switching Circuits," the RCX P-Brick's *input/output* (I/O) can be enhanced by adding electronic circuits. Additional visual and audio circuits can be provided to give the robot a simple means of indicating that a specific task is completed. Diagnostics can be added that enable the robot to perform a pretest before performing its main function.

Another task that the robot can perform is a controlled output source for a timed-delay power driver circuit. The system block diagram in Figure 7-5

Figure 7-3
Bottom view of the Creature Mover robot

Figure 7-4
Completed Creature Mover robot chassis

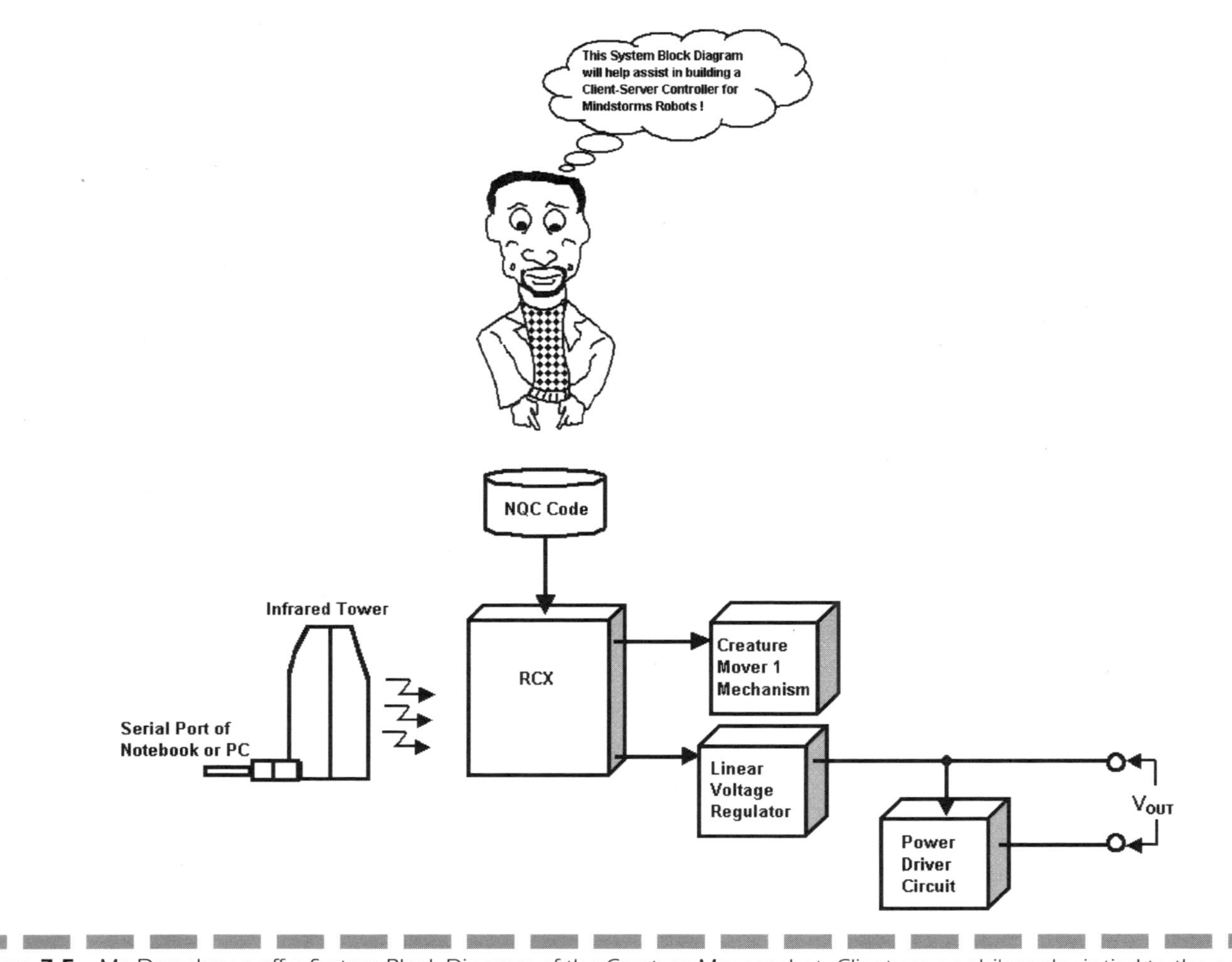

Figure 7-5 Mr. Don shows off a System Block Diagram of the Creature Mover robot. Client-server philosophy is tied to the system block diagram as well.

shows the components of a controlled timed-delay driver. An RCX P-Brick's output is turned on upon completing a specific task. By pushing a momentary switch, the timed-delay driver's output is on for a predetermined time set by a *resistor capacitor* (RC) timing network. To initiate the output timing sequence, the momentary switch must be pressed again. The circuit in Figure 7-6 meets the requirements for driving small loads less than 1 *ampere* (A). The delayed time is the product of the RC network being connected to the load-side driver's *field effect transistor* (FET) gate circuit. The 1-mega-ohm resistor is a variable resistor in parallel with an electrolytic capacitor of 470 microfarads. A microfarad is 1 millionth (1×10^{-6}) of an ampere. The timed delay created by the 1-mega-ohm resistor and the 470-microfarad capacitor can be found using the following equation:

$$\tau(\text{tau}) = R \times C$$

where τ (tau, the Greek letter for *t*) is the time constant, or the total delay time in seconds.

Using the values of 1-mega-ohm for *R* and 470 microfarad for *C*,

$$\tau = \text{1-mega-ohm} \times \text{470 microfarad}$$

$$\tau = \text{470 secs or approximately 8 min}$$

The 8-minute value calculated is the maximum time delay for which the circuit can keep the electrical output active. Therefore, varying the 1-mega-ohm rheostat changes the RC product for creating a time delay.

A method that can be used to help understand the timed-delay power driver circuit is to create a circuit model and run a simulation. This *electronics computer-aided engineering* (ECAE) tool enables the circuit's designer to experiment with electronics networks virtually before building a breadboard version. The parts library has a generous supply of contemporary integrated circuits, transistors, diodes, and test instruments. Components can be wired together using nets by placing the parts on a virtual breadboard.

By using the software's drawing tool, a circuit schematic is created. The component values can be changed and the circuit can be run using the simulator controls. Plots showing *voltage* (V-I) characteristics can be viewed with the help of a *probe*. A probe is an analytical tool that captures electrical parameters at a node and net. The *node* is a single point on the circuit schematic. The net is the line that is joined to the node. Current, voltage, resistance, and power can be displayed within an analysis window using a probe. When conducting a simulation, some components and instruments

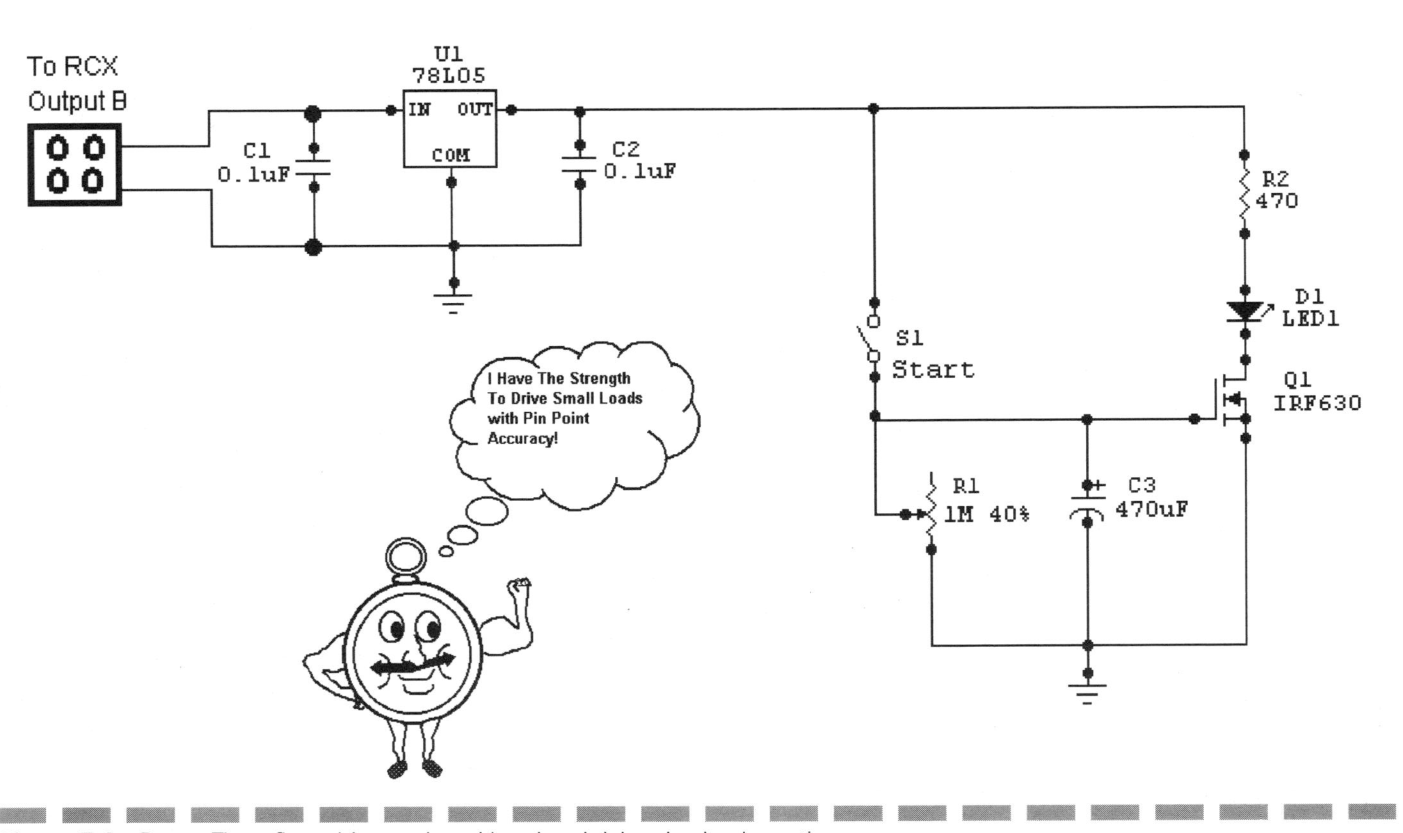

Figure 7-6 Power Timer flexes his muscles with a timed-delay circuit schematic.

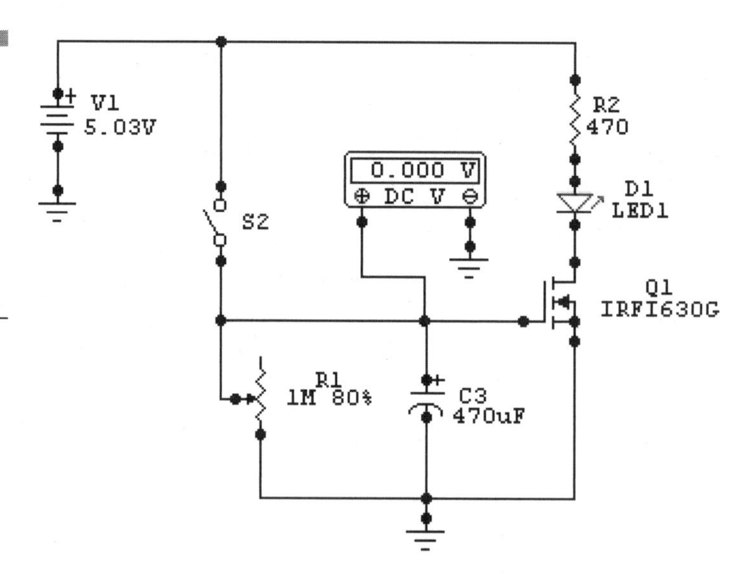

Figure 7-7
An equivalent timed-delay circuit used for simulation in Circuit Designer (student version) software. Notice that a 5.03-volt battery replaces the 7805 IC.

cannot be used in a virtual analysis because the vendor has not yet created the circuit model. Therefore, an equivalent circuit can be built and used with the model in the simulator. Figure 7-7 shows an equivalent model for the timed-delay power driver circuit. The 7805 linear voltage regulator has been replaced with a 5.03-volt battery. The 1-mega-ohm rheostat is used to change the time delay by varying its resistance.

A sample plot of the timed-delay power driver circuit is shown in Figure 7-8. The simulation package used to capture the plot and provide a schematic is called *Circuit Maker*. Circuit Maker makes it easy to build and test circuits with its Windows-based design and analysis tools. Circuit simulation tools enable I/O interfaces to preview possible devices that can be used for Mindstorms robotic development. Circuit simulation also provides testing data that can assist in the creation of Mindstorms robots. Finally, the software acts as a great schematic drawing tool for capturing circuit ideas. The student version (6.0) can be downloaded for free at www.microcode.com.

Building and Testing the RCX-Controlled Timed-Delay Driver Circuit

The RCX-controlled timed-delay driver circuit can be built by breaking the circuit into two device elements: the linear voltage regulator and the timed-delay driver. First, build the 7805 linear voltage regulator shown in Fig-

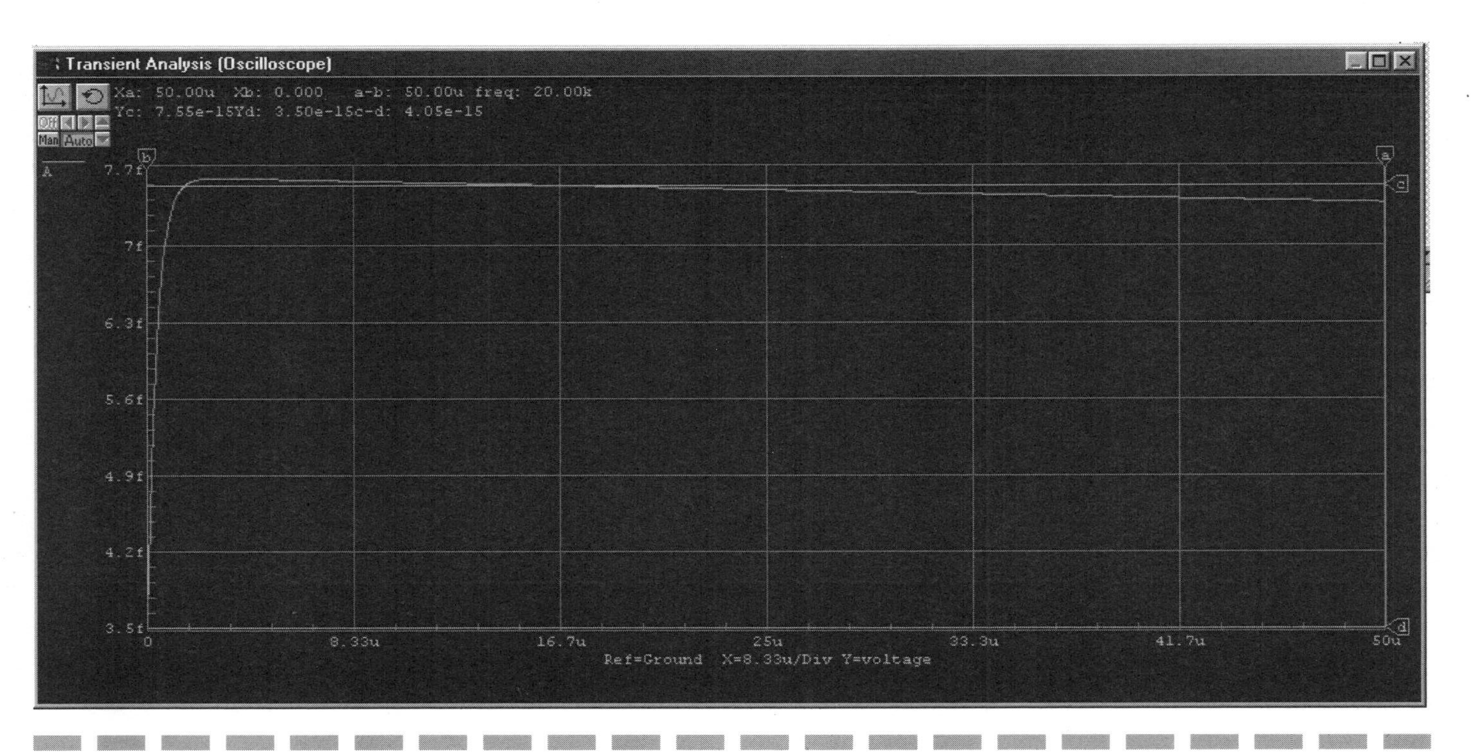

Figure 7-8 Capacitor C3 charging characteristics captured in Circuit Maker 6.0 student version software.

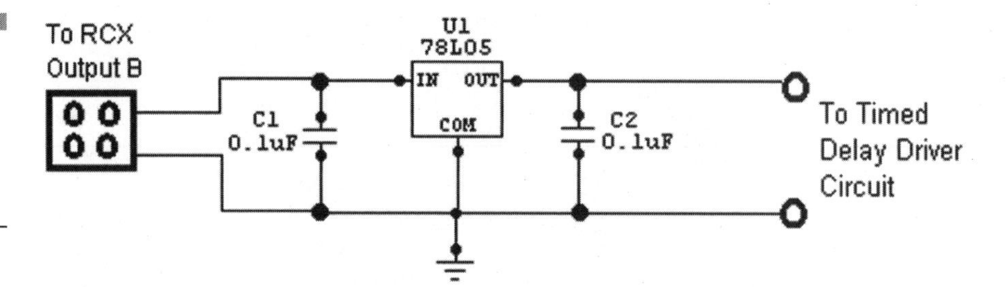

Figure 7-9 Schematic circuit showing the 7805 voltage regulator IC wired to a modified LEGO electric wire

ure 7-9 using a Radio Shack Electronics Learning Lab breadboard or equivalent solderless prototyping system. Place the modified LEGO electric wire onto output B of the RCX P-Brick. Make sure that the wires are pointing up when they are attached to the output of the P-Brick. Once the circuit is built, connect a digital voltmeter across pins 2 and 3 of the 7805 IC. Press and hold the forward B button on the infrared remote control. The digital voltmeter should read 5 volts. If not, recheck the wiring and repeat the measurement.

When the voltage regulator circuit interface and RCX P-Brick work together correctly, the timed-delay driver can be built. Build the circuit using the schematic shown in Figure 7-8. Wire the +5-volt output of the voltage regulator to the timed-delay driver circuit. Check the wiring of the delay circuit prior to testing the completed device. With the infrared remote control, press and hold the forward button and push the momentary *Single Pole-Single Throw* (SPST) switch. The *Light-Emitting Diode* (LED) should stay on for a predetermined time set by the RC network of the circuit and then turn off. Adjust the 1-mega-ohm rheostat for different time-delay values and notice the effect on the LED.

This completes the electrical/electronics interface portion of the project. Next, a brief introduction to the Python language and instructions on building a mini server are provided.

What Is Python Language?

Python is an interpreted *Object-Oriented Programming* (OOP) language that is capable of gluing two dissimilar programs together. Interpreted language provides a vehicle for making the software interactive. Guido Van Rossum, the creator of Python, states the following from the web site www.python.org/doc/essays/blurb.html: "Its high-level built-in data structures, combined with dynamic typing and dynamic binding, make it very attractive for *Rapid Application Development* (RAD) as well as for use as a scripting or glue language to connect existing components together. Python's simple, easy-to-learn syntax emphasizes readability and therefore reduces the cost of program maintenance. Python supports modules and packages, which encourages program modularity and code reuse. The Python interpreter and the extensive standard library are available in source or binary form without charge for all major platforms, and can be freely distributed."

So basically what does this mean to Mindstorms hobbyists and experimentalists? It means that a mini client-server controller can be built using the glue-language principle to share data between two software components: mainly Python and Excel. The Python programming language is interactive; therefore, commands can be typed within the *Integrated Development Environment* (IDE) (for Python, this is called IDLE) *Graphical User Interface* (GUI). As with the IC4 programming language discussed in Chapter 6, code can be tested by typing each line of instruction and the output can be viewed immediately through the Interactive Interpreter of Python. The software is free and can be downloaded from the web site www.python.org. The latest version of Python is 2.2. The mini server discussed in this section is based on this software version; therefore, download the Python 2_2_1.exe file to your hard drive and follow the instructions for extracting the executable Windows application to the appropriate directory on your system.

The following discussion illustrates how the mini server can be built in Python using a key building block known as the *Component Object Model* (COM). Therefore, the win32all-146.exe software must be downloaded and executed from the web site http://starship.python.net/crew/mhammond/win32/Downloads.html.

Building a Mini Server in Python Language

Once the necessary Python components have been installed, a mini server can be built. The mini server provides data for controlling a Mindstorms robot using VBA. The data is passed from a Python GUI and sent to an Excel spreadsheet using COM technology. COM operates under the client-server infrastructure. The data resides in the server put there by the developer of the system. The client is the application that needs the data from the server. The client calls the server using the COM architecture. The COM *application program interface* (API) creates a link for communication that determines the functions that the server provides to the client. If the correct protocol exists between the client and server, they can be written in different languages and glued together. Figure 7-10 shows the network architecture for creating a client-server controller using Python and Excel. COM technology enables hobbyists and experimentalists to create a unique controller using the network architecture shown in Figure 7-10. In order for Python and Excel to communicate, COM objects must be made for the target (which in this case is Excel).

The makepy.py utility must be executed in order to create the COM objects of Excel. Therefore, the directory win32com\client, which was created when the win32all-146.exe file was executed, must be found on the local hard drive. Once this directory is found, double-click the utility. A Windows dialog box appears that has a list of target libraries for COM objects (see Figure 7-11). Select the object for Excel and click OK.

After clicking OK for the Excel COM object, makepy.py utility begins to create a new set of Excel COM objects. Once the objects have been created, Python is ready to connect with Excel.

Open up the IDLE (Python GUI) application and type the following commands using the Interactive Interpreter window of Python:

```
>>> import win32com.client
>>> w=win32com.client.Dispatch ("Excel.Application")
>>> w.Visible=1
>>> w.Workbooks.Add( )
>>> w.Cells(1,1).Value="Hello"
```

As each line is typed, a specific action takes place. When the third line is typed, Excel automatically opens up on the screen. The next line of instruction adds a new workbook and displays Hello at the first cell location of the spreadsheet. Figure 7-12 shows an interactive session using Python as a mini server. By changing the value of w.Cells(1,1).Value, the information is

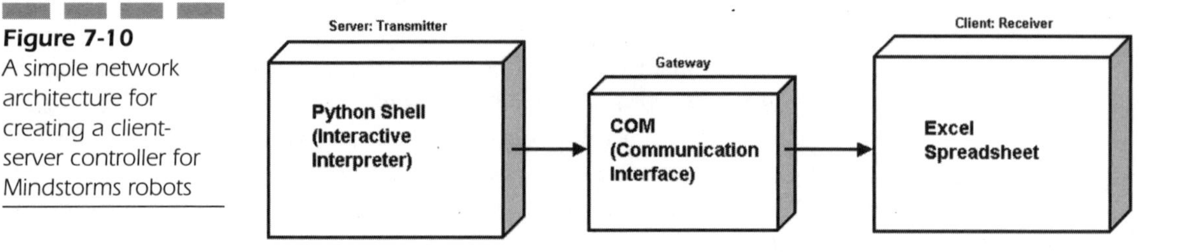

Figure 7-10 A simple network architecture for creating a client-server controller for Mindstorms robots

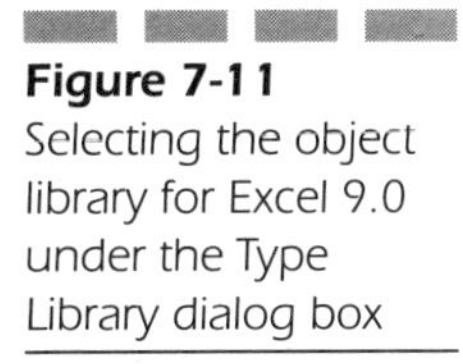

Figure 7-11 Selecting the object library for Excel 9.0 under the Type Library dialog box

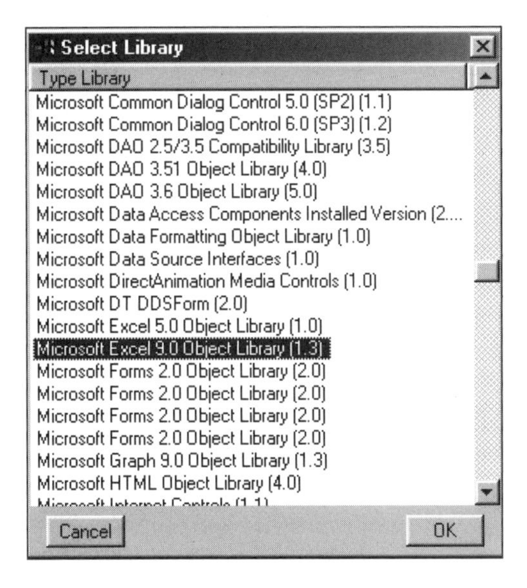

Figure 7-12 A sample interactive session of Python communicating with an Excel spreadsheet

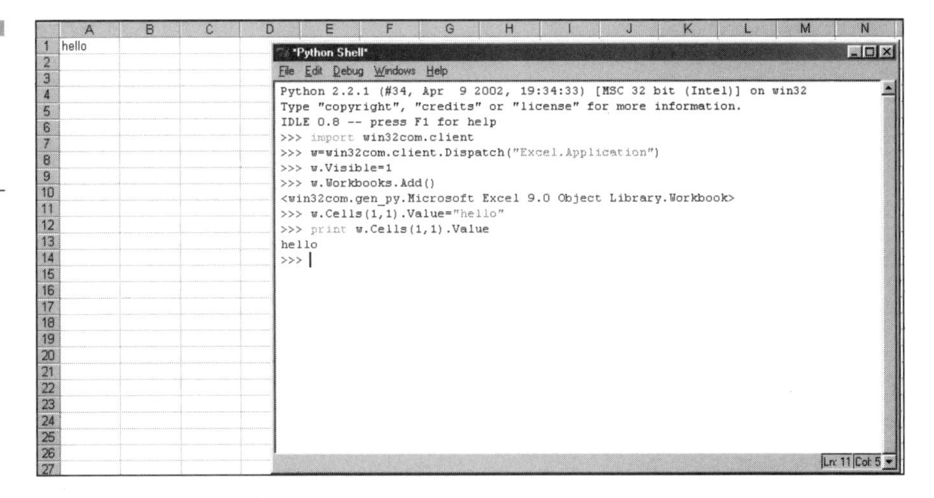

displayed on the worksheet. To capture cell data from Excel and display it within Python, the following command should be used:

```
Print w.Cells(1,1).Value
```

The data at cell location (1,1) is read into w.Cells(1,1).Value and printed on the Python IDLE window. Another way to send data to the spreadsheet is through w.Range("A1").Value. The difference between cells and range is that arrays can be managed by using the cell VBA object, whereas range is unable to perform this type of data management. The first four lines of code in this Python session are key to building a mini server. Therefore, the four lines of code can be reused for future Python applications.

With the ability to send and receive data, the POC of creating a small robot command language becomes feasible. To make the command language easy to understand while remaining in the server domain, the design of the GUI must be based on a window with a list of basic command statements that the user selects with a mouse. After the selection is made, the command is sent to the Excel spreadsheet for further processing to enable the control of a Mindstorms robot. Python has a graphics interface extension program called Tkinter. Tcl (pronounced "tickle") is a programming language developed for Unix-based systems. Tk (pronounced "tick") is also a programming language that enables a graphical interface to exist between the user of the Unix environment and Tcl. The objects used to graphically manipulate data are called *widgets*. Tkinter is a Tk interface for Python because the programming language does not enable the developer to create a user-friendly environment for application-specific customers. The following sample session explains how to use Tkinter within Python to create an application-specific GUI.

Creating a Listbox Using Python—Tkinter

The following POC interactive programming session shows how to create a GUI based on a listbox widget. Once the listbox is created, you will want to have the ability to print the data displayed in the Tkinter widget onto the IDLE window. Type the following code onto the IDLE window:

```
>>> from Tkinter import*
>>> win=Tk()
>>> win.title("Robot Command Language")
''
>>> listbox=Listbox(win)
>>> listbox.insert(END,"List of Commands")
>>> for item in ["Forward","Backward","ON-B","OFF-B"]:
      listbox.insert(END,item)
      listbox.pack()
```

The "from Tkinter import*" enables Python to have complete access to the Tkinter library. If this is typed at the beginning of the application, the developer does not have to type any special commands requesting the Tkinter library. The fourth line shows the response of requesting a string or title to be displayed on the Windows dialog box. The "for item" records the list in an array. If more items are required, add them to the list using the pro-

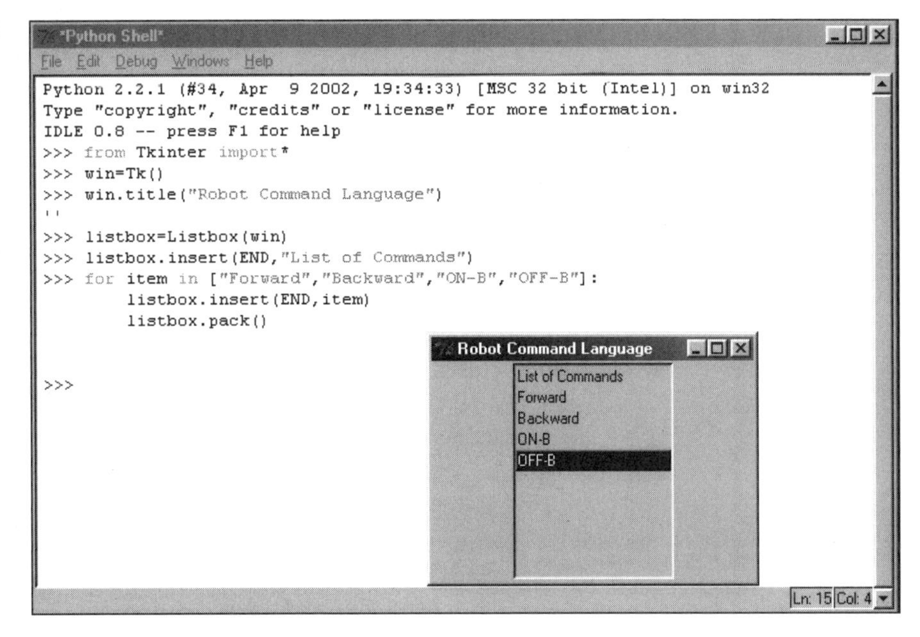

Figure 7-13 Building a Tkinter listbox widget using the Interactive Interpreter of the Python programming language

gramming style shown. The listbox.pack() displays the list on the Windows dialog box. Figure 7-13 shows the Tkinter listbox with the IDLE interactive session. To display the data from the list when selected from the Tkinter widget, use the listbox.get(item) command.

Two other building blocks are required for completing the mini server: initializing communication between Python and Excel and printing selected items from the Tkinter listbox widget to the spreadsheet. Begin building the *Robot Control Language* (RCL) mini server by typing the Python code. Go to the type menu bar of the IDLE GUI for Python, and click File and New Window. A blank window appears on the screen; type in the following lines of code to construct the RCL miniserver application:

```
""" RCL module """
#Setting up communication link to a defined Excel worksheet
import win32com.client
w=win32com.client.Dispatch("Excel.Application")
w.Visible=1
w.Workbooks.Open("C:\Python22\RCL_Panel.xls")

from Tkinter import*
win=Tk()
win.title("Robot Control Language")

listbox=Listbox(win)
listbox.insert(END)
for item in["Forward", "Reverse","Run Task","ON-B","OFF-B"]:
```

```
            listbox.insert(END, item)
           listbox.grid(row=0,column=4)

   def printer():
   part=listbox.get(ACTIVE)
   print part
   w.Range("A1").Value=part

   button=Button(win,text="Select",command=printer)
   button.grid(row=7,column=4)
```

After the code is typed in a module, save the file under the same directory as the Python22 folder on your hard drive as RCL.py. Close the file. Next, open a blank worksheet in Excel and save it as RCL_Panel.xls within the Python22 folder. Close the RCL_Panel.xls worksheet and exit Excel. Now run the RCL.py module under the IDLE GUI by typing import RCL at the prompt. If everything is okay, the RCL_Panel.xls worksheet and the RCL should be displayed on the screen. If not, check the IDLE GUI for error messages. Correct any errors that are present. Exit Python and open the application again. Run the application again. The spreadsheet RCL-panel.xls and the RCL Tkinter listbox should be on the screen. Highlight Forward with the mouse and click Select from the Tkinter listbox widget. The control name Forward should be in cell A1 of the spreadsheet. Highlight the other control names in the list and click Select on the GUI. Notice cell A1. Figure 7-14 shows the RCL Tkinter GUI and the Excel spreadsheet linked together. The cell is overwritten with each control name selection of

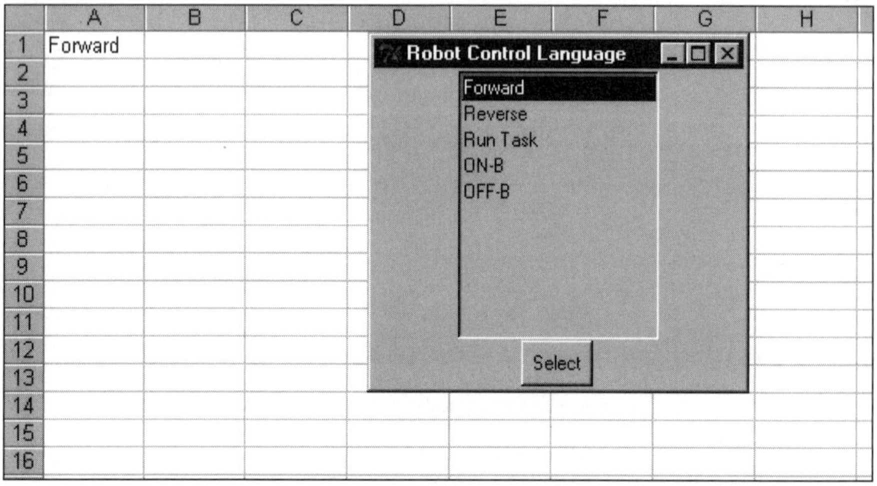

Figure 7-14 A complete Tkinter listbox with the Select button for sending data to an Excel spreadsheet

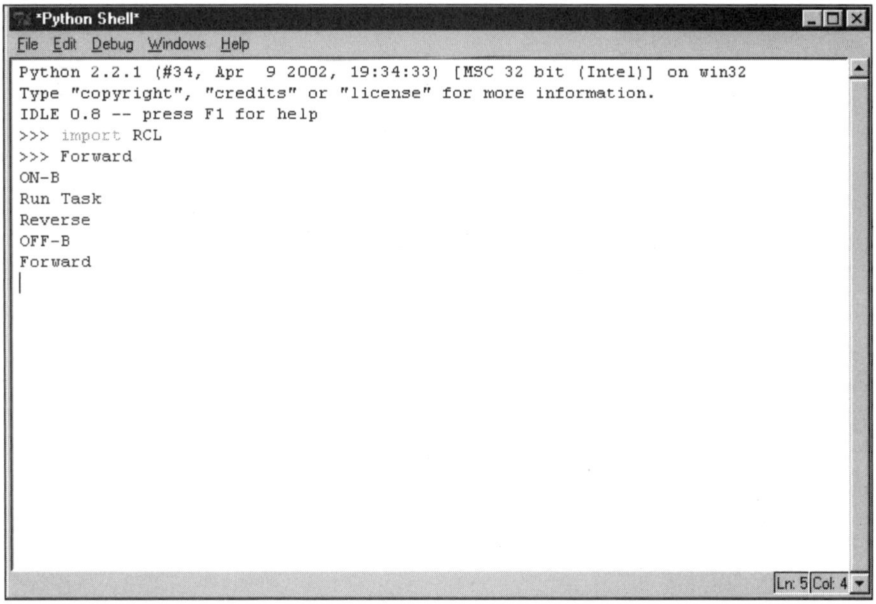

```
*Python Shell*
File Edit Debug Windows Help
Python 2.2.1 (#34, Apr  9 2002, 19:34:33) [MSC 32 bit (Intel)] on win32
Type "copyright", "credits" or "license" for more information.
IDLE 0.8 -- press F1 for help
>>> import RCL
>>> Forward
ON-B
Run Task
Reverse
OFF-B
Forward
|
Ln: 5 Col: 4
```

Figure 7-15
The print statement within the code recording all selected items onto the Python Shell

the button. Also note that the commands are printed to the IDLE GUI, as shown in Figure 7-15.

The program written is quite easy to understand. The first line of code is the name of the program. In Python, program names are called *modules*. A module is an application program where data, behavior, and interface widgets are packaged into a functional unit. The next four lines of code set up a communication link between the Python programming language and the Excel spreadsheet application. In line 5, Python asks the Excel spreadsheet named RCL_Panel.xls to open. It's important that the file is under the same directory as the Python application. If not, an error message is generated that can possibly shut down the Python application. The next eight lines of code create the Tkinter listbox for the Python RCL mini server. The button widget is built using the last two lines of code. The command function printer within the Button Properties instruction enables the selected RCL name to be transferred to the spreadsheet and IDLE GUI. If the control name is not required to be visible on the IDLE GUI, the print part can be removed from the section of code.

As stated in the introduction, Python is an object-oriented language that is self-evident based on how the widgets are built. Using the button widget as an example, consider the following code:

```
button=Button(win,text="Select",command=printer)
```

This line of code creates an instance of the object. This activity involves nothing more than building a template of the object that can be reused for other Python applications. The process of creating an instance of the object is known as *instantiation*. The object creation takes place by defining attributes of button and setting them to the real object of button. Once the button object has been created, the grid property can be defined:

```
button.grid(row=7,column=4)
```

The grid property places the button on the listbox widget using rows and columns. As shown, the button is placed on the listbox widget seven rows down and four columns to the right.

As demonstrated in this example, the ease of using the Python language for developing Windows applications doesn't take hours to learn. The robot control names can be added to the "for item" list statement. By modifying the for item list statement, additional behavior and functions can be used with the mini server. The convenience of this modification comes from its capability to expand the prototype RCL database to suit the various Mindstorms robotic control applications built by hobbyists or experimentalists. As stated earlier in the discussion of the Python programming language, the speed of modifying and testing new lines of code truly demonstrates the RAD capability of the software.

If the mini server functions correctly, the Excel VBA client can be built next. The development of this client software application follows some of the programming style guidelines discussed in Chapter 2, "Developing GUIs: Software Control Basics." We will begin building the client using the RCL_Panel.xls spreadsheet that was used to test the Python mini server.

Building a Mini Client in Excel VBA

In Chapter 2, the POC of using Excel VBA to create GUIs for the software control interfacing of Mindstorms robots was explained using lab projects. The programming style used to write Excel VBA applets and object technology for designing GUI panels was also discussed. Therefore, this discussion focuses on how to control specific functions and tasks of the Mindstorms robot using control names provided by the Python mini server. The important item to note is that the mini server provides the functions, behaviors, and tasks that the client uses to control the target robot.

As shown in Figure 7-5, the specific behavior and/or task is programmed in NQC. NQC provides a lot of flexibility in building behavior- and/or task-based robots. Also, RCX code can be converted into an equivalent NQC application by mapping the specific instructions, tasks, and functions of each programming language. Finally, the ease of programming in NQC makes it a highly regarded software language for novice and experienced Mindstorms robot builders. Let's begin the client-building process by looking at Figure 7-16.

The client environment consists of a spreadsheet, an ActiveX CommandButton, and a UserForm with additional controls for data retrieval and program execution. The client works by retrieving data from cell A1 and displaying it onto the text box located on the UserForm/GUI panel. The information is sent to the RCX P-Brick using the EXECUTE button on the panel. The controls shown on the panel have all been discussed in Chapter 2 with the exception of the image and toggle button ActiveX controls. The image ActiveX control is shown as a line drawing of an RCX Creature Bot.

It is quite easy to place images on a GUI using Excel VBA. The first thing to do is to locate the image ActiveX control on the VBA toolbox. Figure 7-17 shows the image ActiveX control. To use this control, drag it to the location where you want to see it on the UserForm. Select the image control with the mouse and go to the Properties window. Find the Picture property and click inside the box. A little box with three dots should be displayed. Click on the box. A Load Picture dialog box appears on the screen. Locate the file that has the picture you want to load into the image ActiveX control. With the file loaded, the size of the image can be stretched or shrunk by grabbing the

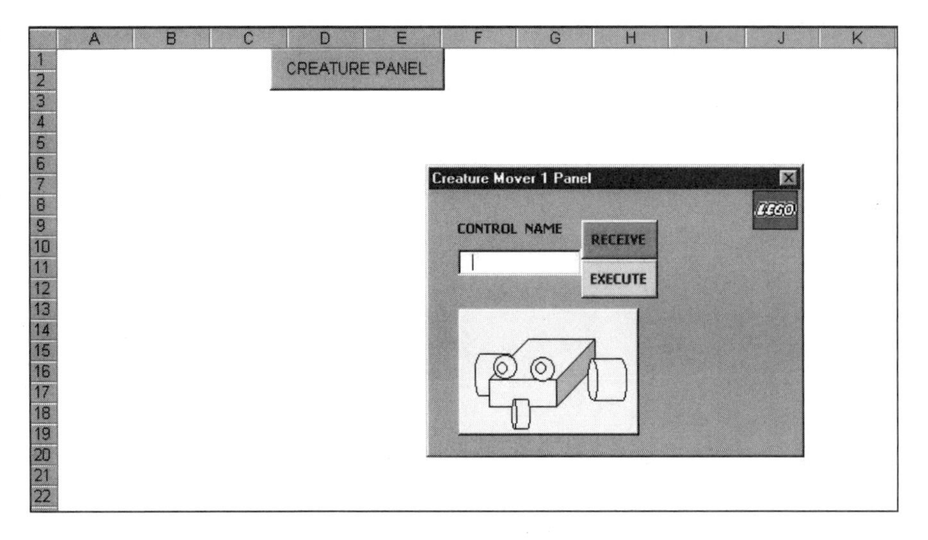

Figure 7-16 Building a client in Excel with the VBA GUI and spreadsheet controls

appropriate handlebars once selected. To create your own images such as the Creature Bot shown in Figure 7-17, use a drawing or paint program. Photos add realism to the panel's appearance and bring out its true functionality. Make sure pictures and photos are saved in bitmap format so VBA processing occurs correctly. Once the picture is completed, check to see if the image is centered correctly with as little white space as possible surrounding it. This helps the image have the best fit within the image ActiveX control. Figure 7-18 shows how to center a hand-drawn picture that will be used with an image control object.

The toggle button ActiveX control shown in Figure 7-19 works like the CommandButton except it has an additional input request. Like the physical electrical switch, the toggle button is a software-latching input control used in applications that require one command but are able to perform two distinct output responses. As seen in Figure 7-19, the toggle button has a raised appearance when the switch is off. The toggle button is then pushed in, or latched closed, when the communication event is requested. Therefore, instead of using two CommandButtons for the ON/OFF control of events, the toggle button provides both control functions using one ActiveX control.

Now that both the controls and spreadsheet have been laid out on the GUI panel, the code used to make them functional is required. The software used to provide client functionality is as follows:

```
Dim MotorA As String
Dim MotorB As String
Private Sub COMMBUTTON_Click()
If CMPanel.COMMBUTTON.Value = True Then
CMPanel.PBrickCtrl.InitComm
CMPanel.SWITCHSTATUS.Caption = "ON"
Else
CMPanel.PBrickCtrl.CloseComm
```

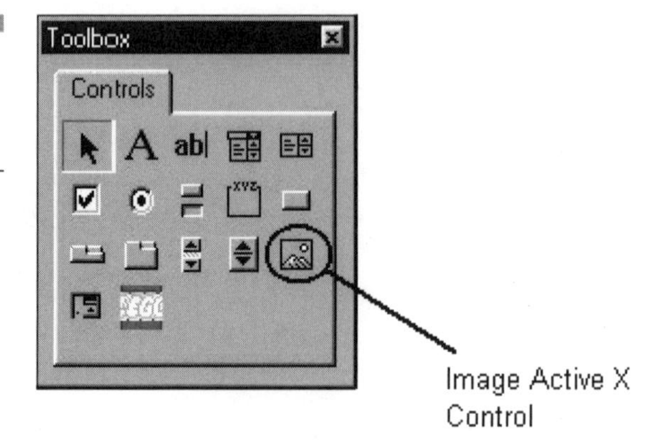

Figure 7-17 The image ActiveX control icon located on the VBA toolbox

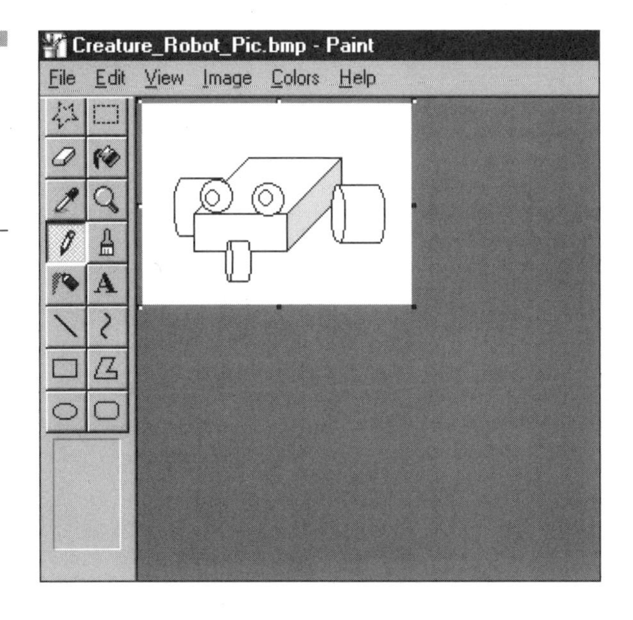

Figure 7-18
Creating a drawing in MS Paint that will be used with the Image ActiveX Control.

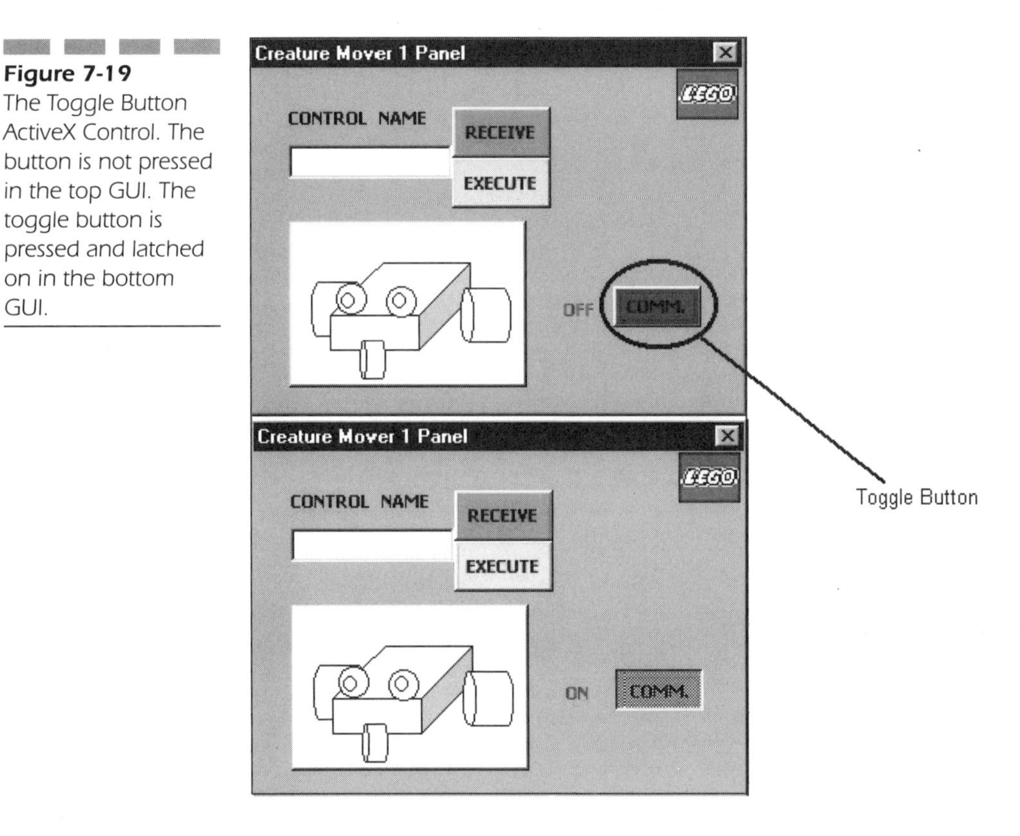

Figure 7-19
The Toggle Button ActiveX Control. The button is not pressed in the top GUI. The toggle button is pressed and latched on in the bottom GUI.

```
CMPanel.SWITCHSTATUS.Caption = "OFF"
End If
End Sub
Private Sub EXECBUTTON_Click()
If CMPanel.DATA.Value = "Forward" Then
MotorA = "0"
CMPanel.PBrickCtrl.SetFwd (MotorA)
CMPanel.PBrickCtrl.On (MotorA)
Else
If CMPanel.DATA.Value = "Reverse" Then
MotorA = "0"
CMPanel.PBrickCtrl.SetRwd (MotorA)
CMPanel.PBrickCtrl.On (MotorA)
Else
If CMPanel.DATA.Value = "Stop" Then
CMPanel.PBrickCtrl.Off (MotorA)
Else
If CMPanel.DATA.Value = "ON-B" Then
MotorB = "1"
CMPanel.PBrickCtrl.On (MotorB)
Else
If CMPanel.DATA.Value = "OFF-B" Then
MotorB = "1"
CMPanel.PBrickCtrl.Off (MotorB)
Else
If CMPanel.DATA.Value = "Run Task" Then
CMPanel.PBrickCtrl.StartTask 0
End If
End If
End If
End If
End If
End If
End Sub
Private Sub RECBUTTON_Click()
CMPanel.DATA.Value = Range("A1").Value
End Sub
```

An important item to note about the programming style of the code is the use of <Parent Object. Child Object. Property> and <Parent Object. Child Object. Method>. This format for naming and decomposing an object is one philosophy of OOP.

This structural approach to programming is used because it provides the ability to encapsulate not only behavior, but also data. For advanced Mindstorms projects such as this one, OOP is a key tool for designing sophisticated controls for LEGO robots. Also, using the OOP convention assists in the debug phase because the controls used in the layout of the GUI panel are identified with meaningful names that relate to the software device function. As shown in the code listing, the OOP names relate to the controls used on the GUI panel. The GUI is named *CMPanel*, which is short for

Creature Mover panel. The name now becomes the Parent Object to all other objects because they are tied to the main object by their associated placement on the VBA UserForm. The associated placed controls are therefore called Child Objects. The properties are the object's physical attributes that can be changed and are usually identified with an equal sign. The equal sign sets the appropriate attribute assigned to the Child Object for the given software application and environment. Methods enable the objects to get things done. For example, consider CMPanel.PBrickCtrl .StartTask 0. Here, the Parent Object is CMPanel and the Child Object is PBrickCtrl. The Method is the StartTask 0 or execute program 0 of the Mindstorms RCX P-Brick. Therefore, when designing code using an OOP language, always think of the relationship between objects as Parent and Child and consider how their attributes can be changed as well as what task the object will accomplish within the program. The code for the command button located on the spreadsheet used to call the CMPanel object is as follows:

```
Private Sub CommandButton1_Click()
CMPanel.Show
End Sub
```

Now that we have all the building blocks available, it is time to put the controller together.

Putting the Controller Together

Once the electrical/electronics controller, the Creature Mover mechanism/drive assembly, and the client and server applets are built, it is time to put them together. The client and server units are software components, requiring the controller to be built virtually rather than physically. The Python Shell must be active in order to run the server. At the prompt, type import RCL. RCL is the name of the Python module acting as a mini server. If the server is running, the client (Excel spreadsheet) will be open. The panel that receives the RCL functions and executes them on the RCX P-Brick can be displayed using the CREATURE PANEL button on the spreadsheet. The function names are selected from the Python listbox widget and sent to the spreadsheet using the Select button. The client obtains the control-function name using the RECEIVE button on the VBA panel.

The function is executed on the RCX P-Brick using the EXECUTE button. Figure 7-20 shows both the VBA and Python control panels on the Excel spreadsheet. The new code for the modified Python control panel is shown as well.

```
""" RCL module """

#Setting up communication link to a defined Excel worksheet
import win32com.client
w=win32com.client.Dispatch("Excel.Application")
w.Visible=1
w.Workbooks.Open("C:\Python22\RCL_Panel.xls")


from Tkinter import*
win=Tk()
win.title("Robot Control Language")


listbox=Listbox(win)
listbox.insert(END)
for item in["Forward", "Reverse","Stop","Run Task","Stop Task","ON-
B","OFF-B"]:
    listbox.insert(END, item)
    listbox.grid(row=0,column=4)

def printer():
    part=listbox.get(ACTIVE)
    print part
 w.Range("A1").Value=part

button=Button(win,text="Select",command=printer)
button.grid(row=7,column=4)
```

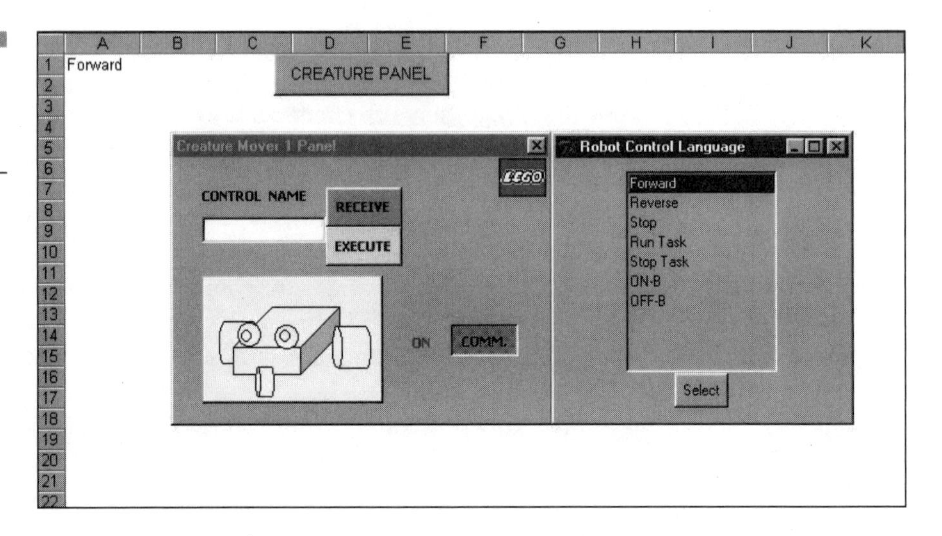

Figure 7-20 The complete client-server GUIs with spreadsheet controls

The Python is a wonderful language to use with Excel VBA because of the seamless applications that can be created. The design and construction of the listbox has the same look as the UserForm and controls for the VBA GUI. As seen in Figure 7-20, the Python GUI is a nice addition to the development of software controllers for Mindstorms robots. Although Python does not directly control the robot, the idea of using a software extension to manipulate the robot is not different than using the Internet to send control data over the Web to a remote machine. In fact, the client-server built in this section can be used to control machines and processes over a data network infrastructure. A Python mini server that has a listbox widget for robot control-function selection is a novel way to program a robot with text-based commands.

The VBA code shown earlier has all of the control-function names embedded within the virtual buttons. The Run Task control is actually two separate programs that work sequentially. The control name is coded in VBA under the "If CMPanel.DATA.Value = Run Task Then" conditional statement. To execute the embedded task residing inside of the RCX P-Brick, the Parent/Child Object CMPanel.PBrickCtrl.StartTask 0 is used to run the control program. The functional objective of the code is to move the Creature Mover back and forth three times. This function is carried out with the NQC programming language. The NQC code is shown in the following listing and in Figure 7-21:

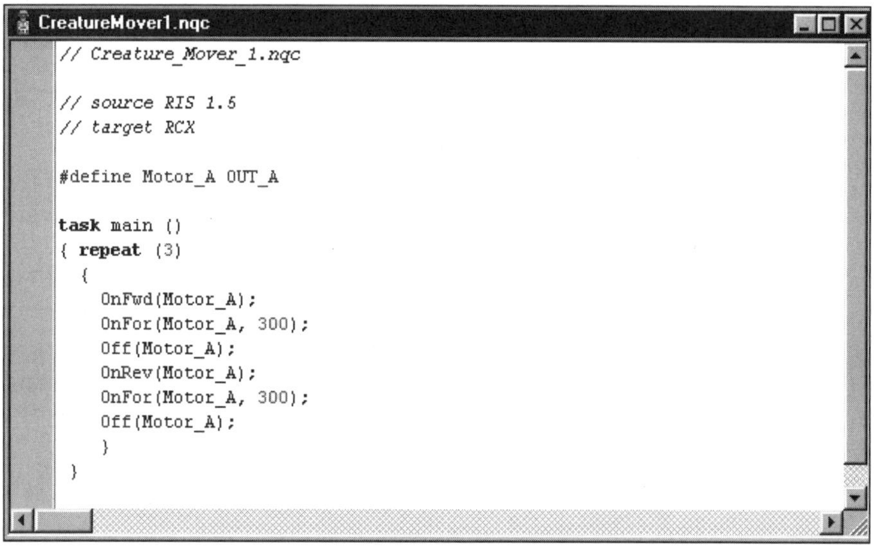

CreatureMover1.nqc

```
// Creature_Mover_1.nqc

// source RIS 1.5
// target RCX

#define Motor_A OUT_A

task main ()
{ repeat (3)
  {
    OnFwd(Motor_A);
    OnFor(Motor_A, 300);
    Off(Motor_A);
    OnRev(Motor_A);
    OnFor(Motor_A, 300);
    Off(Motor_A);
    }
 }
```

Figure 7-21 The NQC code built in the Brick Command Center (BricxCC) editor for the Creature Mover robot

```
// Creature_Mover_1.nqc

// source RIS 1.5
// target RCX

#define Motor_A OUT_A

task main ()
{ repeat (3)
  {
    OnFwd(Motor_A);
    OnFor(Motor_A, 300);
    Off(Motor_A);
    OnRev(Motor_A);
    OnFor(Motor_A, 300);
    Off(Motor_A);
    }
 }
```

The NQC code is very simple and small in size, and enables additional features to be added such as an input switch or sensor. The task main() is linked to the VBA applet using the CMPanel.PBrickCtrl.StartTask 0 Parent/Child Object. If additional tasks were used within the NQC code, they would be identified with the void name() construct. Likewise, StartTask 1 through 9 would correspond to each of the void name() constructs.

The client-server controller has a diagnostics capability that uses the ON-B and OFF-B control-function names. The power timed-delay circuit is initiated by turning on output B using the ON-B control-function name. The timing sequence starts when the momentary pushbutton is pressed. After the timing delay is completed, the low-side driver *Power Metal Oxide Semiconductor Field Effect Transistor* (PMOSFET) turns off. Output B is still on until the OFF-B function is sent to the RCX P-Brick. Figure 7-22 shows the RCX P-Brick wired to the power timed-delay circuit. The LED circuit can be replaced with any small electrical/electronics load under 500 mA. Also, with a slight modification to the wiring, the circuit can turn on immediately when the RCX P-Brick receives the control function from the client. This switching capability can be used to drive an electromechanical relay (see Chapter 3, "Electromechanical Controls Interfacing") that can control a Scout P-Brick. The Scout P-Brick essentially becomes a smart actuator for an advanced Creature Mover robot.

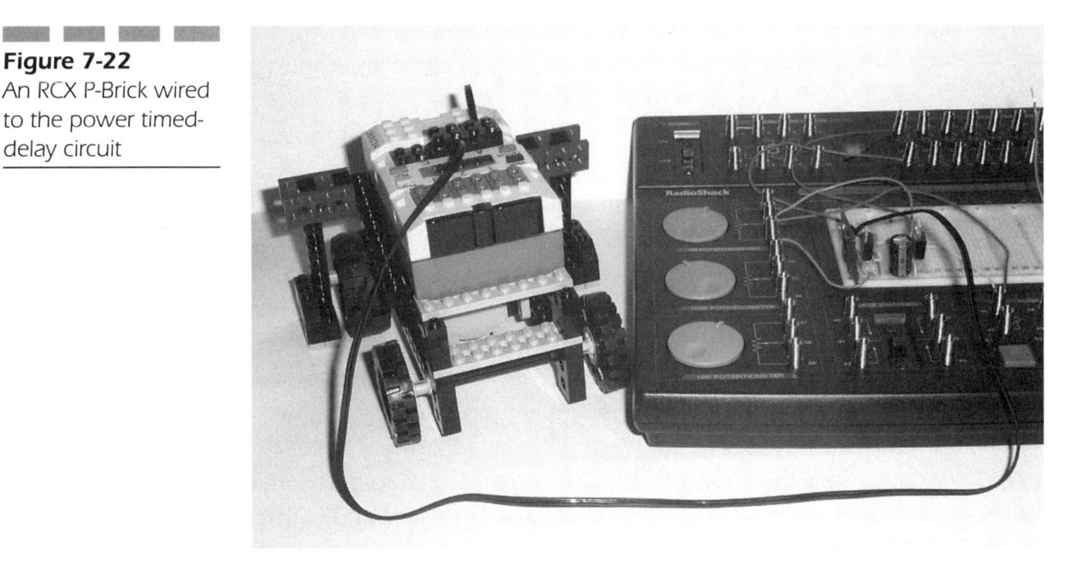

Figure 7-22
An RCX P-Brick wired to the power timed-delay circuit

Final Thoughts and Further Enhancements for the Client-Server Controller

This Mindstorms interfacing project opens the door to the development of many data network architectures that provide robotic control. The Python mini server can be modified to use binary codes that represent specific functions or behaviors for a Mindstorms robot. The codes can be transmitted to an Excel spreadsheet and then converted into their equivalent text-based control-function names. A lookup table in Excel can be built that compares the binary code sent from the mini server to find an equivalent text-based control-function name. The name is then displayed on the spreadsheet for final processing using a VBA GUI. Using some of Excel's functions, the binary data can be converted into the equivalent base 2, 8, and 16 formats and be used to create math-based robots. The numeric-based conversion dictates the type of function or behavior required to execute the correct embedded task programmed inside of the Mindstorms robot.

In terms of hardware projects, the Tkinter widget could have a list showing different time delays. Selecting one time delay and sending it to the Excel spreadsheet can be used as a programming constant that affects the timing function of the Mindstorms robot. The robot can be used as a smart timer for controlling small mechanical assemblies and electrical/electronics controllers at less than 500 mA. The RCX P-Brick could drive seven-segment displays using the Tkinter listbox widget to select specific characters to be created on the unit. The use of silicon diodes to block certain display segments can help to determine the number of LEDs to switch on and off.

The Creature Mover used in the client-server controller project can be enhanced using the Vision Command LEGO Camera. This acts as an electronic eye for maneuvering around obstacles while the robot is being tested. The validation of the Creature Mover and the list of control functions proved to be challenging because the robot and Infrared Tower were located on the kitchen floor and the notebook was placed on top of the counter. With the aid of a camera, several robot crashes into the dishwasher would have been prevented. Actuation of the camera can be accomplished through the same network scheme except it uses different control names with the list. The widget can be built to display control functions like pan, zoom, and tilt. Again, the Tkinter widget is very easy to modify because of the for item in["list of string items"] construct. Control of the robocam follows the same procedure as previously discussed for Creature Mover manipulation. The suggestions presented in this chapter only scratch the surface of the applications that can be built using the client-server controller as the main infrastructure for controlling Mindstorms robots.

The LabView software created by National Instruments is a virtual instrumentation and control application that develops testing and measurement devices for validating industrial consumer products and processes. The technology of the software is based on using graphical icons that represent industrial math processes and functions. The actual configuration of the test system under development is a wired electrical circuit. The interface that the user interacts with is a graphical representation of a test panel with knobs, gauges, and indicator lights. Therefore, the test development GUI is a software interface to the hardware that provides the signals for stimulating the inputs of the electronics controller module that is being tested. The same design approach can be used to develop controls to stimulate the function or behavior of a Mindstorms robot. In the next chapter, a simulator timer control is built using the POC method of Chapters 2 and 7. Animating electromechanical relay-switching contacts are also explored.

CHAPTER 8

Simulator Controls for Mindstorms Robots

Mindstorms robots provide a vast amount of experimental opportunities in the area of automation controls. As discussed in Chapter 3, "Electromechanical Controls Interfacing," electromechanical controls are the basic foundation for operating a Mindstorms robot. By using an electric switch, an electromechanical relay, and software, a simple but powerful robot can be built and controlled quite easily. In Chapter 4, "Electronic Switching Circuits," and Chapter 5, "Sensor-Interfacing Basics," electronics and sensor circuits gave the Mindstorms robot the capability to maneuver around its environment based on the data received and processed by the sensory device.

Simulator controls are a digital manipulative for operating robots with *Graphical User Interfaces* (GUIs). The *Proof Of Concept* (POC) behind using a simulator control is to build a panel with the same capabilities as the physical unit under development—the exception to this is the use of software controls for emulating *Input/Output* (I/O) instead of the physical hardware device. The simulated control enables hobbyists and experimentalists to understand the physical electronic circuit block(s) under development through the execution of a simulation study. The circuit's physical I/O provides information about its behavior and duplicates the electronics-interfacing layer by building a GUI-based controller. The simulator controller can be expanded by adding ActiveX objects to the GUI. The ActiveX control programmed functions give hobbyists and experimentalists complete control over the Mindstorms robot. The key to building a simulator controller is to use an *Electronics Design Automation* (EDA) tool to conduct a simulation experiment on the target circuit. The data received from the simulation exercise can be used to build a virtual controller to operate a Mindstorms robot. A 555 Monostable Timer Circuit is used to build the simulator controller project. The following topics are discussed in this chapter:

- Monostable circuit basics
- Circuit simulation model curves
- Creating a *Virtual Timer User Interface* (VTUI)
- Adding an animated electromechanical relay

Building a simulator controller has the following educational objectives:

- To show how a robot can be controlled using a virtual timer
- To simulate a monostable one-shot circuit using Circuit Designer, Excel *Visual Basic for Applications* (VBA), and Mindstorms *Programmable Brick* (P-Brick) techniques

- To see how animated electronics techniques can be implemented in VBA
- To use a scrollbar ActiveX as a simulated variable resistor

The discussion of a simulator controller begins with understanding the basics of a 555 Monostable Timer Circuit.

Monostable Circuit Basics

Monostable is a class of multivibrator circuit that has the capability to produce a one-stable state output. Another name for a monostable circuit is a *one-shot timer*. In order for the circuit to switch state, an input signal must be applied to it. In a 555 Monostable Timer Circuit topology, the output must be low and the capacitor must be discharged to ground. The trigger input is normally held high using a pull resistor. When a negative pulse that is less than $1/3\ V_{CC}$ [supply voltage applied to the 555 *Integrated Circuit* (IC)] is applied to the trigger input, the comparator inside of the 555 IC is set high, which makes the internal flip-flop's not Q output low. The output signal then becomes high as the capacitor discharges to ground using an NPN transistor. The NPN transistor turns off, enabling the capacitor to charge up using a resistor that is connected to it.

When the voltage across the capacitor reaches $2/3\ V_{CC}$, the internal comparator's output is high, which resets the *Reset-Set* (RS) flip-flop and makes the monostable's output go low. The switching activity of the output occurs when a negative pulse is applied to the trigger input of the 555 IC. The output remains in the high output state based on the value of the external *resistor* (R) and *capacitor* (C). Figure 8-1 shows the circuit schematic of a 555 Monostable Timer Circuit. The equation used to determined the pulse width is as follows:

$$T_{PW} = 1.1RC$$

To properly operate the monostable circuit, the timing resistor should be between 10-kilo-ohm and 14-mega-ohm (one mega-ohm is equal to 1×10^6 ohms) and the timing capacitor should have a capacitance value in the range of 100 pF to 1,000 µF.

The virtual timer that will be built will not have the precise timing values of the actual circuit, but it will simulate the behavior of the timing

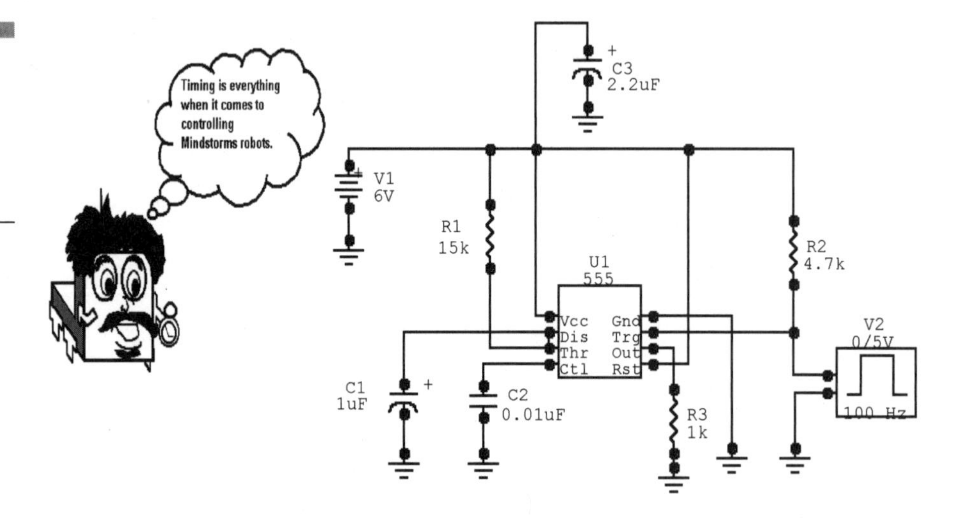

Figure 8-1 Mickey Monos explains the timing philosophy of the 555 Monostable Timer Circuit.

device. The key element of the simulator controller is the behavior of the timing device. The objective of the virtual timer is to develop a flexible way of controlling a Mindstorms robot through a programmable interface. Therefore, by capturing the behavior of the monostable circuit of the GUI, the virtual timer emulates the monostable circuit's input and output timing functions. Some of the unique qualities of developing a virtual timer include the embedded code that lives inside of the GUI and its interaction with the Mindstorms robot's firmware and microcontroller. Using a good software design method ensures that the correct operation is available inside of the GUI-based simulator controller. With a good understanding of how the monostable circuit works, the timing device can be built using an EDA tool to capture the schematic, which is simulated to verify the proper operation of the controller and obtain data from the simulation. The data can be used to build the simulator controller correctly to achieve the proper timing operation of the robot's embedded task(s).

Circuit Simulation Model Curves

The results of running a circuit simulation are best viewed with plots of specific electrical parameters. For the 555 Monostable Timer Circuit, the important electrical parameters are input frequency, output frequency, and the charge and discharge profile. Figures 8-2 through 8-4 show the plots

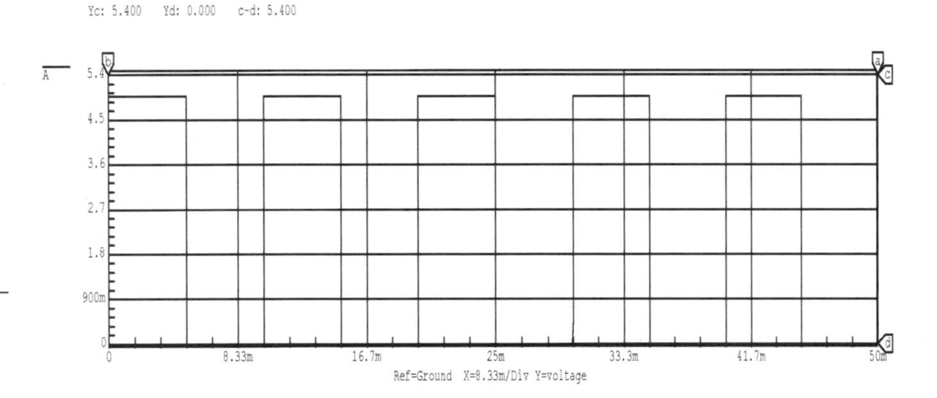

Figure 8-2 Input-switching frequency of 100 Hz plotted with Circuit Maker design software. This signal is used to stimulate the 555 Monostable Timer Circuit model shown in Figure 8-1.

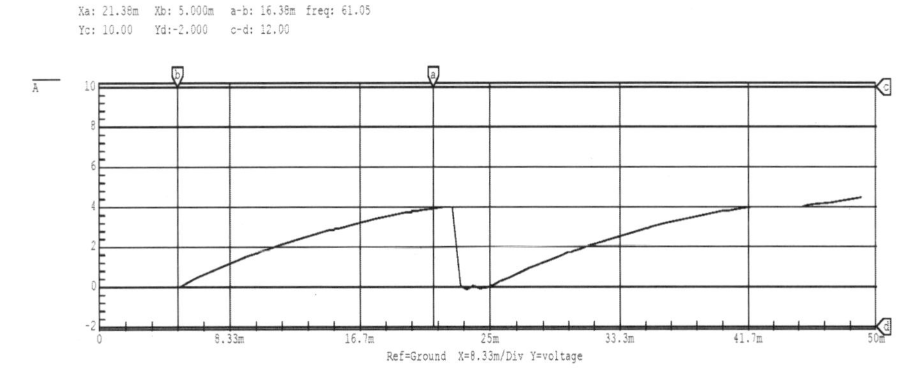

Figure 8-3 Charging and discharging capacitor C1 of the 555 Monostable Timer Circuit. Waveform is plotted with Circuit Maker design software.

obtained from Circuit Maker software (see Chapter 7, "Client-Server Controller for Mindstorms Robots" for details). On the Y-axis of each plot is the voltage measured at the input, C1 capacitor, and the output of the 555 Monostable Timer Circuit. The X-axis represents the time in milliseconds. The plots are obtained by setting up a *Transient Analysis* simulation session of Circuit Maker. A transient analysis looks at an *Alternating Current* (AC) or *Direct Current* (DC) circuit's signals over time. Basically, the signal captured on a plot is a snapshot of the circuit working at a specific time event. The input frequency plot shows the trigger input being switched at a 100 Hz or pulsed at 10-millisecond intervals. The capacitor's charge and discharge plots indicate when a negative pulse applied at the trigger input of

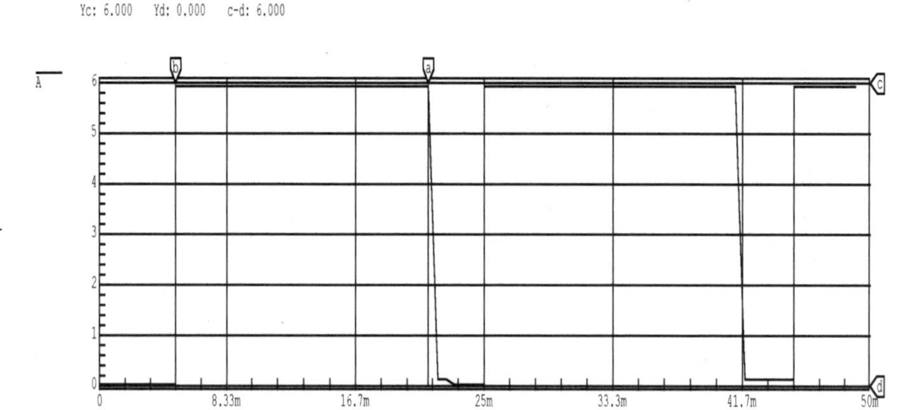

Figure 8-4 Output waveform produced by the 555 Monostable Timer Circuit. The waveform is plotted with Circuit Maker design software.

the IC is less than 1/3 V_{CC} (the capacitor discharges to ground) and when the voltage across the capacitor reaches 2/3 $V_{CC.}$ This plot is typical and is shown in every basic electricity textbook.

The plot shown in Figure 8-4 shows the output switching at a specific frequency driven by the input signal of 100 Hz. The pulse width can be found using the equation $T_{PW} = 1.1RC$. In this example, T_{PW} equals the following:

$$T_{PW} = 1.1 \times 15\ K \times 1\ \mu F$$

$$TPW = 16.5ms$$

To measure the T_{PW} value in Circuit Maker, take the *a* and *b* markers and move them horizontally to where the output waveform's rising and falling edges appear on the plot. A section that lists various measurements should appear just below the title bar. Look for the measurement from *a* to *b*. The measured value should be 16.38 milliseconds. Figure 8-5 shows the output plot being measured using the *a* and *b* markers. Note the difference between the measured value and the calculated one. This measured-versus-calculated validation is very important in circuit analysis because it enables hobbyists and experimentalists to dive deep into the details of basic electricity, physics, and semiconductor behavior. Circuit design practices dictate that analysis should always be followed up with solderless breadboarding. Building the circuit and taking measurements ensures that both the simulation and manual calculations are within 5 percent of each other. Also, after analyzing the charging and discharging profile plot, 2/3 V_{CC}

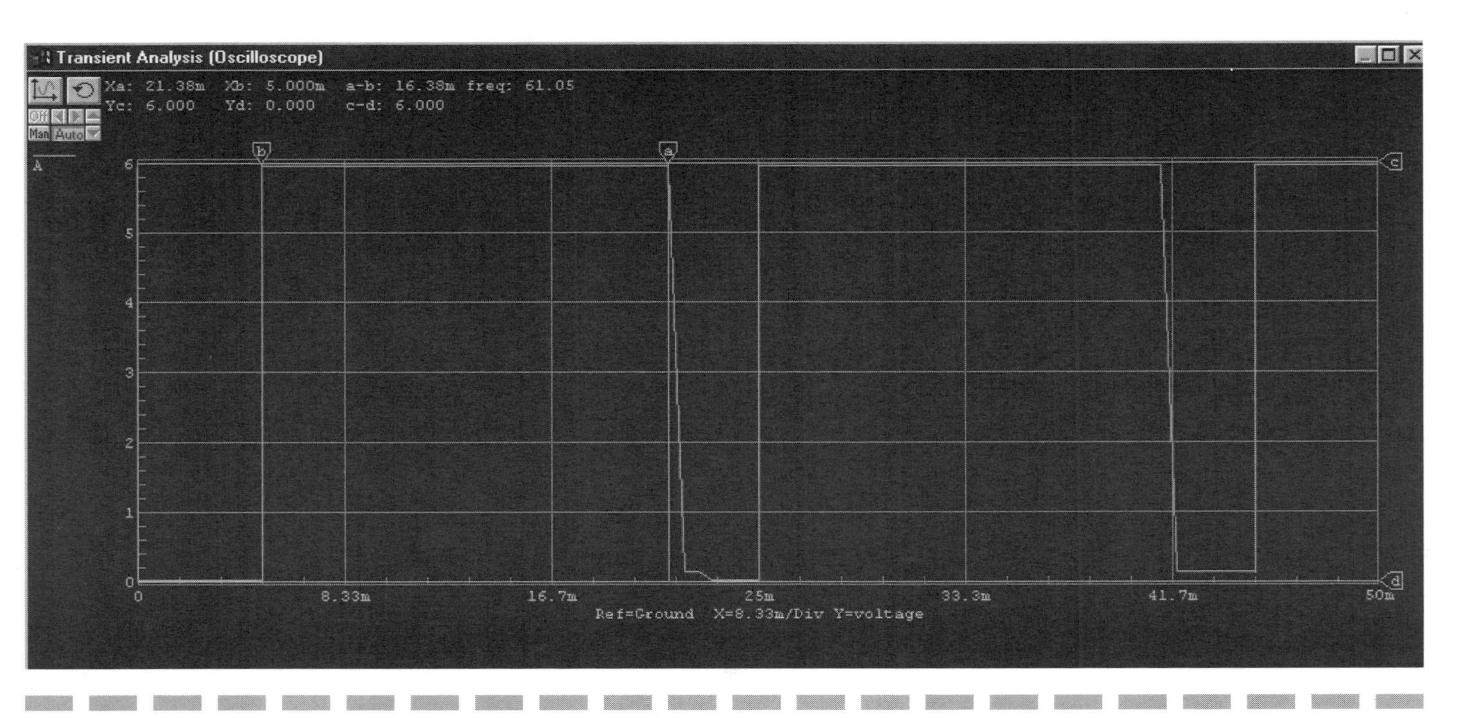

Figure 8-5 The actual screen of the output waveform created by Circuit Maker design software. The pulse width of the timer circuit measures at 16.38 milliseconds using markers a and b.

should equal 4 volts and $\frac{1}{3} V_{CC}$ should equal 2 volts. Therefore, using a simulation tool such as Circuit Maker to build and run circuit models helps hobbyists and experimentalists to understand core designs prior to physically building a simulator controller for Mindstorms robots.

Now that the basics of the 555 Monostable Timer Circuit have been explained, the virtual timer interface can be built. The information presented in the following section aids in the development of the timer interface. The key to the virtual controller's operation is to embed timing functions into the GUI. This data is used to run the robot at the specified time received from the user interface panel of the Excel spreadsheet.

Creating a Virtual Timer User Interface (VTUI)

Building a VTUI is based on art, understanding circuit behavior, and VBA programming. The VTUI relies on using the spreadsheet as the Userform. The simulated circuit from Circuit Maker is placed on the spreadsheet with the appropriate ActiveX controls for changing component values. Create the 555 Monostable Timer Circuit shown in Figure 8-1. Substitute the 15-kilo-ohm resistor with a 1-mega-ohm variable resistor and the 1 µF capacitor with a 100 µF component. To import the circuit diagram from Circuit Maker, it must first be exported to a file. On the main menu bar, highlight File, and select Export and Circuit as Graphic. Save the file with an appropriate name onto your hard drive or 3½-inch diskette. The extension of the file is .wmf.

The circuit is now ready to be inserted onto the spreadsheet. Open a new worksheet and select Insert from the menu bar. Then highlight Picture from the pull-down menu and click From File. Go to the directory that has the circuit and open the file. The circuit diagram should be on the spreadsheet and ready to add controls. Add a text box, CommandButton, and scrollbar controls to the circuit. Place the controls at the designated locations, as shown in Figure 8-6. Adjust the size of the controls to fit the location where they are placed accordingly.

In order for the VBA applet to access the Spirit.ocx library, the LEGO icon must be placed on the spreadsheet. To obtain the Spirit.ocx library, select the Toolbox icon from the Control Toolbox toolbar menu. From the pull-down menu, click the Spirit Control. A crosshair is shown on the spreadsheet. Move it to the appropriate spot, and drag and click until the

LEGO icon (ActiveX control) is the appropriate size. For each of the controls, format the name and caption with meaningful names. Additional objects are added to the spreadsheet such as another CommandButton and formatted B3 cell. As shown in Figure 8-7, these objects are placed at the specific locations on the spreadsheet for visual appeal. To test the functions of the controls and cell location displays, enter the following code into the *Visual Basic Editor* (VBE).

```
Dim a As Double
Dim Time As Double
Private Sub StartButton_Click()
If Range("B3").Value = "3.60437" Then
Range("B6").Value = "It's A Match"
Else
Range("B6").Value = "It's Not A Match"
End If
End Sub
Private Sub VarResistor_Change()
a = VarResistor.Value
ResistorBox.Value = a
Time = 1.1 * a * 0.0001
Range("B3").Value = Time
End Sub
```

The scrollbar control named VarResistor.Value assigns its value to variable *a*, enabling the program to work. The data can be stored in variable *a* and then transferred to the text box control named Resistor. The T_{PW} is calculated using Time as the variable. This calculated value is placed on the

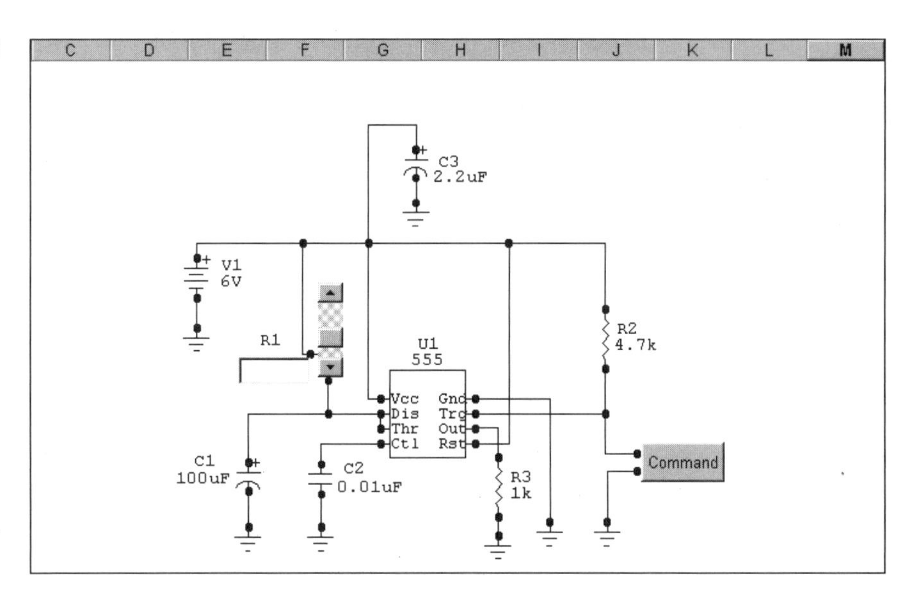

Figure 8-6
The first Monostable Timer Circuit design imported from Circuit Maker to the Excel spreadsheet. The ActiveX controls (scrollbar, text box, and Command-Button) are placed at the appropriate locations of the schematic to provide interactive control of the simulated controller.

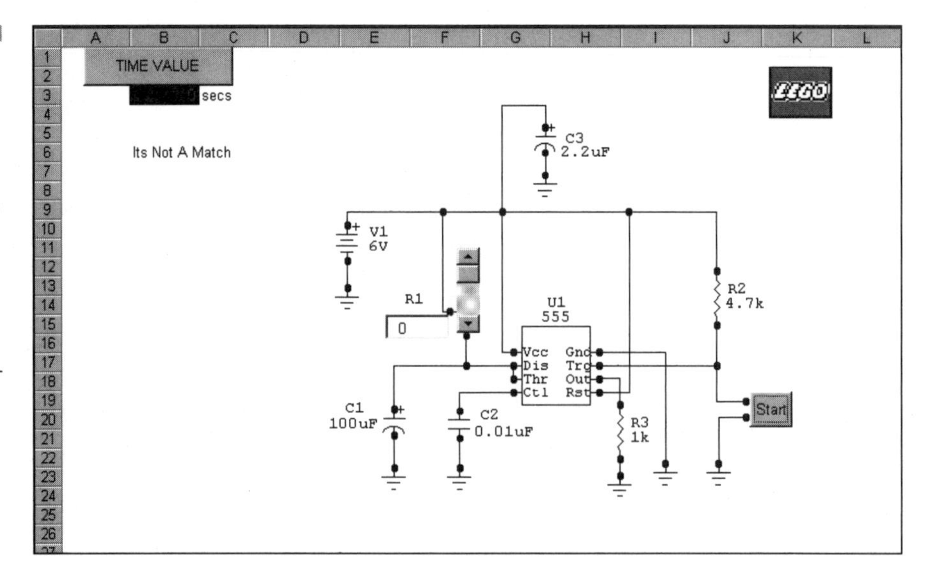

Figure 8-7 The spreadsheet is formatted to provide visual appeal for the simulator controller. LEGO Spirit.ocx (icon) is placed on the spreadsheet, providing access to the complete object library.

spreadsheet at location B3. The VBA Range("Cell Index").Value object is used to place the calculated value of Time at the B3 cell location. A condition is set up where if the value stored in Range("B3").Value is equal to the constant 3.60437, the B6 cell will display the statement "It's A Match." If the Range("B3") is not equal to the constant value, then it prints "It's Not A Match." This test program serves as a template for the actual simulator controller. Therefore, additional modifications are made to the test program to implement the simulator controller. Add the additional ActiveX controls to the spreadsheets, as shown in Figure 8-8. The new code for the modified virtual timer is as follows:

```
Dim a As Double
Dim Time As Double

Private Sub COMMButton_Click()
If Worksheets("Sheet1").COMMButton.Value = True Then
Worksheets("Sheet1").PBrickCtrl.InitComm
Worksheets("Sheet1").Range("E1").Value = "ON"
Else
Worksheets("Sheet1").PBrickCtrl.CloseComm
Worksheets("Sheet1").Range("E1").Value = "OFF"
End If
End Sub
Private Sub StartButton_Click()
Worksheets("Sheet1").PBrickCtrl.StopAllTasks
If Worksheets("Sheet1").Range("B3").Value = "3.60437" Then
Worksheets("Sheet1").Range("B6").Value = "It's A Match"
Worksheets("Sheet1").PBrickCtrl.StartTask 1
Else
```

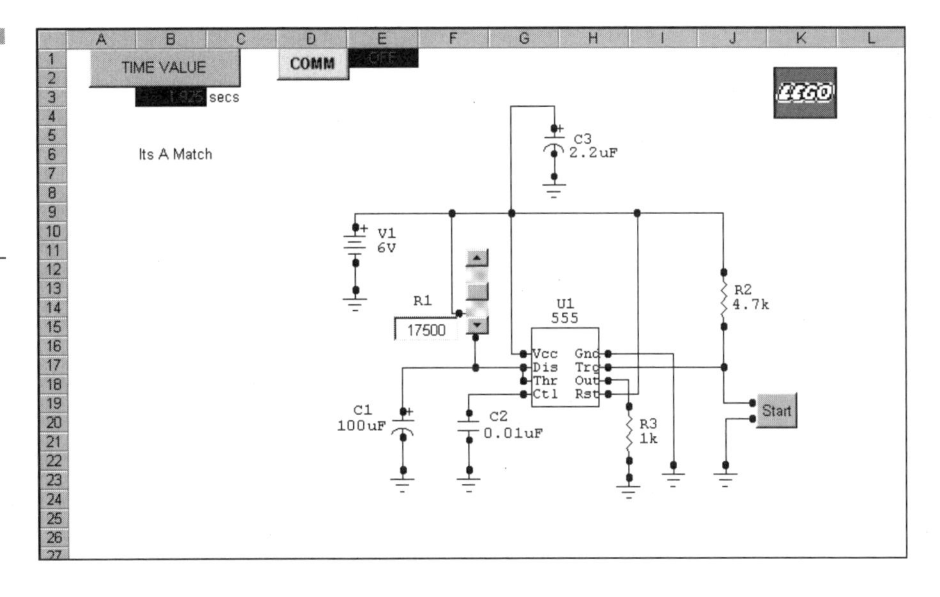

Figure 8-8 Simulator controller with the Toggle Button (COMM) for opening and closing the serial port for the infrared controller

```
Worksheets("Sheet1").Range("B6").Value = "It's Not A Match"
If Worksheets("Sheet1").Range("B3").Value = "2.695" Then
Worksheets("Sheet1").Range("B6").Value = "It's A Match"
Worksheets("Sheet1").PBrickCtrl.StartTask 2
Else
Worksheets("Sheet1").Range("B6").Value = "It's Not A Match"
If Worksheets("Sheet1").Range("B3").Value = "1.925" Then
Worksheets("Sheet1").Range("B6").Value = "It's A Match"
Worksheets("Sheet1").PBrickCtrl.StartTask 3
Else
Worksheets("Sheet1").Range("B6").Value = "It's Not A Match"
End If
End If
End If
End Sub
Private Sub VarResistor_Change()
a = Worksheets("Sheet1").VarResistor.Value
Worksheets("Sheet1").ResistorBox.Value = a
Time = 1.1 * a * 0.0001
Worksheets("Sheet1").Range("B3").Value = Time
End Sub
```

A small conversion factor was embedded in the code for selecting the correct task associated with the time value. The time value is based on the embedded equation Time = $1.1 \times a \times 0.0001$, where *a* is the simulated resistor obtained from the VarResistor.Value object. The constant 0.0001 is the 100 µF capacitor wired to the 555's threshold and discharge pins. The values picked for the variable resistor/scrollbar control are 32767, 24500, and 17500. Using the T_{PW} equation and the three scrollbar control values, the corresponding pulse width values are calculated as 3.60437, 2.695, and

1.925, respectively. These values are added to the code to provide conditions for selecting the appropriate task with the correct timing values. The data was used to write *Not Quite C* (NQC) code to control a Mindstorms robot. The software developed for controlling the Creature Mover robot (discussed in Chapter 7) is as follows:

```
// Virtual_Timer_.nqc

// source RIS 1.5
// target RCX

#define Motor_A OUT_A

task main()
{
  start three;
  start two;
  start one;
}
task three()
{ repeat (3)
  {
    OnFwd(Motor_A);
    OnFor(Motor_A, 36);
    Off(Motor_A);
    OnRev(Motor_A);
    OnFor(Motor_A, 36);
    Off(Motor_A);
    }
 }
  task two()
 { repeat (2)
  {
    OnFwd(Motor_A);
    OnFor(Motor_A, 26);
    Off(Motor_A);
    OnRev(Motor_A);
    OnFor(Motor_A, 26);
    Off(Motor_A);
    }
 }
 task one()
 { repeat (1)
  {
    OnFwd(Motor_A);
    OnFor(Motor_A, 19);
    Off(Motor_A);
    OnRev(Motor_A);
    OnFor(Motor_A, 19);
    Off(Motor_A);
    }
 }
```

As you can see, the NQC code has the base software for providing a reciprocal movement (back-and-forth motion) of the robot's chassis and mechanical drive assembly. Each task has a corresponding time value that is calculated by the virtual timer's worksheet using the T_{PW} equation. Therefore, task three has a time of 3.6 seconds, task two has a time of 2.6 seconds, and task three has a time of 1.9 seconds. If the correct R1 values have been selected, a message also appears that tells the operator whether or not a match occurs. Running the program several times demonstrates how the robot responds to the virtual timer's time values.

If the 555 Monostable Timer Circuit works correctly, the concept of adding an animated electromechanical relay to the VBA applet can be implemented. This feature enhances the functional appearance of a spreadsheet and illustrates a simple animation technique using VBA.

Adding an Animated Electromechanical Relay

To enhance the visual appearance of the virtual timer's spreadsheet controller, a little bit of animation can be helpful. Figure 8-9 shows the new circuit with the timer electromechanical relay added to the output of the 555 Monostable Timer Circuit. The process of creating the schematic is the same as outlined earlier in the section "Creating a Virtual Timer User Interface (VTUI)." Using the circuit that was created and added to the spreadsheet (see Figure 8-9), the electromechanical relay can be built and tested.

The technique for creating the electromechanical relay involves capturing two pictures, which both show the relays in an open or closed position. The pictures for the relays can be hand drawn or modified from a schematic capture program such as Circuit Maker or Electronics Workbench. The pictures are then imported to the Excel spreadsheet using the Image ActiveX Controls. The image controls of each electromechanical relay's contact position are placed on top of an enlarged CommandButton. Placing the images on the CommandButton is critical because the illusion of the electromechanical relay contacts opening and closing must be sequentially correct. Locating the two images on the CommandButton requires some experimenting and a lot of patience. Figure 8-10 shows the steps used to build the electromechanical relay's animated contacts. Once the images have been placed and aligned correctly on the CommandButton, the existing simula-

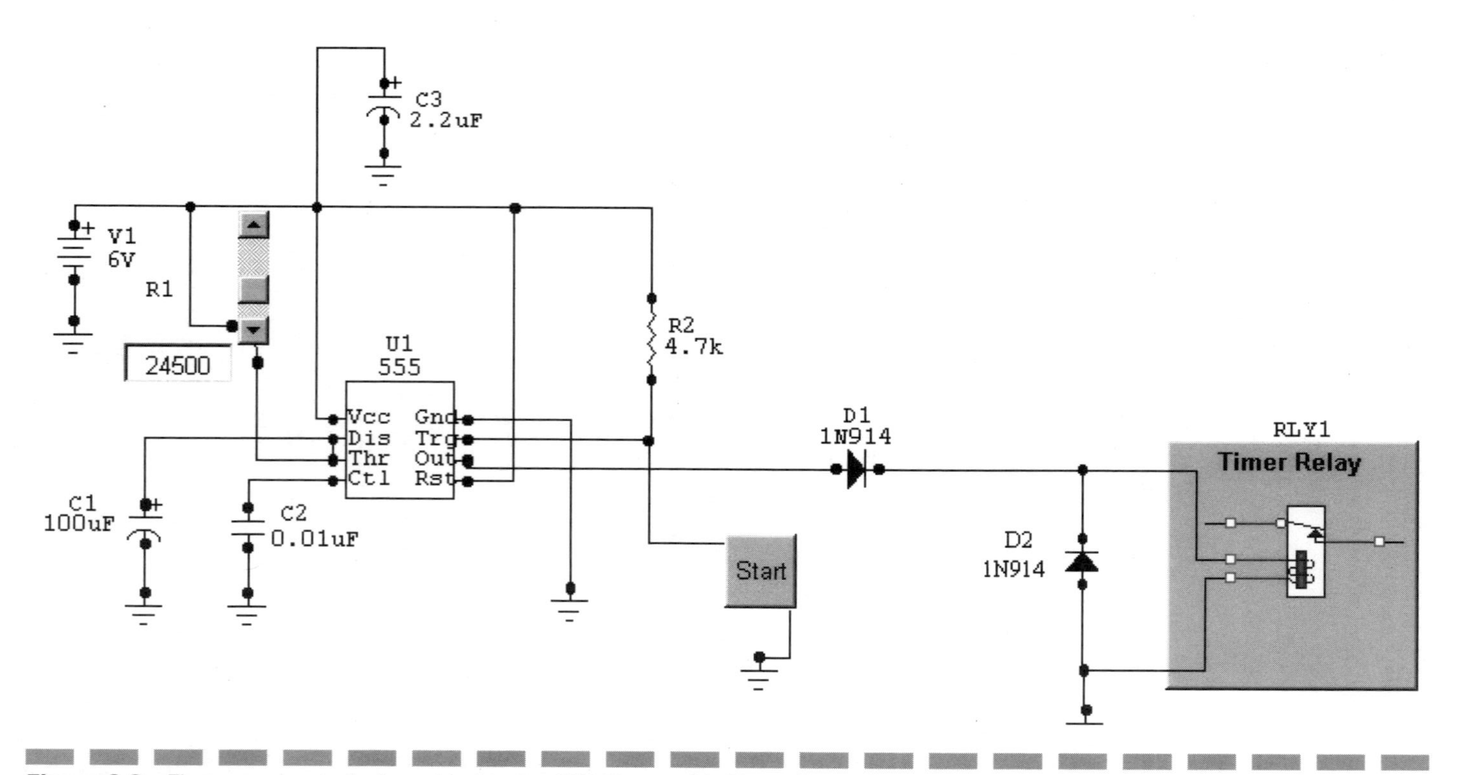

Figure 8-9 Electromechanical relay added to the 555 Monostable Timer Circuit. This design illustrates how simple animation can be used to enhance the visual appearance and function of the controller.

Figure 8-10
The anatomy of an animated electromechanical relay built using VBA. The steps reflect the order in which the components were designed and put together to create a complete animated electromechanical relay package.

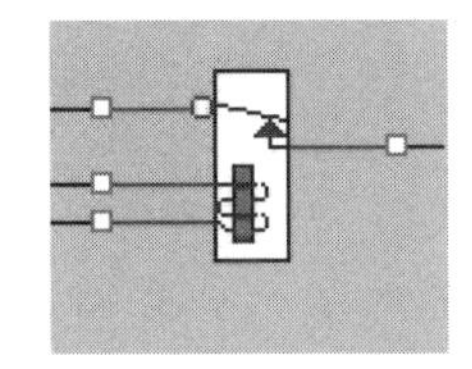

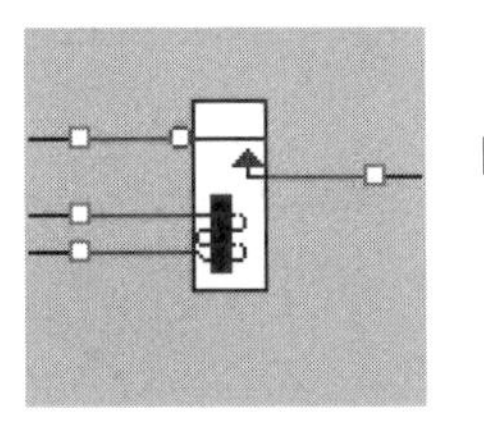

tor control software applet can be modified to affect the visual of the electromechanical relay's contacts. Modify the section of code that starts the time-delay tasks for the Creature Mover robot. The following listing shows the section of code that requires the image control change:

```
Private Sub StartButton_Click()
Worksheets("Sheet1").PBrickCtrl.StopAllTasks
If Worksheets("Sheet1").Range("B3").Value = "3.60437" Then
Worksheets("Sheet1").Range("B6").Value = "It's A Match"
Worksheets("Sheet1").Image1.Visible = True
Worksheets("Sheet1").Image2.Visible = False
Worksheets("Sheet1").PBrickCtrl.StartTask 1
Else
Worksheets("Sheet1").Range("B6").Value = "It's Not A Match"
Worksheets("Sheet1").Image1.Visible = False
Worksheets("Sheet1").Image2.Visible = True
If Worksheets("Sheet1").Range("B3").Value = "2.695" Then
Worksheets("Sheet1").Range("B6").Value = "It's A Match"
Worksheets("Sheet1").Image1.Visible = True
```

```
Worksheets("Sheet1").Image2.Visible = False
Worksheets("Sheet1").PBrickCtrl.StartTask 2
Else
Worksheets("Sheet1").Range("B6").Value = "It's Not A Match"
If Worksheets("Sheet1").Range("B3").Value = "1.925" Then
Worksheets("Sheet1").Range("B6").Value = "It's A Match"
Worksheets("Sheet1").Image1.Visible = True
Worksheets("Sheet1").Image2.Visible = False
Worksheets("Sheet1").PBrickCtrl.StartTask 3
Else
Worksheets("Sheet1").Range("B6").Value = "It's Not A Match"
End If
End If
End If
End Sub
```

The two lines of code that provide the visual for the electromechanical relay contacts are as follows:

```
Worksheets("Sheet1").Image1.Visible = True
Worksheets("Sheet1").Image2.Visible = False
```

The first line of code, Worksheets("Sheet1").Image1.Visible = True, sets the electromechanical relay contacts as closed when the condition is true (if the scrollbar value equals 17500, 24500, and 32762). The second line of code provides the opposite condition where other values except for the three mentioned satisfy a false state. The electromechanical relay contact has its contacts open. Upon running the applet and changing the scrollbar control to the three valid values and nonvalid numbers, the relay's contacts are shown as opening or closing visuals on the CommandButton. This effect provides a nice visual for the virtual timer and opens up a vast array of visual ideas for other simulator controls for Mindstorms robots. The completed code for the revised virtual timer is as follows:

```
Dim a As Double
Dim Time As Double

Private Sub COMMButton_Click()
If Worksheets("Sheet1").COMMButton.Value = True Then
Worksheets("Sheet1").PBrickCtrl.InitComm
Worksheets("Sheet1").Range("E1").Value = "ON"
Else
Worksheets("Sheet1").PBrickCtrl.CloseComm
Worksheets("Sheet1").Range("E1").Value = "OFF"
End If
End Sub

Private Sub StartButton_Click()
Worksheets("Sheet1").PBrickCtrl.StopAllTasks
If Worksheets("Sheet1").Range("B3").Value = "3.60437" Then
Worksheets("Sheet1").Range("B6").Value = "It's A Match"
Worksheets("Sheet1").Image1.Visible = True
```

```
Worksheets("Sheet1").Image2.Visible = False
Worksheets("Sheet1").PBrickCtrl.StartTask 1
Else
Worksheets("Sheet1").Range("B6").Value = "It's Not A Match"
Worksheets("Sheet1").Image1.Visible = False
Worksheets("Sheet1").Image2.Visible = True
If Worksheets("Sheet1").Range("B3").Value = "2.695" Then
Worksheets("Sheet1").Range("B6").Value = "It's A Match"
Worksheets("Sheet1").Image1.Visible = True
Worksheets("Sheet1").Image2.Visible = False
Worksheets("Sheet1").PBrickCtrl.StartTask 2
Else
Worksheets("Sheet1").Range("B6").Value = "It's Not A Match"
If Worksheets("Sheet1").Range("B3").Value = "1.925" Then
Worksheets("Sheet1").Range("B6").Value = "It's A Match"
Worksheets("Sheet1").Image1.Visible = True
Worksheets("Sheet1").Image2.Visible = False
Worksheets("Sheet1").PBrickCtrl.StartTask 3
Else
Worksheets("Sheet1").Range("B6").Value = "It's Not A Match"
End If
End If
End If
End Sub
Private Sub VarResistor_Change()
a = Worksheets("Sheet1").VarResistor.Value
Worksheets("Sheet1").ResistorBox.Value = a
Time = 1.1 * a * 0.0001
Worksheets("Sheet1").Range("B3").Value = Time
End Sub
Dim a As Double
Dim Time As Double

Private Sub COMMButton_Click()
If Worksheets("Sheet1").COMMButton.Value = True Then
Worksheets("Sheet1").PBrickCtrl.InitComm
Worksheets("Sheet1").Range("E1").Value = "ON"
Else
Worksheets("Sheet1").PBrickCtrl.CloseComm
Worksheets("Sheet1").Range("E1").Value = "OFF"
End If
End Sub

Private Sub StartButton_Click()
Worksheets("Sheet1").PBrickCtrl.StopAllTasks
If Worksheets("Sheet1").Range("B3").Value = "3.60437" Then
Worksheets("Sheet1").Range("B6").Value = "It's A Match"
Worksheets("Sheet1").Image1.Visible = True
Worksheets("Sheet1").Image2.Visible = False
Worksheets("Sheet1").PBrickCtrl.StartTask 1
Else
Worksheets("Sheet1").Range("B6").Value = "It's Not A Match"
Worksheets("Sheet1").Image1.Visible = False
Worksheets("Sheet1").Image2.Visible = True
If Worksheets("Sheet1").Range("B3").Value = "2.695" Then
Worksheets("Sheet1").Range("B6").Value = "It's A Match"
Worksheets("Sheet1").Image1.Visible = True
Worksheets("Sheet1").Image2.Visible = False
Worksheets("Sheet1").PBrickCtrl.StartTask 2
```

```
Else
Worksheets("Sheet1").Range("B6").Value = "It's Not A Match"
If Worksheets("Sheet1").Range("B3").Value = "1.925" Then
Worksheets("Sheet1").Range("B6").Value = "It's A Match"
Worksheets("Sheet1").Image1.Visible = True
Worksheets("Sheet1").Image2.Visible = False
Worksheets("Sheet1").PBrickCtrl.StartTask 3
Else
Worksheets("Sheet1").Range("B6").Value = "It's Not A Match"
End If
End If
End If
End Sub
Private Sub VarResistor_Change()
a = Worksheets("Sheet1").VarResistor.Value
Worksheets("Sheet1").ResistorBox.Value = a
Time = 1.1 * a * 0.0001
Worksheets("Sheet1").Range("B3").Value = Time
End Sub
```

Final Thoughts and Further Experimentation

This chapter examines a virtual timer that was developed for controlling the Creature Mover robot built in Chapter 7. The objectives of this discussion on simulator controls are as follows:

- To show how a robot can be controlled using a virtual timer
- To simulate a monostable one-shot circuit using Circuit Maker, Excel VBA, and Mindstorms P-Brick techniques
- To see how animated electronics techniques can be implemented in VBA
- To use a scrollbar ActiveX as a simulated variable resistor

Excel VBA provides interesting opportunities to create simulator controls that not only control a Mindstorms robot, but also enable the GUI to provide a visual appeal to the POC philosophy of product development. This virtual timer project only scratches the surface of simulated controls. Electrical controls using ladder logic can be created using the technique described in this chapter. The relays can be modeled using ActiveX controls, and robot functions can be assigned to a task using the NQC programming language. The robot emulates the actuator identified in the ladder logic output card. Essentially, automation controls for industrial robots can be developed using the simulator controls for manipulating Mindstorms robots. The techniques used in Chapter 2, "Developing GUIs: Software Control Basics," for developing software controls can be implemented to create GUIs for simulator controllers. A Python Tkinter GUI can also serve as a unique method of providing data for target electrical parameters that the simulator can process and use to control a Mindstorms robot. The key is to explore, build, and create.

The *Basic Stamp* (BS) is a favorite tool that can be used to create embedded devices. With its inputs and outputs, the BS can be used to build robots and smart machines. Using the *Silicon Diode Interface Circuit* (SDIC), which was discussed in Chapter 5, the BS interface's with *Robot Command Explorer* (RCX) P-Brick to develop a wealth of embedded devices and machines. In Chapter 9, "Remote-Control Techniques," a master-slave controller is built using these innovative embedded programming tools. The next chapter also discusses Distributive Systems using a host controller and smart load.

CHAPTER 9

Remote-Control Techniques

In Chapter 4, "Electronic Switching Circuits," several switching circuits that can be used to control Mindstorms robots were explored. The electromechanical relay is the core component used to switch the robot on and off based on the embedded program providing the function or behavior to the intelligent machine. The definition of *remote control* from a roboticist's perspective is the ability to operate a robot at a distance from the smart machine. Remote control can be accomplished by using a wire or electromagnetic energy. The LEGO handheld controller is an example of using electromagnetic energy because of the infrared beam that is emitted from the unit. Infrared is an element of the electromagnetic wave spectrum. The electromechanical controls used in Chapter 4 have wires between the robot and its switching contacts. This is an example of remote control by using wire. The remote-control-by-wire design is also known as a *master-slave control*.

In this chapter, the master-slave design is implemented using two smart devices: a Basic Stamp and *Robot Command Explorer* (RCX) *programmable brick* (P-Brick). The Basic Stamp is a great programming device for creating toys, robots, kinetic art, and unique inventions. The P-Basic programming language makes it easy to develop gadgets and inventions within an hour of receiving instruction. The RCX P-Brick is ideal for creating robots, crazy gadgets, and inventions. The P-Brick can be programmed in RCX code, *Not Quite C* (NQC), Java, and *Interactive C Version 4.0* (IC4) languages. In addition, the RCX P-Brick gives the Basic Stamp a mechatronics appeal because of the smart mechanics that can be built with the *Robotics Invention System* (RIS). Instead of wiring a motor, a *smart motor* can be built because of the embedded processor power of the RCX P-Brick associated with the gearing in the electromechanical unit. Therefore, this chapter explores the master-slave topology for the remote control of Mindstorms robots using a Basic Stamp and RCX P-Brick. The objective of this project is to have a Basic Stamp control the RCX P-Brick using a electronics interface between the two embedded devices to translate signals shared by the programmable units. This chapter investigates the following topics:

- Building simple Basic Stamp master controller hardware
- Programming the Basic Stamp master controller in *Parallax Basic* (P-Basic) language
- Developing the Distributive Load System for Mindstorms robots
- Biology-based robots
- Programming the RCX P-Brick using IC4 language
- Enhancing IC4 and P-Basic programs for remote-controlled robots
- Developmental thoughts: Basic Stamp, Mindstorms P-Bricks, and robots

Building Simple Basic Stamp Master Controller Hardware

The master-slave controller topology is used to conveniently control an electrical/electronic apparatus from a distance. Figure 9-1 shows two master-slave block diagram topologies (geometry or configuration). In each of the block diagrams, a voltage converter is used to correct the *Input/Output* (I/O) signal for both the Basic Stamp and RCX P-Brick.

Figure 9-1a shows the topology for the Basic Stamp (master controller) feeding a hardwire signal to the RCX P-Brick (slave controller) with the aide of the transistor relay driver (which is discussed in Chapter 4). The output produced by the Basic Stamp can be a 0- to +5-volt *Direct Current* (DC) or a *Pulse-Width Modulation* (PWM) signal. The transistor relay driver provides the correct signal interface for the RCX P-Brick to process. The IC4 code written for the RCX P-Brick processes the signal received from the Basic Stamp's output. The Basic Stamp uses a programming language called P-Basic. P-Basic was written to enable embedded projects to be built without the hassle of learning assembly or C language for hours. The IC4 programming language is capable of reading PWM or analog or digital input signals. The ability to read various input signals makes it easy to interface the Basic Stamp with the RCX P-Brick.

In Figure 9-1b, the RCX P-Brick is the master controller and the Basic Stamp is the slave device. The output voltage of the RCX P-Brick is higher than Basic Stamp can handle. A 0- to 5-volt input signal is required for the Basic Stamp. To reduce the output voltage of the RCX P-Brick to a lower signal value that the Basic Stamp can use safely, a 7805 voltage regulator circuit can be used to match the two programmable devices. The output voltage of 7 volts from the RCX P-Brick is reduced to 5 volts DC, which the Basic Stamp can process safely.

The same programming languages used in the topology of Figure 9-1a can be applied to reading input values, processing incoming data, and turning on the appropriate output devices. The circuit topology used in the following project is the same one that appears in Figure 9-1a. Building a master controller using the Basic Stamp is quite easy. The most basic controller to build is a one-input, one-output device. The embedded software reads the incoming signal and, based on the voltage level (equivalent digital value), the microcontroller (Parallax Custom PIC16C57) provides the appropriate output response. Because the digital(int p) in the IC4 programming language is looking for a passive switch input, the transistor relay driver is used to provide the correct switching event. As discussed in

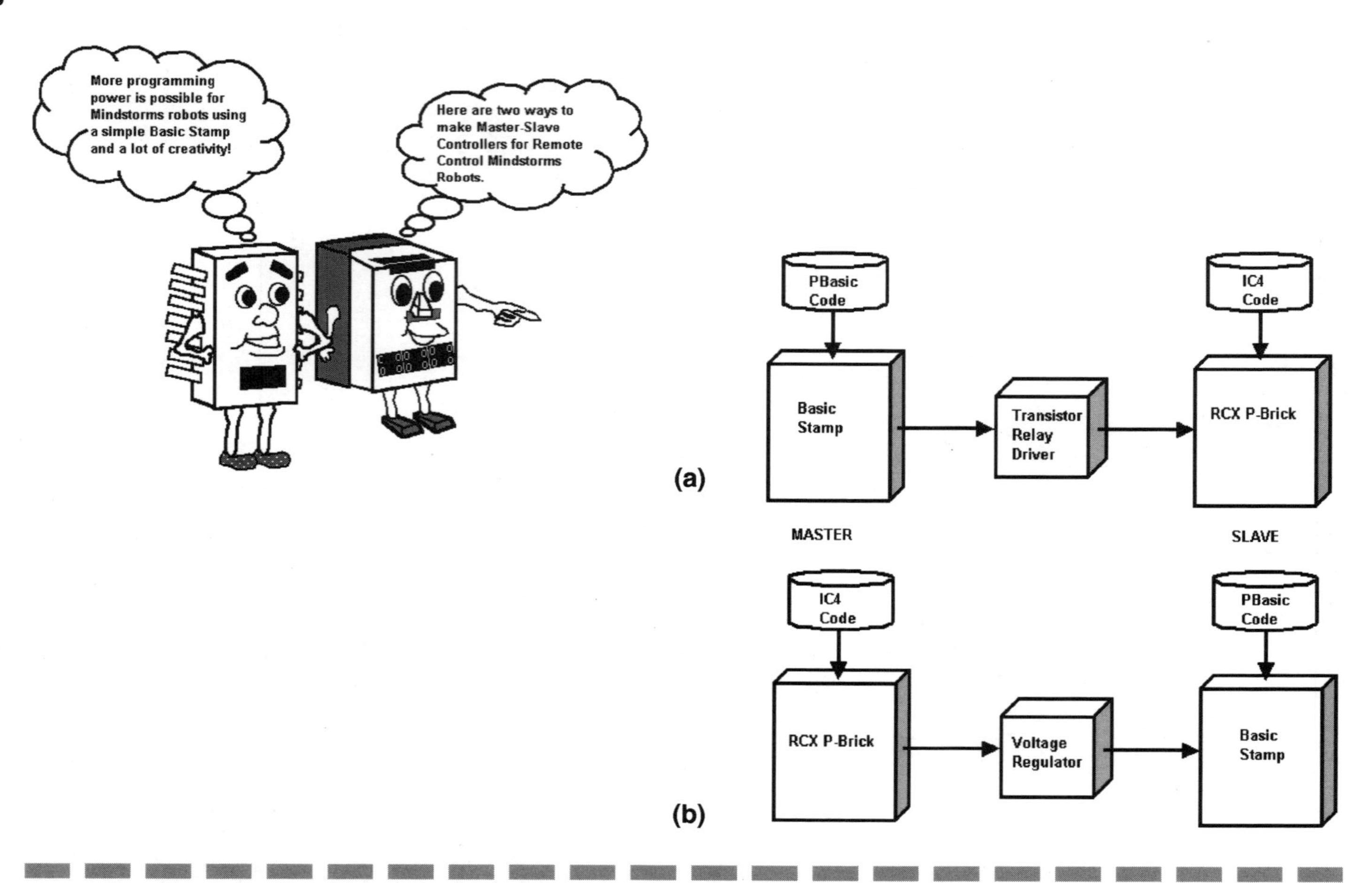

Figure 9-1 Two master-slave topologies that can be used to remotely control Mindstorms robots. Chip and RCX discuss remote-control topologies.

Chapter 3, "Electromechanical Controls Interfacing," and Chapter 4, "Electronic Switching Circuits," the relay contacts provide the digital data (1 or 0) the RCX P-Brick will process. The programmable brick will make the decision to turn the appropriate output on or off. Figure 9-2 shows the circuit schematic of the Basic Stamp master controller. The circuit can be built quite easily using a solderless breadboard and a modified LEGO electric wire.

To build the controller, the Basic Stamp master controller circuit must be breadboarded first. After checking the wiring, software written in P-Basic is downloaded to the programmable device. Traditionally, downloading a program to the Basic Stamp requires the use of DOS. The ALT-R keys are used to download the program to Basic Stamp for program execution and storage. There is currently a Windows-based text editor for the Basic Stamp 2 that can be downloaded for free from the Parallaxinc web site www.parallaxinc.com/html_files/downloads/downloads.htm. Unfortunately, the software only works for Basic Stamp 2, 2e, 2sx, and 2p devices. Once this is loaded onto the PC or notebook computer, running the application is quite easy.

The file is an executable Windows application; therefore, it can be placed on the desktop of the PC or notebook computer. With the Basic Stamp Windows text editor opened, the following code can be typed:

```
btnWK  var  byte
btnWK=0
```

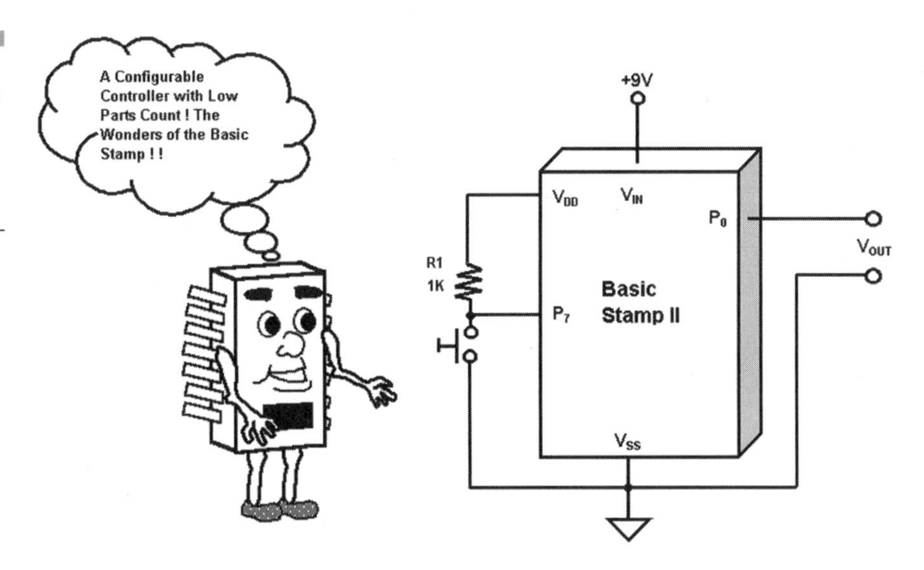

Figure 9-2 Circuit schematic of a Basic Stamp master controller—an up-close-and-personal observation by Chip

```
LOOP:
    BUTTON 7,0,255,250,btnWK,0,noPress
     debug "This is a test"
     debug "-"
     TOGGLE 0
noPress:goto loop
```

This code is written for the Basic Stamp 2 computer, but should also work for the Basic Stamp 1 computer with a few slight modifications. Pages 373 through 448 of the *Basic Stamp Programming Manual* (version 1.9) show the Basic Stamp 1 and 2 conversion programming. The code examples in this chapter are written for Basic Stamp 2; therefore, consult the manual to modify the programs to run on Basic Stamp 1.

Figure 9-3 shows the code for the Basic Stamp master controller entered into the Windows text editor. Check the code typed into the editor for any errors. If there are no errors, click Run at the menu bar. The program is checked for syntax errors before it is downloaded into the Basic Stamp. Once it is loaded into the device, without power, the code remains there for the life span of the unit. The 24LC16B is an *Electrically Erasable Programmable Read-Only Memory* (EEPROM) *Integrated Circuit* (IC) that is responsible for retaining the program in memory without power. This gives

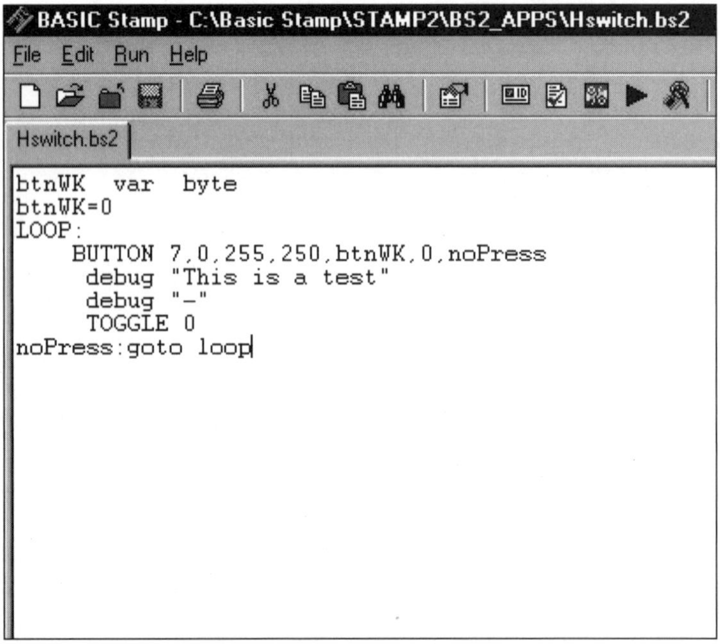

Figure 9-3 P-Basic code for the Basic Stamp master controller. The program is entered into a Windows-based text editor for a Basic Stamp 2 family of devices.

the Basic Stamp great flexibility to store the master controller software without worrying that the code will be lost in a power outage.

Take a digital voltmeter and connect it to pin P0 of the stamp with the negative lead attached to ground. Apply +9 volts to the Basic Stamp. Push and release the switch wired at P7 to ground. The output should read approximately +5 volts. Pin P0 remains in that voltage state until another switch event is requested. This toggle or latching output is wired to the transistor relay driver to provide the passive digital input data. Wire the Basic Stamp master controller to the transistor relay driver by connecting pin P0 to the 470-ohm base series resistor. See Figure 9-4. Take an ohmmeter and place the test leads across the relay contacts. Push and release the switch. The ohmmeter should read 0 ohms. This resistance measurement signifies that the correct relay contacts (*normally open* [NO]) have been selected. If a 0-ohm reading is not displayed on the ohmmeter, use the next pin of the relay. Repeat the measurement to ensure that the 0-ohm value is present when the relay contacts close. Push and release the switch several times while examining the ohmmeter reading.

If the circuit is working correctly, the RCX P-Brick can be wired to the Basic Stamp master controller using a modified LEGO electric wire. Wire the LEGO modified electric wire to the transistor relay driver's NO contacts using a 1-kilo-ohm resistor as the whetting current component for the passive digital switching circuit (see Chapter 3). Download an IC4 program into the RCX P-Brick. Usc thc following IC4 code to test the master-slave control function:

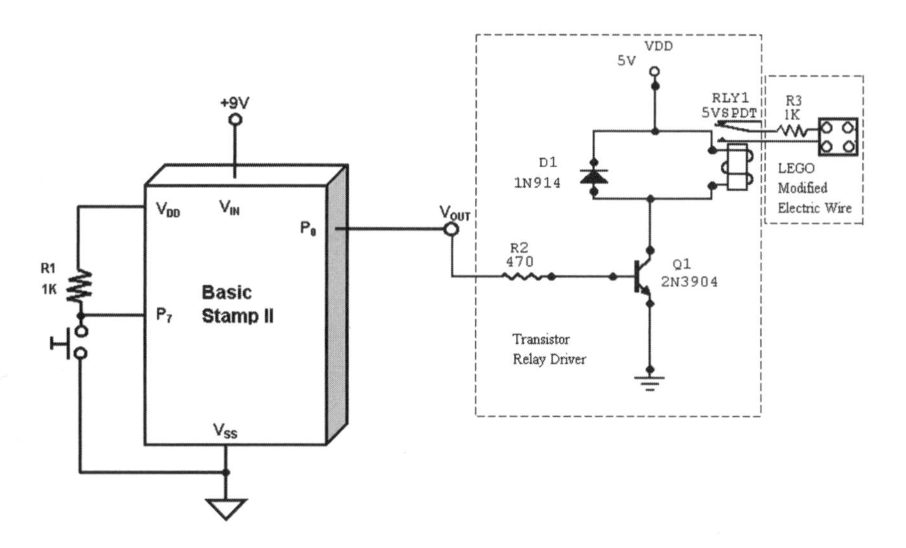

Figure 9-4 The complete circuit schematic of Basic Stamp master controller with a transistor relay driver

```
void main()
{     while (1){
        if  (digital(1)==1){
            printf("1");
            fd(1);
            sleep(5.0);
            off(1);
        }
        else {
            printf("0");
            off(1);
        }
    }
}
```

Apply the +9 volts to the master controller and attach a LEGO small motor to output A. Turn on the RCX P-Brick and push Run to execute the embedded test code. Push and release the Master Controller button. The *Liquid Crystal Display* (LCD) should display a 1 while the motor is running. After 5 seconds, the motor stops and the LCD displays a 0 on the mini screen. If the program does not work, check the code for the correct port assignments for the input and output of the RCX P-Brick. Run the program again, looking for the correct LCD and output A response. The RCX P-Brick can be replaced with the Scout programmable unit as well. The Basic Stamp is an incredible software tool for Mindstorms robot development. Now, let's take a closer look at programming the master controller using the P-Basic language.

Programming the Basic Stamp Master Controller in P-Basic Language

Writing code for the Basic Stamp is pretty straightforward. P-Basic was developed to help nonprogrammers write software for the Parallax Custom PIC16C57 microcontroller without the heavy burden of using assembly language. P-Basic is like the standard Basic language with the exception of having the capability to interface with the outside world through the microcontroller's I/O. The *Basic Stamp Programming Manual* (version 1.9) that is supplied with the Basic Stamp 2 device is a comprehensive manual that contains application notes and conversion information for Basic Stamp 1 and 2 components. The best way to learn about programming is to dive into a project. The Basic Stamp master controller is a very good project to use to introduce the topic of software writing using the P-Basic programming language. Also, this software/hardware tool increases the capabilities of RCX P-Brick robot development by providing more inputs for the Mindstorms machine, which enables more sophisticated robots to be built.

The code used to check out the hardwired interface between the Basic Stamp 2 and RCX P-Brick can be found on page 250 of the *Basic Stamp Programming Manual* (version 1.9). The following code is taken from page 250:

```
btnWK  var  byte
btnWK=0
LOOP:
    BUTTON 7,0,255,250,btnWK,0,noPress
     debug "*"
noPress:goto loop
```

The BUTTON instruction is used to debounce an electromechanical switch's contacts, perform auto repeat, and GOTO, or branch, to the address when the BUTTON's request state has been reached. The BUTTON instruction can receive data from an Active HI or LO switching circuit. The P-Basic BUTTON instruction has the following syntax format: BUTTON pin, downstate, delay, rate, byte variable, target, and address. The pin is a variable or constant (0 through 15) that specifies which I/O pin to use. The pin is usually made to behave as input when the BUTTON instruction is used. The downstate is a variable/constant that specifies an Active HI (1) or an Active LO (0) logic state when the BUTTON is pressed. Delay indicates how long the BUTTON must be pressed before auto-repeat can occur. It can also be a variable or constant with a numeric data set of 0 through 255. Rate is a variable or constant that specifies the number of cycles between auto-repeat. Its numeric data set is also 0 through 255. The byte variable is the workspace for BUTTON. To use this workspace, it must be set to zero, clearing it for the BUTTON instruction. The target state indicates the state of the BUTTON or switch being pressed (1). Upon knowing the state of the BUTTON, a branch or jump to another part of the program occurs. The address is a label used during a branching event if the BUTTON is in the target state.

The BUTTON instruction is a key parameter for developing remote-control interfaces for Mindstorms robots. The code shown previously uses the BUTTON instruction to detect an Active LO (0) digital input signal when the switch is pressed. The DEBUG instruction prints an asterisk inside of a Windows dialog box. If the switch is not pressed, a polling or monitoring loop becomes active.

The programming style for this P-Basic example needs to be refined whereby the organization of variables/constants and the description of the program, filename, and code enable the Mindstorms robot application to be

```
'==========================================================================================
' File: DIGITAL_SW.BS2
'
' Provides a digital input for the RCX/Scout P-Bricks. Transistor
' Relay Driver Circuit used for digital input is connected to P0.
' =========================================================================================

  TRDpin     CON      0          ' Transistor Relay Driver connected to Pin 0

  SWpin      CON      7          ' Switch connected to Pin 7

  DlyVal     CON      255        ' Delay Value

  ALow       CON      0          ' Active Low switch state

  SWrate     CON      255        ' Switch cycles between autorepeats

  btnWK      var      byte       ' Workspace for BUTTON instruction

  btnWK=0                        ' Clear the workspace variable

'__________________________________________________________________________________________

Loop:
   BUTTON SWpin,ALow,DlyVal,SWrate,btnWk,TRDpin,noPress  ' Goto to noPress
   DEBUG "*"                                             ' Print asterisks on the Debug window
noPress: GOTO Loop                                       ' Repeat endlessely.
```

Figure 9-5 Modified code of the BUTTON instruction using the programming style convention

managed effectively. Also, future modifications will be easy due to this software management technique and data organization.

When using this programming style, use names that have meaning to the function that it will perform. Figure 9-5 shows a programming style convention and refinement of the sample P-Basic code. The software starts off with the name of the file. The name should be related to the function of the software. A description of the program is very important because after years of working on the code, a reminder of how the program works helps during the modification process for the present application.

Next, a list of constants and variables identifies the key parameters used in storing data. The main body of the program uses these constants and variables to manipulate and execute subroutines. The programming style is very important in capturing I/O requirements for software design and development. The naming convention of the I/O signals helps to develop code because the programming convention enables you to capture the names through the list of constants and variables. A system block diagram (see Chapter 2, "Developing GUIs: Software Control Basics") can assist in identifying the I/O that can be used to create the list. The programming style can also help to debug code because breakpoints can be added using constants and variables to monitor key data read/writes. Processing and jumps/branches can be checked during program execution using a specifed programming style as well.

Using Figure 9-5, enter the code into the Basic Stamp 2 Windows version text editor. Check the code for syntax errors by selecting Run and Check Syntax on the menu bar. If there are no errors, select Run and run it again. A debug terminal dialog box appears on the screen. If the BUTTON is pressed, the debug terminal box shows an asterisk. Any string variable or constant can be displayed using the DEBUG instruction. This instruction provides a method of displaying information about the Basic Stamp's interface with the robot as well as the sensors and switches that are wired to it.

The transistor relay driver can easily be latched on and turned off by using the TOGGLE instruction. The TOGGLE instruction flip-flops the specified output port of the Basic Stamp. The specified output latches on or off in the state when the TOGGLE instruction is executed. For example, if input P7 of the Basic Stamp notices that the switch is depressed, the TOGGLE instruction is initiated at output P0. The transistor relay driver is turned on, energizing the relay's coil to switch its contact position. The transistor relay driver is latched until a second switch activation at pin P7 occurs. The TOGGLE instruction is very useful for testing the outputs of an embedded controller. Each output latches on or off when the instruction is called by switch activation at an input port of the Basic Stamp. The TOG-

```
'=====================================================================================
' File: DIGITAL_SW.BS2
'
' Provides a digital input for the RCX/Scout P-Bricks. Transistor
' Relay Driver Circuit used for digital input is connected to P0.
' =====================================================================================

 TRDpin     CON       0          ' Transistor Relay Driver connected to Pin 0

 SWpin      CON       7          ' Switch connected to Pin 7

 DlyVal     CON       255        ' Delay Value

 ALow       CON       0          ' Active Low switch state

 SWrate     CON       255        ' Switch cycles between autorepeats

 btnWK      var       byte       ' Workspace for BUTTON instruction

 btnWK=0                         ' Clear the workspace variable

'-------------------------------------------------------------------------------------

Loop:
   BUTTON SWpin,ALow,DlyVal,SWrate,btnWk,TRDpin,noPress   ' Goto to noPress
   DEBUG "*"                                             ' Print asterisks on the Debug window
   TOGGLE TRDpin                                         ' Toggle pin 7
noPress: GOTO Loop                                       ' Repeat endlessely.
```

Figure 9-6 The complete code for the Basic Stamp master controller. The TOGGLE instruction provides an embedded latching feature for momentary electric switches.

GLE instruction provides a physical means for controlling Mindstorms robots. Although not shown in Figure 9-5, this instruction can easily be added to the code. Enter TOGGLE TRDpin on the line between the DEBUG and noPress instructions. The final code should look like the following:

```
BUTTON SWpin,ALow,DlyVal,SWrate,btnWk,TRDpin,noPress
   DEBUG "*"
   TOGGLE TRDpin
noPress: GOTO Loop
```

Figure 9-6 shows how the code looks using the programming style convention discussed in this section.

The GOTO instruction is a traditional Basic language instruction that is used to branch or jump to another part of the program based on the execution of a relational condition. The label noPress is the branch location if the electric switch is not activated. If the switch is pressed, the DEBUG instruction prints an asterisk.

The P-Basic programming language is quite easy to use, as illustrated in the software code for the Basic Stamp master controller. This concept is best understood through experience. By entering code into the text editor and downloading it to the Basic Stamp, you can quickly learn how to use this product with the RCX P-Brick. Using the sample code discussed in this section, try changing the asterisk of the DEBUG instruction to display the message "Switch has been pushed." This feature enables a mini diagnostics tool to be embedded within the application. With the P-Basic programming language, the prototyping of embedded master controller systems can produce the end result quick and effectively, which in this case is a hardwired remote-control box for Mindstorms robots.

Distributive Load System for Mindstorms Robots

The *Proof Of Concept* (POC) that is developed using a Basic Stamp computer and the RCX P-Brick is a *Distributive Load System*. A distributive load system is based on the master-slave method of controlling electric devices remotely (shown earlier in Figure 9-1). The key element of a distributive load system is that the receiving device must have a microcontroller and an understanding of the communications protocol. The transmission of data sent by the master controller is routed to the slave unit using a hardwired communication network. The protocol is usually software based, and the master and slave units are capable of sending and receiving data along the hardwired network using the messages identified within the communications documentation. The master unit is responsible for directing all communication messages and electrical signals to the slave unit over the hardwired network. With an understanding of the bus protocol and software interface layer, the slave unit can receive the data sent by the master controller over the communications network or bus and process it based on the embedded microcontroller's physical I/O.

The key to a distributive load system is to have a smart load on the receiving end of the communication network or bus. A smart load is an electrical device that has an embedded microcontroller that is used for making decisions based on the input data it receives. The smart load is typically located at a distance from the host controller, and its output response is determined by the embedded microcontroller processing the data it receives over a communication bus. A smart load helps to reduce the number of hardwired connections, reducing the cost of the remote-controlled system. Communication networks can have either one wire or four wires depending on the application and the amount of data being sent and received by both embedded devices. The master-slave controller architecture for robotic control has two wires.

A simple distributive load system can be easily built using the Basic Stamp and RCX P-Brick. Figure 9-7 shows the distributive load system architecture of a remote-controlled Mindstorms robot. The switching contacts of the transistor relay driver provide the hardwired interface that enables the master-slave communication between the two units. The data that is received by the slave unit (RCX P-Brick) is binary 1 or 0. The RCX P-Brick's embedded software makes a decision based on the data to turn the motor on or off.

The communication bus is a physical wiring connection that enables master-slave controllers to share the binary data for controlling the robot's motorized drive mechanism. To tie the two microcontroller-based devices together on one conductive plane, the LEGO Electric 2×8 plate can be used as a communication bus for a distributive load network for master-slave-controlled Mindstorms robots. Figure 9-8 shows the LEGO Electric 2×8 plate establishing a communication bus for networking the Basic Stamp and RCX P-Brick-embedded devices.

The Scout P-Brick can be a direct substitute for the RCX P-Brick because it contains the internal 10-kilo-ohm resistor (see Chapter 3) circuit that interfaces with the external inputs, enabling a seamless match for distributive load systems. To control the Scout P-Brick using this hardware network architecture, either *Stand-Alone Code* (SAC) (see Chapter 1, "Wireless

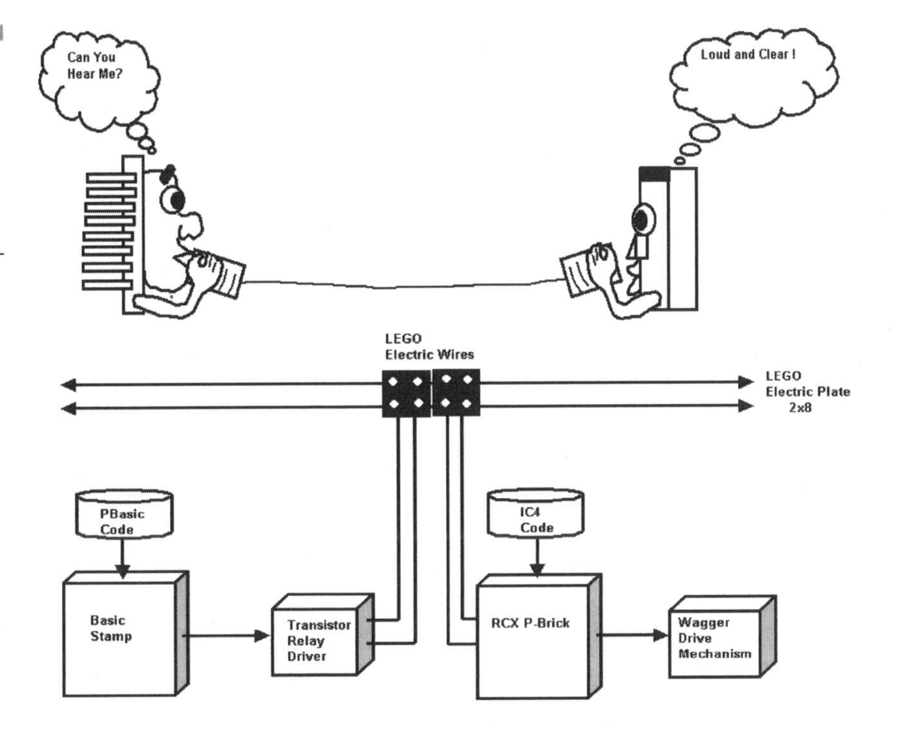

Figure 9-7 A distributive load system using the Basic Stamp, RCX P-Brick, and the LEGO Technic Electric 2×8 plate

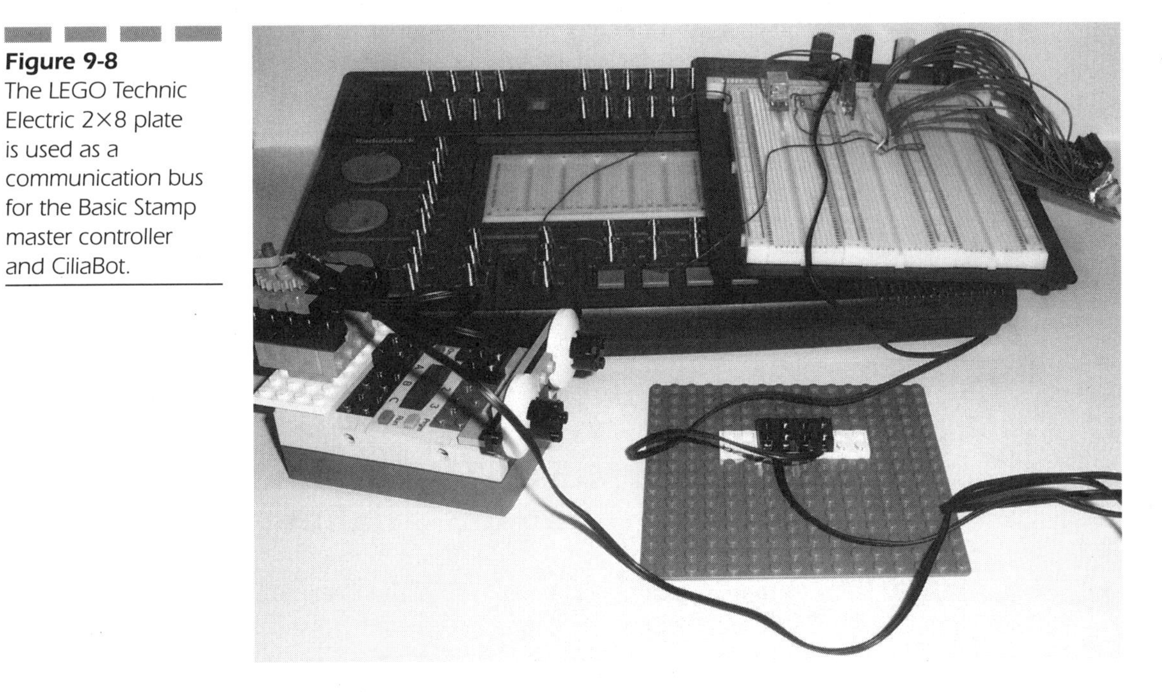

Figure 9-8 The LEGO Technic Electric 2×8 plate is used as a communication bus for the Basic Stamp master controller and CiliaBot.

Basics,") or *LEGO Assembly* (LASM) (see Chapter 6, "Using Procedural Languages for Mindstorms Robot Control,") can be programmed into the P-Brick to read a passive binary digit from the Basic Stamp. The output response is dependent on the programming language used and the conditions set in the embedded code of the Scout P-Brick. The Bug Constructopedia would make a good distributive load system project using the Scout P-Brick. The LEGO Electric 2×8 plate makes it convenient and easy to add other LEGO Mindstorms robots. To make a longer communication bus, add and snap the additional LEGO Electric 2×8 plates together in a sequential chain.

Biology-Based Robots

Software design for robots is based on programming tasks or behaviors that the machine will perform using mechanics as the physical extension to the coded commands. Biology is a good source for robot design because of the behaviors in plants and animals as well as the physical attributes they exhibit. Software can be designed to capture these attributes, establishing a base for building the mechanics for the robot. The LEGO Extreme Creatures is an expansion set that enables hobbyists and experimentalists to explore physical and behavioral designs of hybrid animals. It's because of this concept that CiliaBot was designed. CiliaBot is based on the basic paramecium design of a creature known as a *ciliate*.

Ciliates [pronounced "SIHL-ee-uhts"] are complex animal-like protists. A protist is a one-celled organism that has a clear membrane surrounding its nuclear material. These creatures move by using hairlike projections called *cilia* [pronounced "SIHL-ee-uh"]. The paramecium is the best-known ciliate. It is found in both freshwater and seawater. It has a definite shape—some say it is shaped like the sole of a slipper. The ciliate owes its shape to a stiff but flexible outside covering called a *pellicle* [pronounced "PEHL-ih-kuh"]. Figure 9-9 shows a sketch of a typical ciliate.

How Ciliates Move—The Mechanical Design Inspiration for CiliaBot

Ciliates move about by means of cilia. Cilia are like flagella in structure (a whiplike thread of protoplasm on a cell that causes the cell to move by the flagella's motion), but they are much shorter and more numerous than flagella. Cilia cover the whole surface of most ciliates. The movement of the

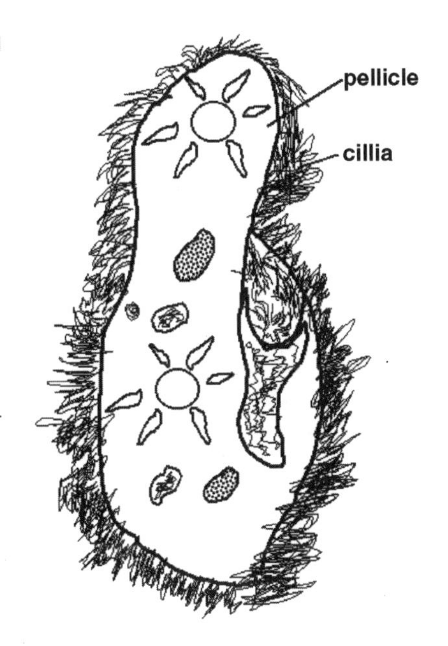

Figure 9-9
A photo showing a ciliate. This paramecium provided inspiration for the CiliaBot project discussed in this chapter. Photo courtesy of http://www.cytochemistry.net/Cell-biology/cilia.htm.

CiliaBot is based on a clubbed wagging tail mechanism. As the tail wags back and forth, the CiliaBot body moves in a similar fashion. The construction of the wagging tail is based on the *wagger* mechanical design shown on pages 21 through 23 of the Extreme Creatures Constructopedia. The wagger mechanism is remotely controlled using the Basic Stamp master controller and RCX P-Brick. By pushing and releasing the momentary switch, P0 output is approximately +5 volts DC, which turns on the transistor relay driver. The RCX P-Brick reads the binary digital data and turns on output A. The wagger mechanism begins to move, propelling the CiliaBot slowly across the top surface of a table.

Figure 9-10 shows the system block diagram of a master-slave control applet for CiliaBot. A small club on the end of the LEGO Electric 2×8 plate is added to the wagger mechanism. The club helps propel the RCX P-Brick in small sideways and circular motions across a smooth surface.

The wagger mechanism design is based on the LEGO Technic Electric 2×8 plate rotational axis being off center. A Technic lift arm (1×5 half beam) and camera are the primary mechanical components used to create the wagger's movement. The other LEGO components provide a sound assembly and structure for mounting the motor and supporting the lift arm and camera parts. Figures 9-11 through 9-15 show the construction of the CiliaBot. Bringing this small robot to life requires embedded software. The IC4 programming language provides the necessary behavior of movement for CiliaBot.

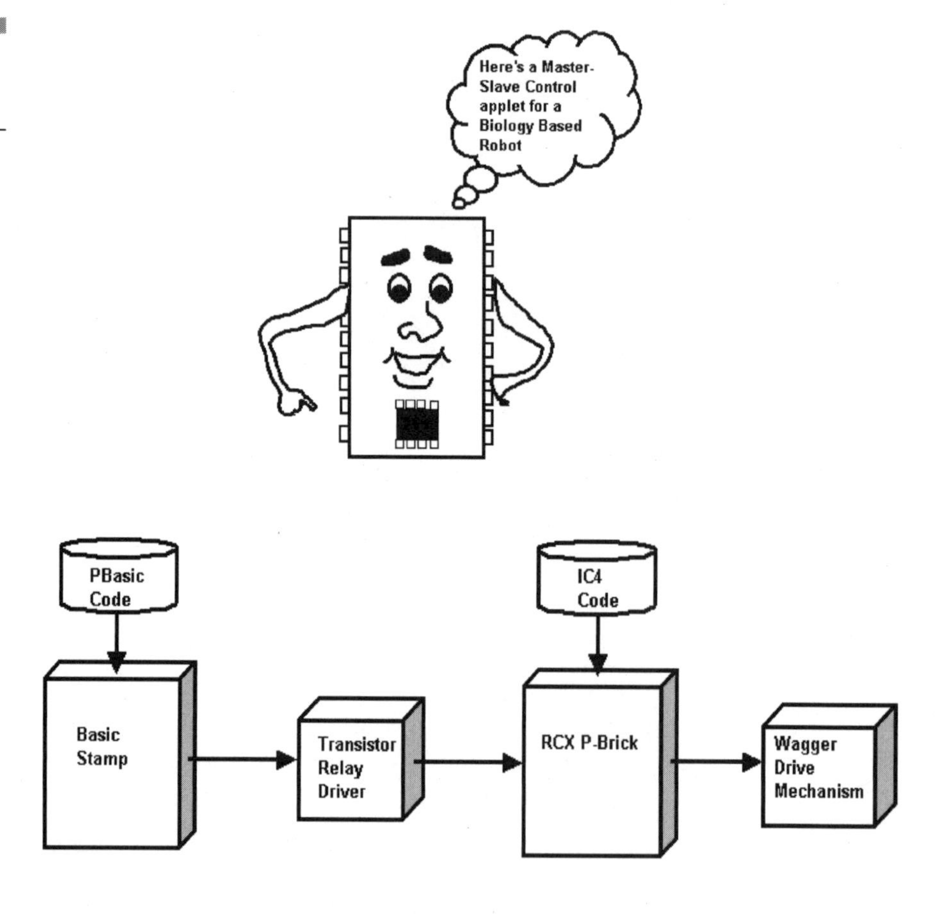

Figure 9-10
A System Block Diagram for CiliaBot

Figure 9-11
Top or plane view of the wagger mechanism

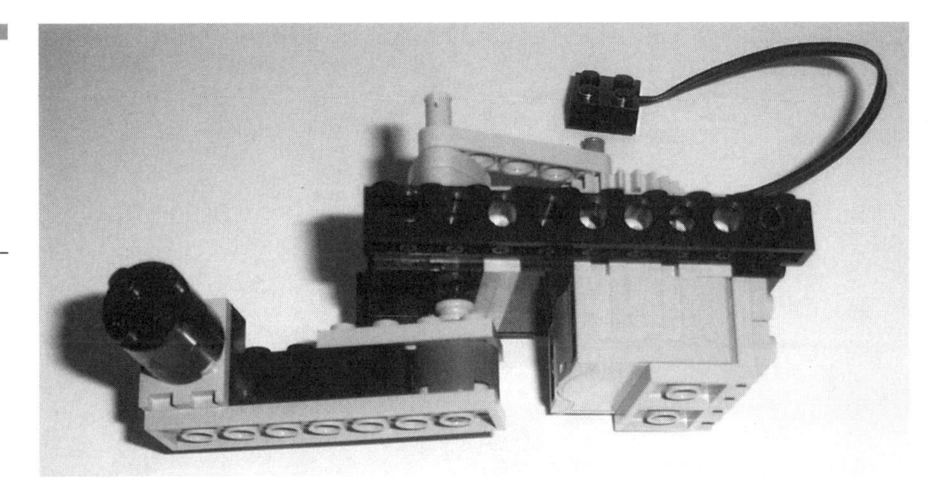

Figure 9-12 Side view of the wagger mechanism showing a small LEGO 9-volt motor and the construction of the clubbed tail

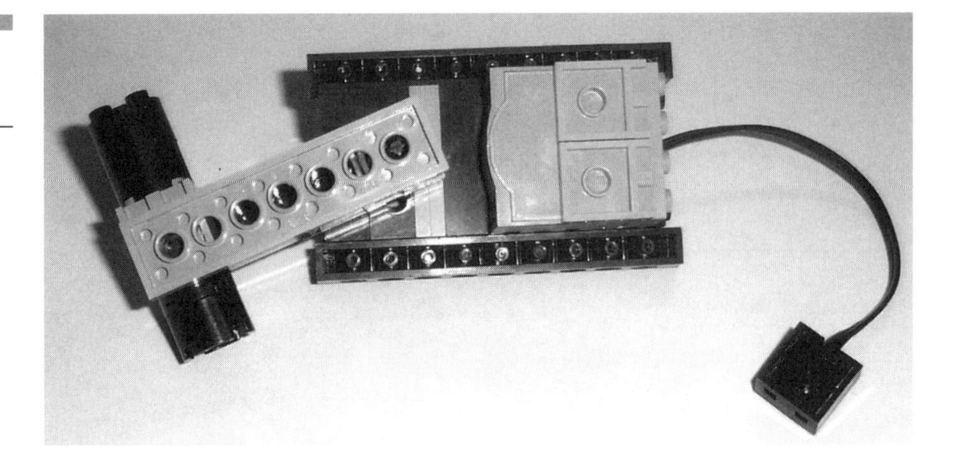

Figure 9-13 Bottom view of the wagger mechanism

Programming the CiliaBot Using IC4 Language

IC4 (see Chapter 6) is an excellent programming language to use with the RCX P-Brick. The language is very easy to use, and it enables small messages to be displayed on the LCD using the printf(int) command. The Cilia Bot's movement behavior is programmed in IC4 because of these two

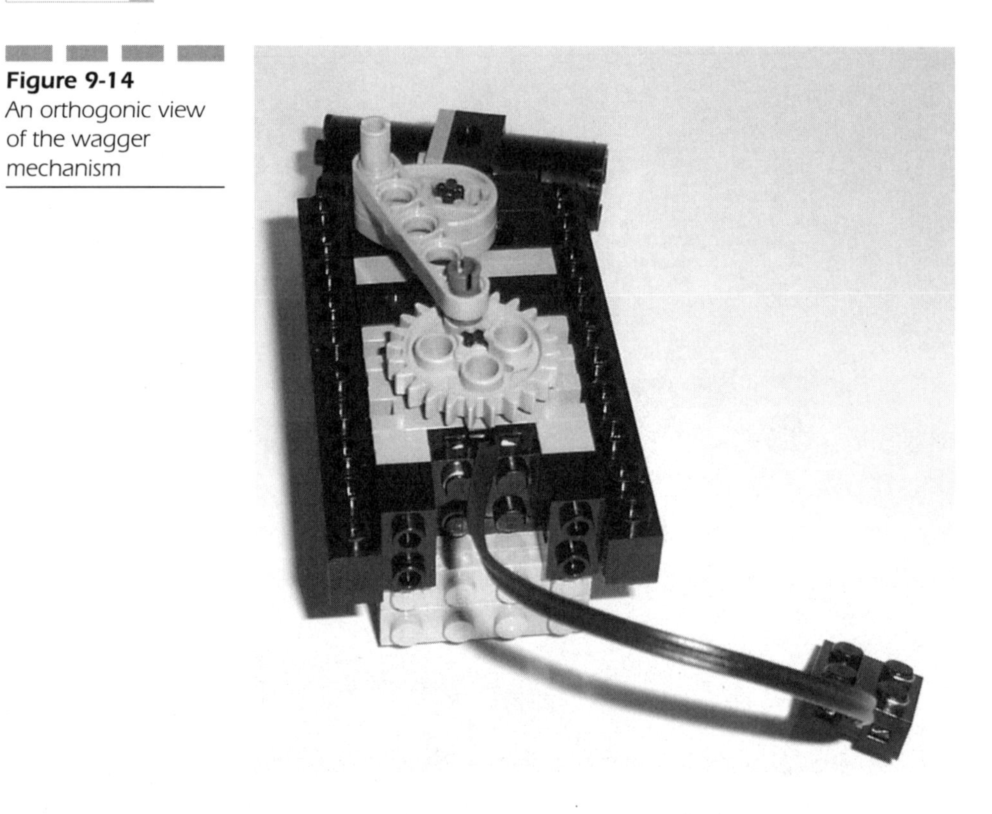

Figure 9-14
An orthogonic view of the wagger mechanism

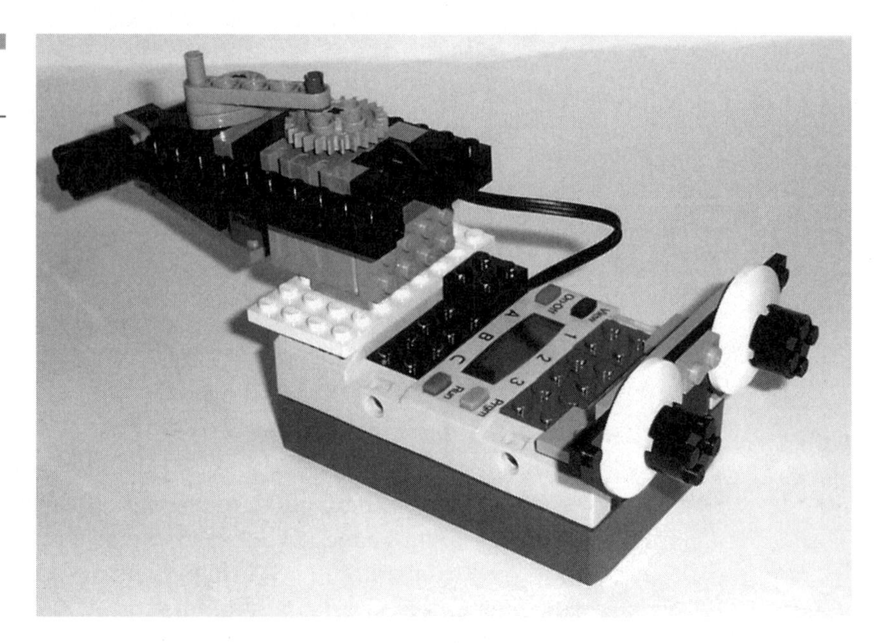

Figure 9-15
A completed CiliaBot

factors. The LCD can be used as a simple switch status indicator for monitoring input data received from the master controller. The IC4 code for controlling the wagger mechanism is as follows:

```
void main()
{     while (1){
        if  (digital(1)==1){
            printf("1");
            fd(1);
            sleep(5.0);
            off(1);
        }
        else {
            printf("0");
            off(1);
        }
    }
}
```

To change the information that will be displayed on the LCD of the RCX P-Brick, the printf("0") or printf("1") can be modified easily. Instead of using binary digits 1 or 0, text can be used. For example, try using printf("HI") and printf("LO") on the LCD. The following lists the complete code with the modified printf instructions using HI and LO to indicate the status of Basic Stamp master controller output:

```
void main()
{     while (1){
        if  (digital(1)==1){
            printf("HI");
            fd(1);
            sleep(5.0);
            off(1);
        }
        else {
            printf("LO");
            off(1);
        }
    }
}
```

To check for the two text messages, type in the commands using the editor and notice the LCD and the information that is displayed. An alpha-character data table can be made using two columns for letter generation. One column for the actual letter viewed on RCX LCD and the other for the desired character to display on the P-Brick. The columns will be compiled and used when designiing LCD messages for diagnostics or functional indicators of Mindstorms robots.

Enhancing IC4 and P-Basic Programs for Remote-Controlled Robots

The embedded programming languages of IC4 and P-Basic are wonderful software development tools for building remote-controlled robots. The master-slave control topology discussed in this chapter illustrates some of the basic programming techniques that can be used to build a programmable remote-control device for manipulating a Mindstorms robot. The sample codes used in the Basic Stamp and RCX P-Brick enable basic programming building instructions to be used as an educational tool for constructing a unique Mindstorms robot with a front-end Basic Stamp master controller. Each language also has some instructions that can enhance the performance of a Mindstorms robot.

The SLEEP mode enables the Basic Stamp to reduce its quiescent current to about 50 μA by turning itself on and off for a specific time in seconds. The length of the SLEEP mode can range from 2.3 seconds to 18 hours. The output pins of the Basic Stamp 2 are polled every 2.3 seconds due to the Parallax Custom PIC16C57 IC that manages time. If the application is driving loads during SLEEP mode (sourcing or sinking current through output-high or output-low pins), the current flowing through at that time is interrupted. The interruption occurs because the watchdog timer resets the awoken Basic Stamp. The I/O gets scrambled for approximately 18 milliseconds. When the P-Basic language regains control of the microcontroller's interpreter firmware, the I/O is restored prior to the watchdog resets. To avoid the watchdog resets, add resistors to the wired loads for glitch management. The P-Basic program written for the master controller in this chapter can be modified to include the SLEEP mode:

```
BUTTON SWpin,ALow,DlyVal,SWrate,btnWk,TRDpin,noPress
   DEBUG "*"
   TOGGLE TRDpin
   SLEEP 10       ' Sleep for 10 seconds
noPress: GOTO Loop
```

When the program is running, the output at P0 toggles and the Basic Stamp goes to sleep for 10 seconds.

Two other P-Basic commands that are comparable to the SLEEP mode are END and NAP. The NAP mode enters SLEEP mode for a short period of time. The quiescent current during this low-current mode is 50 μA. This P-Basic instruction is quite easy to use:

```
BUTTON SWpin,ALow,DlyVal,SWrate,btnWk,TRDpin,noPress
   DEBUG "*"
   TOGGLE TRDpin
```

```
    Snooze:
        NAP 4          'Nap for 288mS
GOTO snooze;
```

During the NAP interval, output P0 is on, but it turns off for a fraction of a second. The 288-millisecond time is calculated by using the following equation:

$$\text{Length of Nap} = 2^{\text{Period}} \times 18\text{mS}$$

The period for this calculation equals 4. Substitute the numeric value of 4 for the period into the equation:

$$\text{Length of Nap} = 2^{4} \times 18\text{mS}$$
$$\text{Length of Nap} = 0.288\text{secs or } 288 \text{ mS}$$

On page 285 of the *Basic Stamp Programming Manual* (version 1.9), the length of the NAP times can been calculated for those interested in using this instruction to reduce the I/O quiescent current of a Basic Stamp remote-control application for Mindstorms robots.

The END puts the Basic Stamp into an inactive, low-power mode. The Basic Stamp's current draw is approximately 50 µA.

The key enhancement to using the END, NAP, and SLEEP instructions is that you can extend the life span of a 9-volt battery by allowing the Basic Stamp's I/O quiescent current to be low. The master controller used to drive the RCX P-Brick should not only be able to retain its embedded program, but it should also supply voltage. Occasionally, the Basic Stamp might have to sit on the shelf or benchtop for data acquisition or remote-control devices. Therefore, to ensure that the battery supply voltage is sufficient for such applications, techniques that reduce I/O quiescent current must be used. The LEGO P-Brick has a SLEEP mode scheme embedded inside of its firmware, enabling the battery to retain its supply voltage over a period of weeks while sitting on a workshop or lab bench. These three P-Basic instructions are definitely enhancements that can be used in all Basic Stamp applications, especially remote-control devices for Mindstorms robots.

An indirect battery-saver function of the IC4 programming language that is somewhat similar to the P-Basic instructions is brake(int p). The gearing mechanism used in LEGO DC motors provides sufficient speed and torque for the simple and complex mechanical assemblies used in Mindstorms robots. The power that is supplied to the motors over a period of time takes its toll on the batteries inside of the P-Brick. The electromotive force is the back feed voltage generated by rotating the shaft of the internal DC motor. This physical phenomenon is also known as the *generator action* of the motor. In robotic applications, the generator action is a direct response to the motor floating to stop because of the internal gearing built inside of the gray housing surrounding these mechanical components. Depending on the critical nature of the robotic application, floating motors are not acceptable, especially in competition where stop accuracy is required when picking up objects or maneuvering obstacle courses. The brake(int p) function not only disables the power to the motor's terminals, but it also provides a path of least resistance, shorting the output circuit to ground. The result is a dynamic braking scheme that stops the motor shaft rotation instantaneously. The int p refers to outputs 1 through 3 of the RCX P-Brick. In order to use the function in a Mindstorms application, the IC4 code used to control the wagging tail of CiliaBot has been modified with the brake(int p) function:

```
void main()
{     while (1){
          if  (digital(1)==1){
              printf("HI");
              fd(1);
              sleep(5.0);
              brake(1);
          }
          else {
              printf("LO");
              brake(1);
          }
      }
}
```

The original code used the off(1) instruction to stop the motor attached to output A of the RCX P-Brick. The motor floats to a stop using the instruction, enabling a small quiescent current to flow within the RCX P-Brick. The brake(1) function immediately stops the output A motor's shaft from further rotating or floating. Another brake(int p) function is alloff(). When executed, this IC4 function turns off all motors immediately. Using the same code for CiliaBot, the function is as follows:

```
void main()
{     while (1){
          if  (digital(1)==1){
              printf("HI");
              fd(1);
              sleep(5.0);
              alloff( );
          }
          else {
              printf("LO");
              alloff( );
          }
      }
}
```

Again, when the lines with the alloff() function are reached, the motors stop immediately.

The IC4 programming language is an alternative to the NQC platform. Both software languages have the same instruction and function set as well as programming style. One of the enhancement features that IC4 language contributes to the remote-control applications is the direct RCX P-Brick control response of the motors attached to its outputs. Like the END, NAP, and SLEEP instructions of the P-Basic programming language, the brake(int p) and alloff(int p) functions of IC4 battery saver capability can be used.

Developmental Thoughts: Basic Stamp, Mindstorms P-Bricks, and Robots

The use of electric motors and *Light-Emitting Diodes* (LEDs) as output electrical loads of a Basic Stamp leaves very little to the imagination. The distributive system concept discussed earlier enables the RCX and Scout P-Bricks to perform in a smart load capacity because of the embedded microcontroller packaged inside of the programmable units. By creating robots, the P-Brick can be adapted to give the Basic Stamp's output the physical makeup of the actual unit. In conventional lab projects, the device under control is described in the text, but the use of an LED or motor as a surrogate does not convey the system concept very well. The RCX and Scout P-Bricks enable hobbyists and experimentalists to build a functional model that looks and performs like the intended component. The RCX and Scout P-Bricks enable POC development to take on new meaning because of the immediate creation of engineering tabletop models due to the seamless physical rapid prototyping inherent with using these two programmable platforms. Basic Stamp and Mindstorms robotics provide a new dimension in creativity and innovation because paper designs can be validated in a matter of minutes at a very low cost.

The robot concepts found in books such as *Basic Robotic Concepts*[1] by John Holland and *The Complete Handbook of Robotics*[2] by Edward Safford can be modified quite easily using the Basic Stamp and RCX and Scout P-Brick devices.

---

[1]*Basic Robotic Concepts*, Howard W. Sams & Co, Inc., ISBN 0-672-21952-2, copyright 1983.
[2]*The Complete Handbook of Robotics*, TAB Books, Inc., ISBN 0-8306-9872-8, copyright 1978.

Mobility designs, robot brains, interfaces to a computer, motors, and methods can easily have a face-lift using the Basic Stamp and RCX and Scout P-Bricks. The Basic Stamp and P-Bricks are small computers with as much computing power as a desktop. The major difference between the Basic Stamp or P-Bricks and a desktop is the small size of the Basic Stamp and P-Bricks. Combining the Basic Stamp with the RCX and Scout P-Brick gives more flexibility in small tabletop robotic designs and programmable remote-control applications. The Basic Stamp and RCX and Scout P-Brick programmable platforms offer a world of unlimited remote-control applications that is only limited by one's imagination.

Further Thoughts and Experimentation

The POC for remote-control techniques discussed in this chapter is only a small bridge to more complex intelligent machines and manipulative schemes that can be developed using the Basic Stamp and Mindstorms P-Bricks. The primary objective of this chapter is to show how different programmable units can be used to develop a basic remote-control device for Mindstorms robots. The techniques used are a compilation of the previous chapters discussed in this book on hardware and software interfacing to LEGO robots. The previous chapters provide the building blocks needed to build the Basic Stamp master controller for operating the biology-based robot CiliaBot.

In the next chapter, virtual prototyping and control using advanced instrumentation ActiveX controls are explained. Software design methods for developing advanced *Graphical User Interfaces* (GUIs) are illustrated using the LEGO Mindstorms RCX P-Brick. Chapter 2 provides the necessary background information on software controls and GUI development that will assist in building panels with instrumentation ActiveX controls. An interface design methodology is also explained in the next chapter.

CHAPTER 10

Virtual Prototyping and Control Using ActiveX Controls

Chapter 2, "Developing GUIs: Software Control Basics," explores how a control panel based on a *graphical user interface* (GUI) can be used to operate a Mindstorms robot. The ActiveX component is an object-based technology where events, properties, and methods can be manipulated through human interaction. The control panels created in Chapter 2 enable a robot to move with the touch of a button. In subsequent chapters, the control panel grew in sophistication to where data could be recorded and then printed on a spreadsheet for further analysis. The control panel also became part of the virtual network that enabled data to be processed and obtained from an open-source language GUI, which was ultimately used to control a robot. The ActiveX controls used in the previous chapters to build the control panels were the standard objects provided by the desktop Windows operating system environment.

Designing GUIs for control panels is definitely more an art than a science, and the type of object used in the layout helps the developer to work inside the robot and inside the controls used to manipulate its behavior. Building a hardware version of a control panel for a LEGO Mindstorms robot may not be cost effective, and physical changes may also slow down the software development phase of the robot. Virtual prototyping is a low-cost and creative way of developing sophisticated and customized controls for Mindstorms robot operation. Sometimes a small test panel may be required to test diagnostics written in the *Interactive C Version 4.0* (IC4) or *Not Quite C* (NQC) programming language, or to investigate a new task or behavioral model. Instead of wiring up a physical *black box* for such investigations, a virtual unit can be built in a manner of minutes.

This chapter explores software design methods for creating advanced GUIs using instrumentation ActiveX controls. Three instrumentation test panels that interface with the *Robot Command Explorer* (RCX) *programmable brick* (P-Brick) are demonstrated using a Basic Stamp and other supporting electronic components. The following topics are discussed in this chapter:

- Real-Time ActiveX instrumentation controls
- A basic *real-time diagnostics and control* (RTDC) circuit
- Using *Light-Emitting Diode* (LED) indicator switches for Mindstorms robot controls
- Master/smart-slave controller technical philosophy and musing
- Real-Time Control interface design methods

Real-Time ActiveX Instrumentation Controls

ActiveX controls are traditionally used for Windows business applications such as data entry, menu selection, and report generation. The standard ActiveX controls that are packaged with the Windows operating system are migrating from business applications to technical applets (small applications) because of the adequate visual appeal for the virtual prototyping of control panels. Virtual control panels for operating robots should have the appearance of the physical unit. The basic panel built in Chapter 2 for V-Bot uses the standard ActiveX controls to operate the robot. The CommandButton is the primary user interface control used to generate an event when the object is clicked. For a novice learning how to design GUIs, the CommandButton is quite adequate for building a robot control panel. The number of robot functions and behaviors that must be controlled requires an individual CommandButton to execute each embedded task. The overall design of the panel using the CommandButton would look bulky and too robotized.

Real-Time ActiveX instrumentation controls enable you to design and lay out a GUI that looks professional and has the virtual feel of a physical control. The controls that can be used include rocker switches with LEDs, rotary switches, analog gauges, seven-segment displays, toggle switches, valves, tanks, and xy plotters. The controls are configured like standard ActiveX objects and provide easy control over LEGO Mindstorms robots. These ActiveX objects are built by the company Iocomp Software. You can obtain a free evaluation pack of the Real-Time Controls from their web site at www.iocomp.com. Go to the company's homepage and click Downloads. The Downloads web page will be displayed on the screen. Under the Quick Download section, using the Select Download Type down arrow, click Evaluation Downloads from the pull-down menu. Next, select Iocomp Components 2.1.0 SP5 ActiveX Evaluation from the And Then Select Your Download listbox. Click Download and save the IocompComponentsEV-ActiveX.exe file onto your hard drive. Figure 10-1 shows the proper menu items to select from the listbox. An important note about the ActiveX evaluation pack is that an occasional "Evaluation 10-minute limit" message appears with each Real-Time Control. This message pops up every 10 minutes and is a reminder to buy the full software package for $300+. To eliminate this message, get out of the window or development task and return back into the activity.

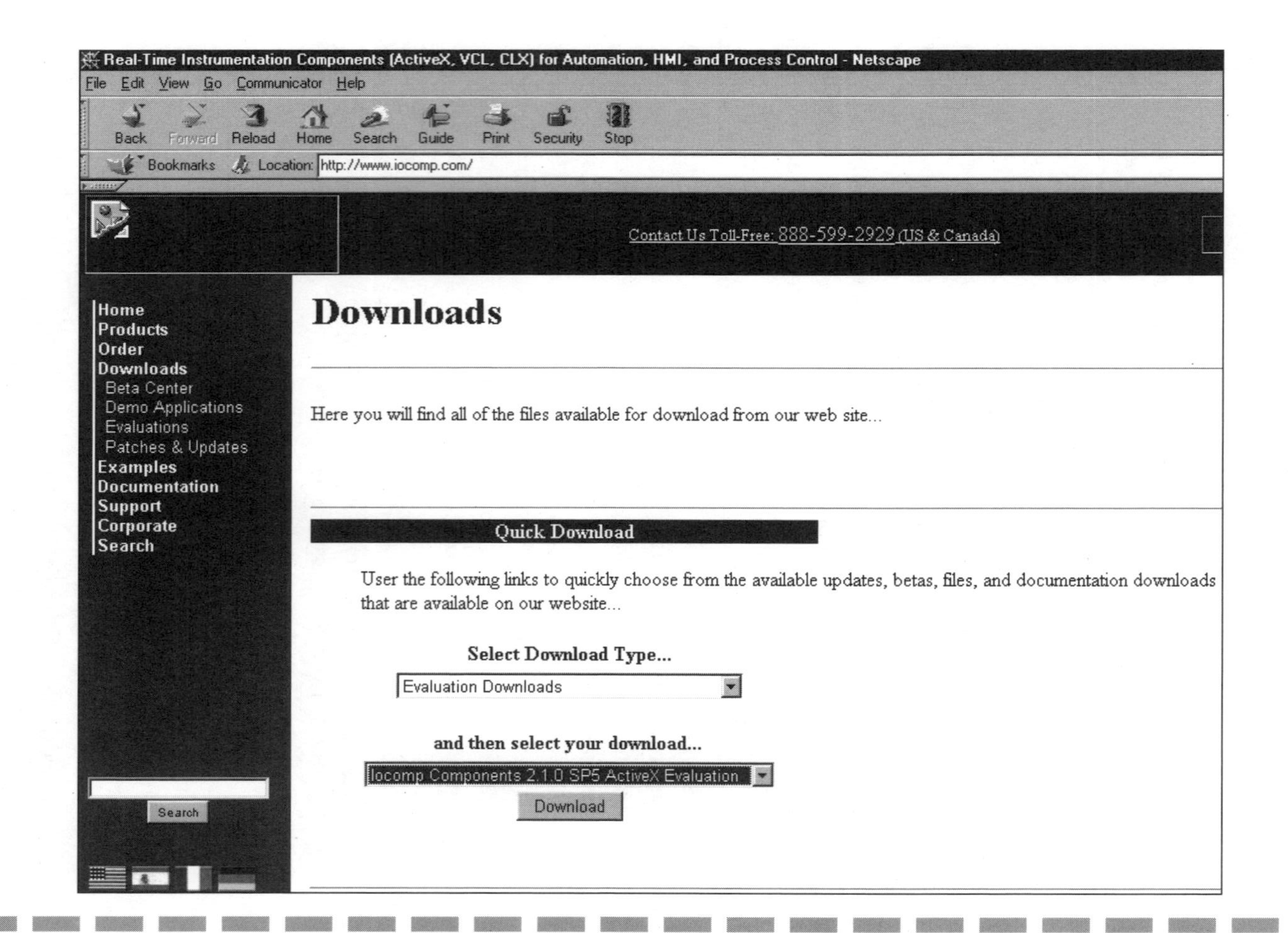

Figure 10-1 Iocomp's web site: gaining access to the real-time control evaluation software

For educational purposes, the evaluation copy with the occasional popup message is still a good deal for hobbyists and experimentalists. Install the software by following the installation instructions from the executable file. Also, a copy of Iocomp's "Getting Started Guide" is highly recommended and can be obtained from the Documentations web page. To use the Real-Time Controls, the References must be added to software development area, which in this case is Excel. Follow the procedure outlined in Chapter 2 in the section "Subprocedure for Adding the LEGO ActiveX Control to the Toolbox." The following .ocx files must be referenced before the controls can be added to the toolbox: iPlotLibrary.ocx, iProfessionalLibrary.ocx, isAnalogLibrary.ocx, isDigitalLibrary.ocx, and iStripChartXControl.ocx. Once the Real-Time Controls and their references are added to Excel's *Visual Basic for Applications* (VBA) environment, a professional-looking virtual control panel can be built.

A Switch Lever with a Seven-Segment Display Indicator Control Panel

To start off the virtual prototyping experimentation, a basic control panel using the Real-Time Controls must be built. The GUI's control function indicates when the toggle switch is in the up position. It also shows a binary 1 on the LED 7-segment display as well. When the switch is in the down or home position, the 7-segment display must show a 0. The following controls are used to build the required control panel:

- iSwitchLeverX A binary switch displayed as a lever. To use the switch position in a logic conditional statement, the Active property is used. The OnChangeUser event can be used to respond to changes in the Active property. The VBA syntax is as follows:

```
iSwitchLeverX.Active=True
```

- iSevenSegmentCharacterX Use the VBA Value property to get or change the display character using the enumerated property type. The enumerated property types are as follows:

```
type SevenSegmentCharacter(ssc) = (ssc0 = 0, ssc1 = 1, ssc2 =2,
ssc3 = 3, ssc4 = 4, ssc5 = 5, ssc6 = 6, ssc7 =7, ssc8 = 8, ssc9 =
9, sscA = 10, sscB =11, sscC = 12, sscD = 13, sscE = 14, sscF =
15, sscPeriod = 16, sscColon = 17, sscPlus = 18, sscMinus = 19,
sscBlank = 20, sscLine = 21, sscUpArrow = 22, sscDownArrow = 23)
```

The VBA syntax is as follows:

```
"iSevenSegmentCharacterX.Value" =ssc
```

An important item to note is that the AutoSize property should equal False. If it does not, the software automatically adjusts the size of the display in accordance with the alphanumeric that is set using the ssc constant. What is drawn on the UserForm will look different when the application is run.

The UserForm is built using these two main controls along with the other ActiveX devices, as shown in Figure 10-2. The other controls consist of two labels and a frame ActiveX control. Once the panel has been built (as shown in Figure 10-2), VBA software can be written. The following shows the VBA code that can be used with the software GUI controls:

```
Private Sub iSwitchLeverX1_OnChangeUser()
If DigPnl.iSwitchLeverX1.Active = True Then
DigPnl.iSevenSegmentCharacterX1.Value = ssc1
Range("A1").Value = DigPnl.iSevenSegmentCharacterX1.Character
Else
DigPnl.iSevenSegmentCharacterX1.Value = ssc0
```

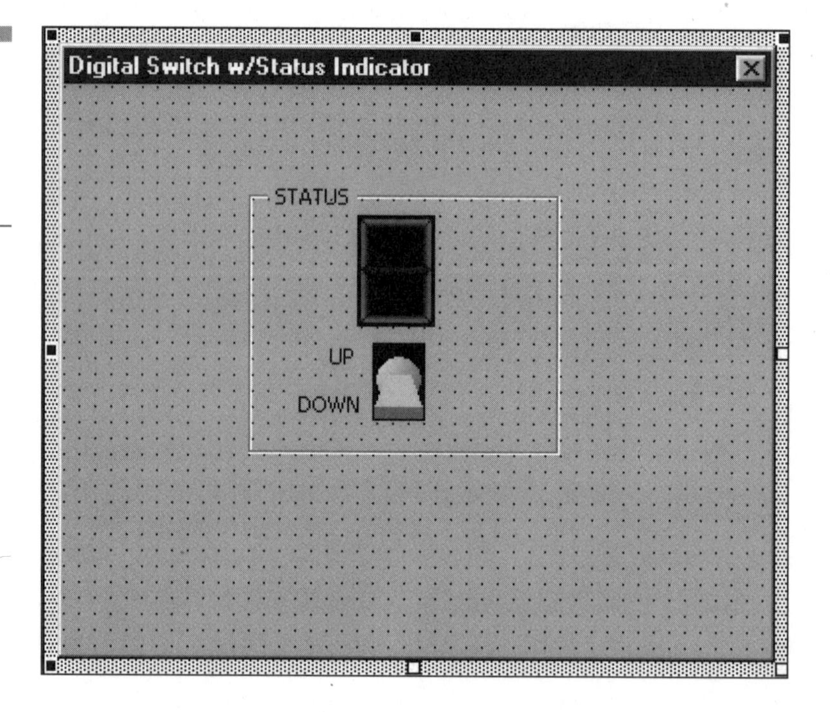

Figure 10-2
A simple GUI using the lever switch and LED seven-segment display ActiveX controls

```
Range("A1").Value = DigPnl.iSevenSegmentCharacterX1.Character
End If
End Sub
```

This VBA code is simple in logic and enables you to easily change the character shown on the seven-segment display. If the contacts of the switch are closed (up position), then DigPnl.iSevenSegmentCharacterX1.Value = ssc1 displays a 1. The equivalent value of ssc1 is displayed in cell A1 with the VBA instruction Range("A1").Value = DigPnl.iSevenSegmentCharacterX1.Character. If the lever switch contacts are open (down position), the VBA code displays a 0 using DigPnl.iSevenSegmentCharacterX1.Value = ssc0. This data is also shown in cell A1 using the Range("A1") object.

The size of the segment can be changed by selecting the object and editing the Segment Size property to a value of 2. Figure 10-3 shows the new size for the 7-segment display with a value of 2. After the code is built, a CommandButton can be added to the spreadsheet. To add an ActiveX control from the Tools menu bar, design mode must be initiated. When a design-mode session is running, ActiveX controls from the Tools menu bar can be placed on the spreadsheet. Changes can be made and code can be added to the controls in design mode. Right-clicking the controls brings up the *Visual Basic Editor* (VBE). This ActiveX control calls the GUI control

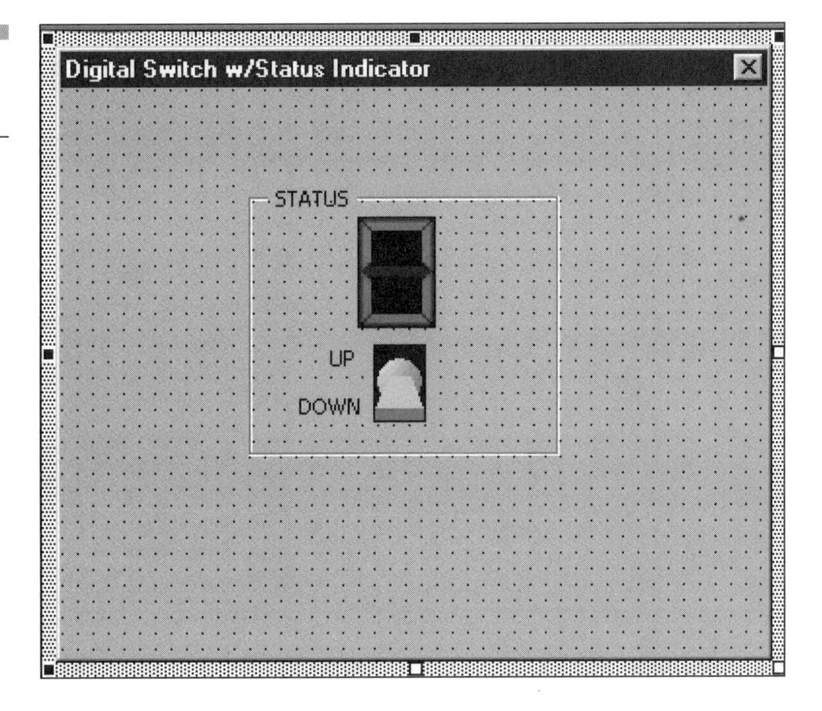

Figure 10-3 Modifying the LED 7-segment size to 2

panel each time the button is pressed. The VBA code used to call the GUI control panel is as follows:

```
Private Sub CommandButton1_Click()
DigPnl.Show
End Sub
```

In both the Sub CommandButton1 and iSwitchLeverX1 code applets, the UserForm's name was changed to "DigPnl." The name change is accomplished using the UserForm1's property list. This name also carries over to the iSwitchLeverX1 code applet in order to maintain proper consistency and eliminate a runtime error. Figure 10-4 shows the Real-Time Control displaying a logic 1. As stated earlier, the evaluation version is fully functional, except that the controls time out every 10 minutes. Remove the annoying register reminder by closing the application and running it again. Figure 10-5 shows the 10-minute limit message splashed all over the controls.

A GO-NO-GO Test of the Digital Switch

Once the spreadsheet is designed and the GUI control panel is built, a quick Go-No-Go test can be carried out. By clicking the Digital Panel, the Digital Switch w/Status Indicator panel should appear on the spreadsheet. Move

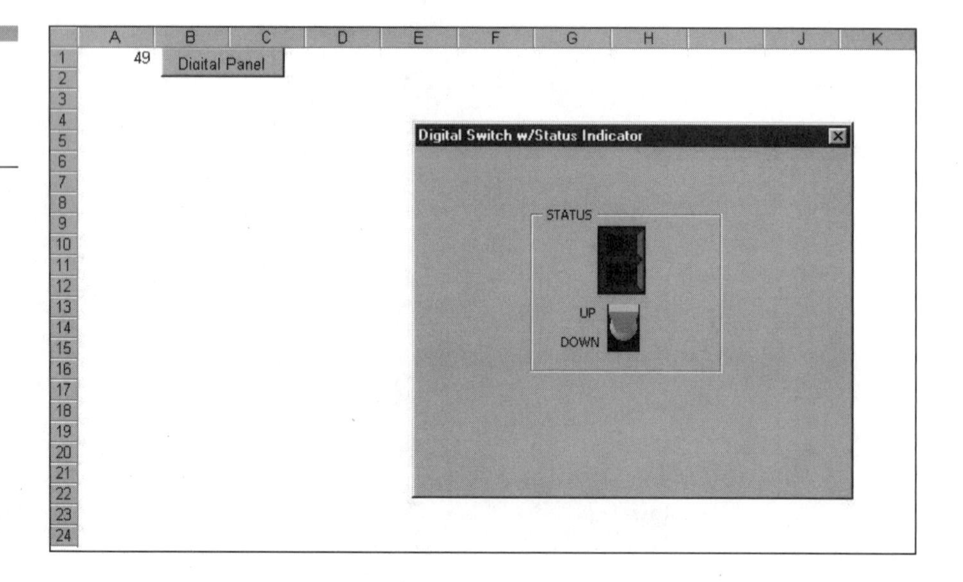

Figure 10-4 Digital switch GUI controls placed on the spreadsheet

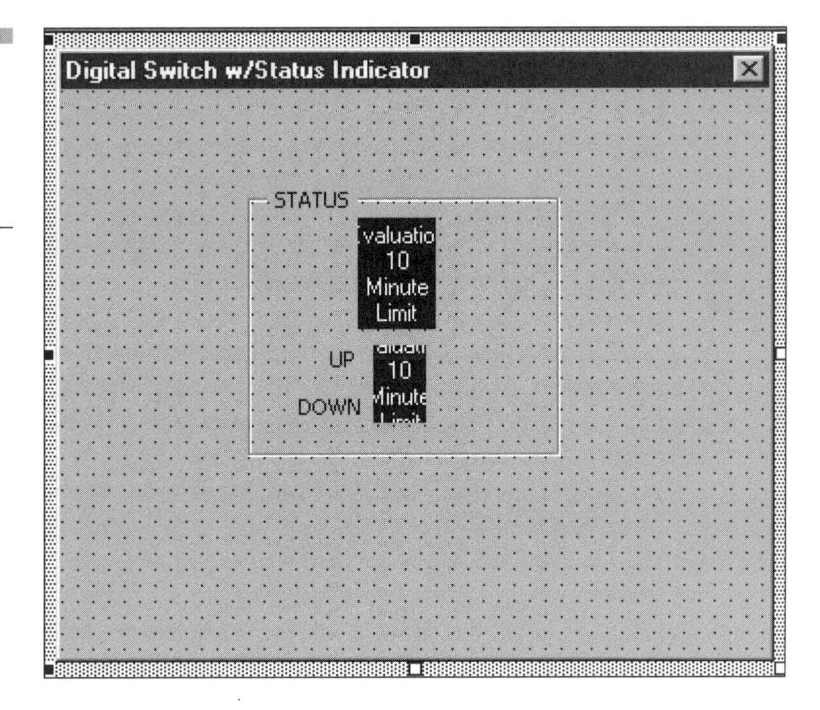

Figure 10-5 The 10-minute timeout of the evaluation software placed on each control object

the lever switch by clicking it. The switch should be shown in the up position and a digit of 1 should be displayed. The number 49 in cell A1 represents ssc1. Clicking the switch again puts it in the down position and a digit of 0 should be displayed. The number 48 in cell A1 represents ssc0. Click the switch several times and notice how the seven-segment display keeps track of the virtual switching events. The design of the virtual panel is very simple and performs the digital display of binary digits 1 and 0 precisely. The spacebar on the keyboard can toggle the switch as well. By pressing the spacebar once, the lever switch flips up. If the spacebar is hit again, the switch flips down. The control panel is now ready for software controls to be added to physically communicate with an RCX P-Brick.

A Virtual Master/Physical Slave Controller

A virtual master/physical slave controller can be built using a software GUI control panel and an RCX P-Brick. The *Proof Of Concept* (POC) study consists of using the built Digital Switch control panel as the virtual master controller and the RCX P-Brick as the physical slave. This concept is validated in Chapter 2 with V-Bot. The only difference between the two master controllers is the GUI. The Real-Time Controls provide the visual appearance of a robot control panel. Also, the virtual design can be turned into a physical unit by selecting its equivalent real components and wiring them to operate like their software counterparts.

Figure 10-6 shows a system block diagram for a virtual master/physical slave controller. To make this diagram a working control panel, a few additional building blocks are required. On the master controls side, a virtual switch is needed to enable and disable the serial port for outside communication with the RCX P-Brick. The Infrared Tower is the data communications medium between the virtual panel and the RCX P-Brick. The virtual switch and Infrared Tower provide a vital link for establishing a gateway to the physical controls of the RCX P-Brick. The other building block controls the outputs of the RCX P-Brick. In Chapter 2, the PBrickCtrl.On and PBrickCtrl.Off objects enable outputs to be controlled using VBA software. The same objects are also used with the Real-Time Controls panel. The virtual switch used to open and close the serial communication port is iSwitchLEDx. The iSwitchLEDx is a binary switch with a rectangle LED indicator. If the switch is clicked once, the LED is on and its Active property is True. If the switch is clicked again, the LED is off and its Active property is False. To make the switch provide more status information to the operator of the GUI control panel, ON and OFF captions are displayed on the body of the switch. These captions are visible when the switch is clicked two times. The syntax used to program the ActiveX object is as follows:

```
iSwitchLEDX.Caption = string
```

where string equals ON or OFF.

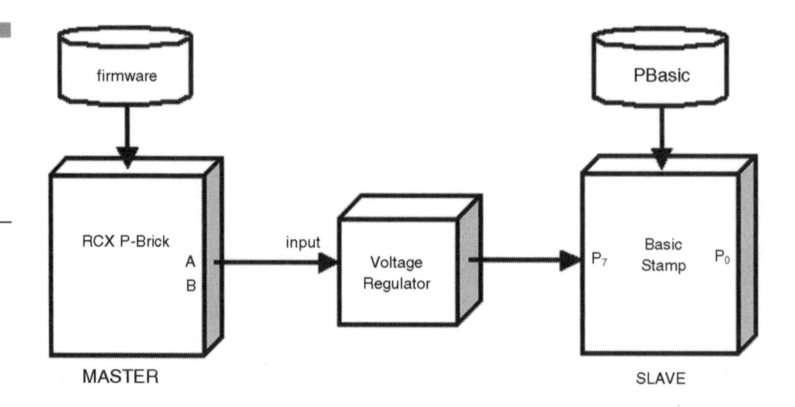

Figure 10-6 The system block diagram of the master (RCX P-Brick) and slave (Basic Stamp) controller

After adding the LED indicator switch to the panel (as shown in Figure 10-7), software can be added to provide serial communication control of the panel as well as control of output A on the RCX P-Brick. The following software can be added to the control panel to operate output A of the P-Brick:

```
Dim MotorA As String
Private Sub iSwitchLedX1_OnChange()
If DigPnl.iSwitchLedX1.Active = True Then
DigPnl.iSwitchLedX1.Caption = "ON"
DigPnl.PBrickCtrl.InitComm
Else
DigPnl.iSwitchLedX1.Caption = "OFF"
DigPnl.PBrickCtrl.CloseComm
End If
End Sub

Private Sub iSwitchLeverX1_OnChangeUser()
If DigPnl.iSwitchLeverX1.Active = True Then
DigPnl.iSevenSegmentCharacterX1.Value = ssc1
MotorA = "0"
DigPnl.PBrickCtrl.On (MotorA)
Range("A1").Value = DigPnl.iSevenSegmentCharacterX1.Character
Else
DigPnl.iSevenSegmentCharacterX1.Value = ssc0
MotorA = "0"
DigPnl.PBrickCtrl.Off (MotorA)
Range("A1").Value = DigPnl.iSevenSegmentCharacterX1.Character
End If
End Sub
```

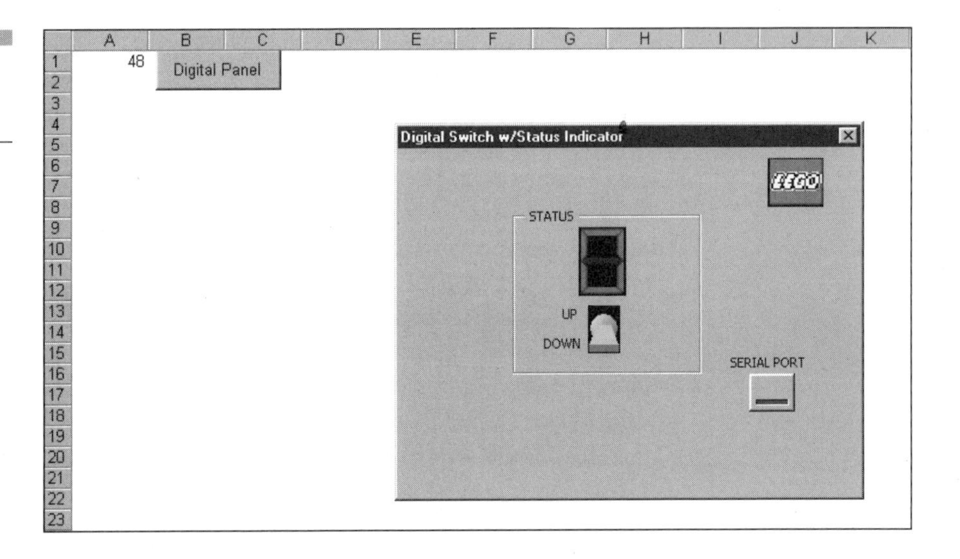

Figure 10-7 Master panel with serial port control

The DigPnl.iSwitchLedX1.Caption = “ON” object is responsible for showing the state of the switch with text and a visual indicator. The OFF caption is used with the same object, except the LED is not lit, signifying that the serial port is closed. The serial port is open, enabling the PC or desktop application software to communicate with the Mindstorms robot. The VBA line of instruction responsible for opening the serial communication port is DigPnl.PBrickCtrl.InitComm. The DigPnl.PBrickCtrl.CloseComm is used to close the serial communication port.

Once the code is added to the control panel, the software is ready to be tested. Click the Digital Panel button on the spreadsheet. The Digital Switch panel should be visible on the spreadsheet. Click the Serial Communication button, and the green LED, along with the caption ON, should be visible. Add a LEGO small 9-volt motor to output A using the electric wire to the RCX P-Brick. Turn the RCX P-Brick and click the lever switch. A logic 1 should appear on the 7-segment display. The LEGO small 9-volt motor and the 7-segment display will be on. Click the lever switch again and the 7-segment display should show a logic 0. The LEGO small 9-volt motor turns off at this point.

A Basic Real-Time Diagnostics and Control (RTDC) Circuit

To build an embedded diagnostics tool for a Mindstorms robot, an actual seven-segment display driver (as discussed in the section “A DCU Diagnostics Tool Lab Procedure” in Chapter 4, “Electronic Switching Circuits”) can be wired to the RCX P-Brick for a basic RTDC.[2] Robot diagnostics are residential miniprograms in control systems that can be activated automatically when electrical power is turned on or manually initiated with a readout box. The purpose for the miniprograms is to isolate a fault or validate the specific functions of the robot. When a failed or passed condition is detected, a visual or audible indicator is given to the operator. The RTDC is an experimental tool that validates the instrumentation GUI panel and its physical interface to the RCX P-Brick. The gateway or hardware link to the physical domain of the RCX P-Brick is through the Infrared Tower. A visual indicator for the RTDC would be to hardwire the 7-segment display to show a logic 1. This binary digit is viewed when the virtual logic switch is in the up position. When the lever switch is in the down position, the seven-segment display is off.

Figure 10-8 shows the system block diagram for the RTDC. The RTDC uses the one of the three outputs of the RCX P-Brick to perform a Go-No-Go test. If the 7-segment display is hardwired to display a binary 1 (meaning a HI state), this provides a simple tool for validating software programs that use the outputs of the RCX P-Brick for robot drive control. The 7805 voltage regulator *Integrated Circuit* (IC) discussed in Chapter 4 in the section "An Audible Tone Generator Lab Project Procedure" provides a +5-volt *Direct Current* (DC) supply for the 7-segment display. The voltage regulator circuit also acts as an electronic switching circuit when the RCX P-Brick's output A turns on by the virtual control of the Real-Time VBA GUI panel. The circuit schematic for the voltage regulator driver electronics is shown in Figure 10-9.

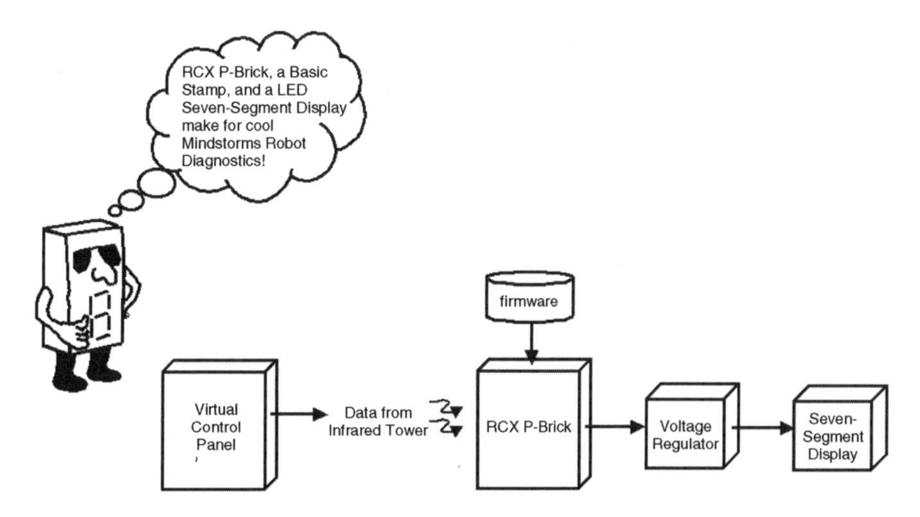

Figure 10-8 Led Seven introduces The System Block Diagram of the RCX P-Brick hardware interface circuit for controlling the LED seven-segment display.

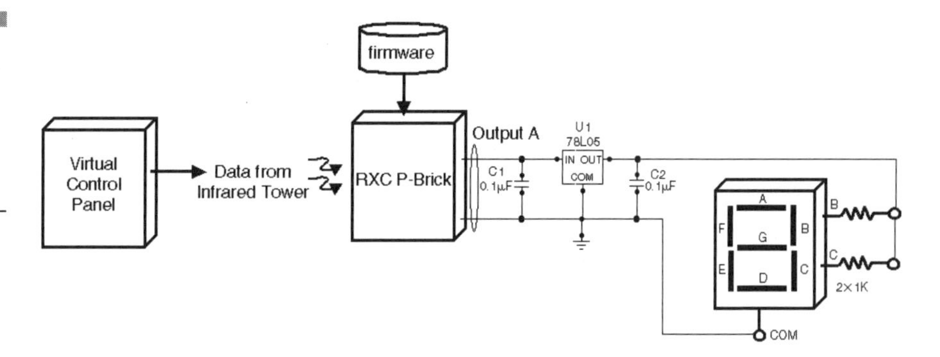

Figure 10-9 The circuit schematic of the hardware interface for the LED seven-segment display

The voltage regulator circuit is built on a solderless breadboard such as the Radio Shack Electronics Learning Lab or an equivalent whiteboard. The modified LEGO electric wire is attached across the 0.1 µF capacitor. The other end of the electric wire is attached to output A of the RCX P-Brick. The wire attached to output A should be dressed or pointed up for proper circuit interfacing. The other 0.1 µF capacitor is wired across the output pin of the 7805 IC and ground. The 7805 circuit is tested by using the VBA Real-Time Control panel and a *Digital Multimeter* (DMM) connected across the 0.1 µF output capacitor and ground. If the RCX P-Brick is turned on and the Infrared Tower faces the transceiver of the programmable unit, the lever switch on the GUI is flipped to the up position. The DMM should read +5.01 volts DC. If the output voltage reading is different, turn off the RCX P-Brick and check the wiring. If the output voltage reads correctly, the seven-segment display can be wired. The display is hardwired for a binary 1 if the virtual logic switch is in the up or closed position. The LEGO small 9-volt motor can be wired to the input circuit of the voltage regulator IC by taking a standard electric wire and attaching it between the motor and output A. Therefore, output A should have two electric wires attached to it: one for the voltage regulator circuit and one for the small 9-volt motor. With each lever switch actuation, the 7-segment display and small 9-volt motor will turn on and off at the same. Also, there is no battery drain on output A's internal DC supply voltage.

Almost simultaneously, the binary digits are displayed and the motor switches on or off. The Real-Time GUI control panel makes it quite easy, educational, and fun to prototype Mindstorms robots. The two instrumentation controls used in the GUI panel only scratch the surface of virtual prototyping. This RTDC can be expanded whereby the +5-volt signal from the 7805 voltage regulator circuit can provide a logic-level input to the Basic Stamp. The GUI is modified to accommodate a rocker switch with an integral LED indicator.

Using an LED Indicator Switch for Mindstorms Robot Control

The GUI control panel shown in Figure 10-7 can be modified to use a rocker switch with an LED indicator. The rocker switch has the same latching capability as the lever switch. The Real-Time Control VBA syntax for programming the rocker switch is as follows:

```
"iSwitchRockerX1.Active=True"
```

The iSwitchRockerX is a binary switch with a rocker actuator for latching digital input request. To get or set the switch position, use the Active property. To change the active color of the LED indicator, the property IndicatorActiveColor is used. The VBA syntax for programming the LED indicator is as follows:

```
"iSwitchRockerX1.IndicatorActiveColor=vb(color)"
```

When the color is brown, red, green, and so on, the rocker switch replaces the lever device as the main switching element on the GUI control panel. Also, the RCX P-Brick and 7805 voltage regulator provide a +5-volt input signal to the Basic Stamp, thus forming a master-slave controller, which is discussed in Chapter 9, "Remote-Control Techniques."

Building the modified panel consists of reusing code (see Chapter 4) and changing it to meet the new design requirements. An expedient way of modifying reusable VBA code is to export/import the existing software. The following section shows the procedure discussed in Chapter 4 to create or modify a GUI control using an existing UserForm.

Developing a New GUI Using an Existing UserForm Procedure

The following steps outline the procedure for building a new GUI using an existing UserForm design:

1. Open the spreadsheet with the appropriate UserForm.
2. Go to the VBE by clicking the icon with the toolbar.
3. Under VBAProject window, find the target UserForm.
4. Select and right-click it.
5. Under the pull-down menu select Export File . . . .
6. Put the UserForm (name.frm) into the appropriate directory.
7. Close out the worksheet and open an new .xls file.

8. Go to VBE.
9. Select File and click Import File . . . .
10. Go to the directory where the .frm will be placed.
11. Double-click the UserForm. A folder named Forms should be visible within the VBA project.
12. Open the folder; the UserForm should be there.
13. Change the name of the UserForm to DigPnlA under the Properties window.
14. Highlight the name DigPnlA and right-click View Code.
15. The VBA code for this applet should be displayed in the VBE.

With the old DigPnl on the screen, the rocker switch ActiveX Real-Time Control can replace the existing lever device. Replace the lever switch with the rocker device, as shown in Figure 10-10. Remember that in order to use the Real-Time Controls, the References must be added to the software development area, which in this case is Excel. Follow the procedure outlined in Chapter 2 in the section "Subprocedure for Adding the LEGO ActiveX Control to the Toolbox." The following .ocx files must be referenced before

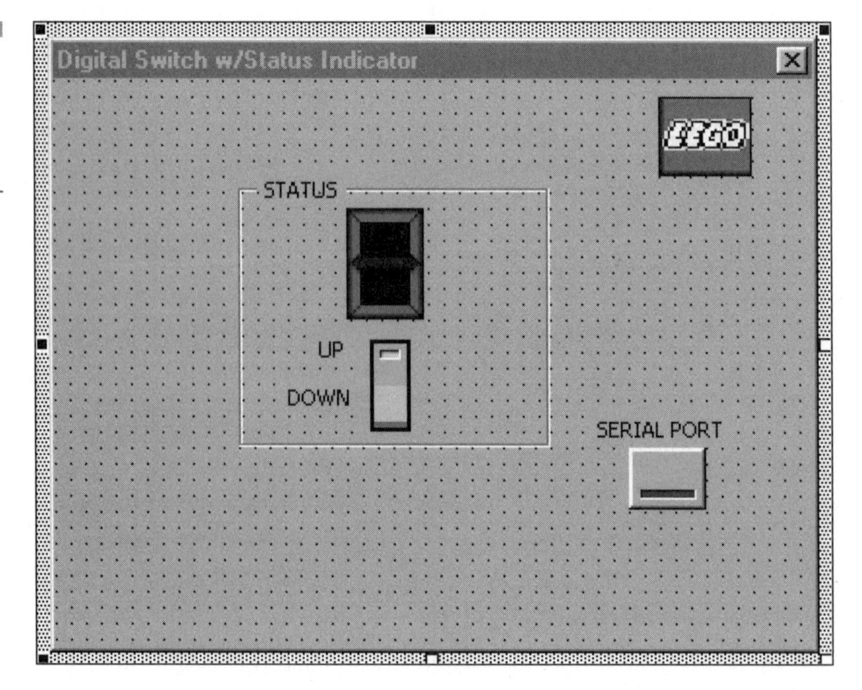

Figure 10-10 GUI panel modified with the rocker switch LED indicator control

the controls can be added to the toolbox: iPlotLibrary.ocx, iProfessionalLibrary.ocx, isAnalogLibrary.ocx, isDigitalLibrary.ocx, and iStripChartXControl.ocx. If these objects are not registered, they will not work in the prescribed application.

Once the panel is modified using the rocker switch LED indicator, software can be written for the RCX P-Brick master controller. The following VBA program can be used for this applet. Enter the following code into the VBE:

```
Dim MotorA As String

Private Sub iSwitchLedX1_OnChange()
If DigPnlA.iSwitchLedX1.Active = True Then
DigPnlA.iSwitchLedX1.Caption = "ON"
DigPnlA.PBrickCtrl.InitComm
Else
DigPnlA.iSwitchLedX1.Caption = "OFF"
DigPnlA.PBrickCtrl.CloseComm
End If
End Sub

   Private Sub iSwitchRockerX1_OnChangeUser()
   If DigPnlA.iSwitchRockerX1.Active = True Then
   DigPnlA.iSwitchRockerX1.IndicatorActiveColor = vbRed
    DigPnlA.iSevenSegmentCharacterX1.Value = ssc1
MotorA = "0"
DigPnlA.PBrickCtrl.On (MotorA)
Range("A1").Value = DigPnlA.iSevenSegmentCharacterX1.Character
Else
DigPnlA.iSevenSegmentCharacterX1.Value = ssc0
MotorA = "0"
DigPnlA.PBrickCtrl.Off (MotorA)
Range("A1").Value = DigPnlA.iSevenSegmentCharacterX1.Character
End If
End Sub
```

The code is very similar to the first lever switch panel with the exception of the following software changes:

```
If DigPnlA.iSwitchRockerX1.Active = True Then
DigPnlA.iSwitchRockerX1.IndicatorActiveColor = vbRed
```

As explained earlier, the first line of code is used as a condition for checking the rocker switch's logic status. The second VBA line of code is used to set the rocker's LED indicator color to red. The rest of the applet functions like the lever switch device previously discussed in the section "A Switch with a Seven Segment Display Indicator Control Panel" on page 309 of this chapter. To test the code, turn on the RCX P-Brick and have the new GUI displayed on the spreadsheet. Click the Serial Port button and notice that the green LED is on. Next, click the rocker switch in the up position. The seven-segment display should have a logic 1. The physical LED display shows an illuminated 1. Click the switch again for the down position. The display should have a logic 0. The breadboard device will be turned off, signifying logic 0. The Basic Stamp circuit from Chapter 9 is wired to the RCX P-Brick's output A. Figure 10-11 shows the complete wiring schematic of the new controller. For testing purposes, the original code used to validate the Basic Stamp's *Input/Output* (I/O) function can be entered into the Basic Stamp text editor, as shown in the following code:

```
btnWK  var  byte
btnWK=0
LOOP:
    BUTTON 7,0,255,250,btnWK,0,noPress
     debug "This is a test"
     debug "-"
     TOGGLE 0
noPress:goto loop
```

Download the code into the Basic Stamp. Place an audible tone continuity tester or ohmmeter across the *Normally Open* (NO) contacts of the +5-volt relay. With the power supplied to the Basic Stamp using a 9-volt DC source, click the ActiveX rocker switch on the GUI panel. The GUI LED 7-segment display should show a binary 1. A binary 1 should also appear on the real 7-segment display. If it does not, check the wiring and try it again. Click the rocker switch and a binary 0 should appear on the optoelectronic device. The hardwired component should be blank. The audible tone conti-

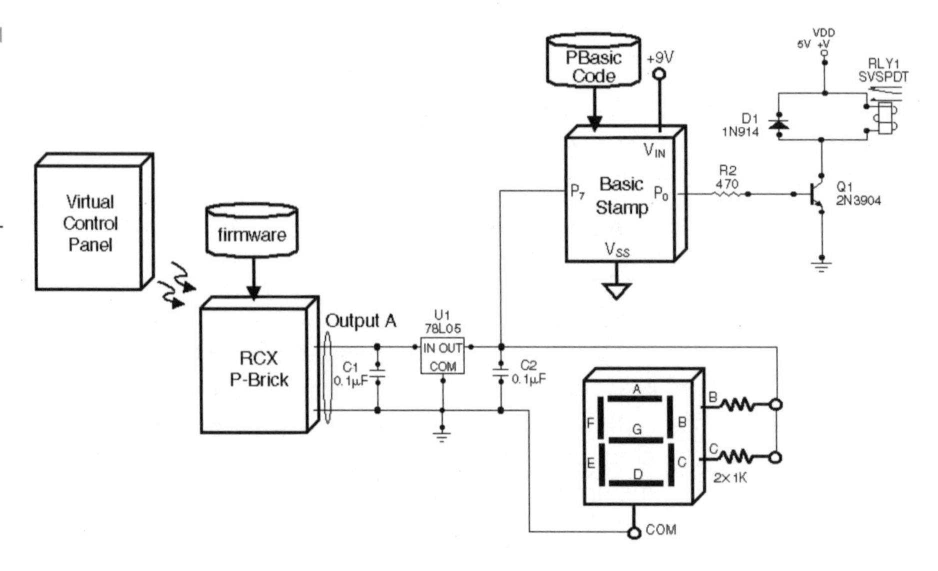

Figure 10-11 The circuit schematic showing RCX P-Brick to Basic Stamp transistor relay driver interface

nuity tester or ohmmeter should sound off, signifying that the relay contacts are closed.

The effect of operating in the Active Low state is that the correct input request and output response of the Basic Stamp can be determined. The output response is based on the TOGGLE instruction latching the relay on a single input request. Another rocker switch closure opens the relay contacts, putting the output in an unlatched state. Two additional virtual input requests can be added using ActiveX rocker switches and VBA code. The relay contacts can be wired to an input of the RCX P-Brick for feedback and the control of a LEGO small 9-volt motor. Therefore, not only can the robot be put into motion, but its output can also be monitored and controlled using a feedback signal. The mechatronics known for controlling and monitoring the output with feedback is called a closed-loop system. Advanced robotics has a closed-loop system of some sort for controlling. If a small error is induced in the system because of mechanical dynamics, a control signal is sent to the servomechanism to make a correction. Using the master/smart-slave controller, a Mindstorms robot's mechanical assembly for motor drive can be used to correct errors in the machine's dynamics. This closed-loop technique can be used to turn on output B upon receiving a request at input 1. Some VBA code could be added where this value would automatically be displayed on the spreadsheet. The unique item of this master/smart-slave controller is that each programmable device controls the other using hardwired points. Figure 10-12 shows the complete system block diagram of all the interconnected components.

Figure 10-12 System Block Diagram of master/smart-slave controller

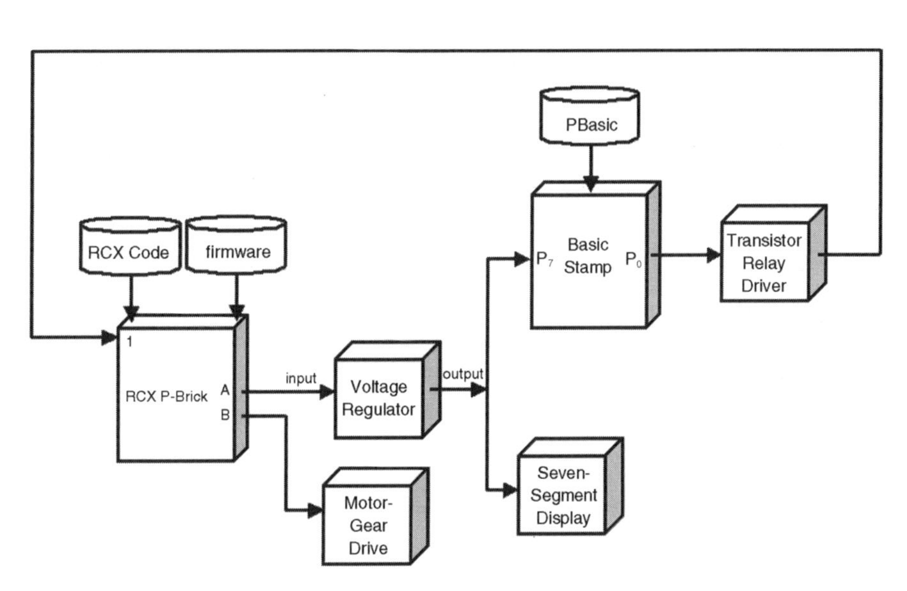

To build the system shown in Figure 10-12, the relay contacts must be wired to input 1. The modified LEGO electric wires connect the Basic Stamp's PO output. The transistor relay driver contacts are the input request signal for initiating the embedded task. To ensure a correct switch voltage level for input 1, add the series 1-kilo-ohm resistor to either relay contact. See the section "Understanding the Touch Sensor RCX/Scout Interface" in Chapter 3, "Electromechanical Controls Interfacing," for a review on the purpose of the series 1-kilo-ohm resistor. The embedded program reads the signal at input 1 and turns output B on or off if the input request is correct. Wiring the two programmable devices physically can be accomplished using the LEGO Technic Electric 2×8 plate and additional modified electric wire. A distributive load system (see Chapter 9) is a good way of electrically connecting the RCX P-Brick to the Basic Stamp when building the master/smart-slave controller.

The program can be written easily in RCX code where pressing a set of contacts turns output B on and releasing the contacts at input 1 turns output B off. The key VBA instruction used to initiate a particular preprogram function is DigPnlA.PBrickCtrl.StartTask 0. The StartTask command as defined on page 48 of the LEGO Mindstorms *Software Development Kit* (SDK) is the "execution of P-Brick's task [number]. Tasks always start from the beginning of the task (that is, the very first program line in the task). If the task [number] was already running, it is stopped and then restarted (from the very first program line in the task). If no task was running before issuing this command, the little man in the *Liquid Crystal Display* (LCD) screen on the P-Brick will start running." Basically, StartTask enables a particular function to be executed when that task is called within a computer program. Automatic execution of the embedded task is carried out when the correct request from an input event occurs. The modified and complete VBA code for the rocker switch with LED indicator GUI controls using the StartTask command appears in the following code:

```
Dim MotorA As String

Private Sub iSwitchLedX1_OnChange()
If DigPnlA.iSwitchLedX1.Active = True Then
DigPnlA.iSwitchLedX1.Caption = "ON"
DigPnlA.PBrickCtrl.InitComm
Else
DigPnlA.iSwitchLedX1.Caption = "OFF"
DigPnlA.PBrickCtrl.CloseComm
End If
End Sub
```

```
Private Sub iSwitchRockerX1_OnChangeUser()
If DigPnlA.iSwitchRockerX1.Active = True Then
DigPnlA.iSwitchRockerX1.IndicatorActiveColor = vbRed
DigPnlA.iSevenSegmentCharacterX1.Value = ssc1
MotorA = "0"
DigPnlA.PBrickCtrl.On (MotorA)
DigPnlA.PBrickCtrl.StartTask 0
Range("A1").Value = DigPnlA.iSevenSegmentCharacterX1.Character
Else
DigPnlA.iSevenSegmentCharacterX1.Value = ssc0
MotorA = "0"
DigPnlA.PBrickCtrl.Off (MotorA)
DigPnlA.PBrickCtrl.StopTask 0
Range("A1").Value = DigPnlA.iSevenSegmentCharacterX1.Character
End If
End Sub
```

Again, the software development practice of code reuse (see Chapter 2) is put into good practice through this program. The StopTask command "stops execution of P-Bricks task [Number]. If all tasks are stopped, the little man in the display on the P-Brick will stop running." See page 49 of the LEGO Mindstorms SDK. Therefore, when the code reaches the line DigPnlA.PBrickCtrl.StopTask 0, execution of the RCX code stops. The electrical load at output B stops when the StopTask command is executed.

When the control panel is used to turn the LED seven-segment display on, output B remains off until a second request of the rocker switch is initiated. The sequence of events for this software applet is as follows:

1. With the rocker switch in the down position, the binary data displayed on the LED 7-segment display is a 0. The internal program (StartTask) is not running and output B is off.
2. With the rocker switch in the up position, the binary data displayed on the LED 7-segment display is a 1. The hardware circuit display is also a binary 1. The internal program (StartTask) is running and output B is off.
3. With the rocker switch in the down position, the binary data displayed on the LED 7-segment display is a 0. The hardware circuit display is blank (off). The internal program (StopTask) is not running and output B is on.

Repeating Steps 2 and 3 causes output B to turn off. To see the actual logic data being applied at input 1, press the View button once. This places a small cursor (arrow) to input 1, enabling a binary 1 value to be displayed on the LCD.

Master/Smart-Slave Controller Technical Philosophy and Musing

The master/smart-slave controller is quite unique because of its *closed-loop* method to control another output on the RCX P-Brick. The Basic Stamp can also provide additional outputs for the RCX P-Brick by using a multiplex scheme similar to the master/smart-slave controller topology. By using one output and wiring it to an input of the Basic Stamp, the customized PIC microcontroller can switch the additional outputs that it can provide on or off using the *Parallax Basic* (P-Basic) programming language. The design requirements define what behavior or task the robot will perform. The I/O electrical interface of the RCX P-Brick and Basic Stamp can be designed with these functional objectives.

A system block diagram proves valuable in defining the software and physical layers (see Chapter 2) of the target Mindstorms robot. The embedded tasks or behaviors of the RCX P-Brick can use the IC4, NQC, or RCX code programming languages. RCX code was used in the master/smart-slave controller exercise because an output can be turned off using one input requirement. Also, the RCX code is a good programming language choice for quickly testing an idea for Mindstorms robot control. Software design doesn't have to be so complicated as to distract hobbyists and experimentalists from the goal of designing an experimental robot that not only has good design features, but also has a technical education benefit.

The interesting aspect of the VBA GUI control panel is that its functional control capability to manipulate Mindstorms robots can be easily expanded. The two Real-Time Control panels built enable you to move from one control feature design to another one seamlessly. The second panel was upgraded from a lever switch to a rocker with an LED indicator. The second panel also has the capability to not only provide diagnostics, but also to drive the other two available outputs based on a single input provided by the Basic Stamp. Therefore, think of this hardwired input to the RCX P-Brick as a command control signal (see the section "Electromechanical Relays for RCX/Scout Input Control" in Chapter 3). This expanded hardware interface mechanism is a viable tool for building advanced Mindstorms robots using a distributive load system topology.

One last item to note about the VBA GUI control panel is that its capability to add a front-end system with user inputs to a non-Windows control object gives the programmable device an extension in configurable design. Configurable design is an object's capability to change its I/O function using a GUI-based control panel. The RCX P-Brick is a configurable design because a control panel can be built that can manipulate its I/O function through a Windows-based GUI (RCX code). The Basic Stamp does not provide a mechanism that enables the I/O of the programmable unit to be changed using a Windows GUI. When building a VBA GUI controller using Real-Time instrumentation ActiveX or standard control objects and linking the panel to the RCX P-Brick, the Basic Stamp's I/Os can be affected through hardwired connections. As demonstrated in the two Real-Time Control projects, the Basic Stamp becomes a slave to the RCX P-Brick. Therefore, the software resources available to the RCX P-Brick through a VBA-built GUI are shared by electrically hardwired connections to its interfacing input layer. Thus, a programmable front end has been established for the Basic Stamp.

Real-Time Control Interface Design Methods

The two panels discussed in this chapter are quite simple in design as well as function. They perform direct control of the Mindstorms I/O physical layer (see Chapter 2). The two panels built are successful designs because they lack additional controls and software logic confusion. The Real-Time Control interface design method is based on how important an ActiveX object is for manipulating a Mindstorms robot. The lever and rocker switches provide direct control of a Mindstorms robot. Additional controls can complicate the use of a GUI-based panel to a point where it becomes useless. If additional controls are desired, they should be selected on the basis of importance to the overall control of the robot. The controls used on the GUI panel should directly relate to the robot's function or behavior. The rocker switch VBA panel indicates that the ActiveX object not only controls the P-Brick's output, but it also starts the embedded robot task. Output B can be considered the command control signal for the Basic Stamp's input. The Active-Low signal is used by the Basic Stamp to turn the transistor relay driver circuit off. The following code is responsible for enabling the RCX P-Brick to carry out the dual robot functions:

```
Private Sub iSwitchRockerX1_OnChangeUser()
If DigPnlA.iSwitchRockerX1.Active = True Then
DigPnlA.iSwitchRockerX1.IndicatorActiveColor = vbRed
DigPnlA.iSevenSegmentCharacterX1.Value = ssc1
MotorA = "0"
DigPnlA.PBrickCtrl.On (MotorA)
DigPnlA.PBrickCtrl.StartTask 0
```

The dual functions are executed if the rocker switch is active or turned on. The following two lines of instruction are responsible for the dual robot functions:

```
DigPnlA.PBrickCtrl.On (MotorA)
DigPnlA.PBrickCtrl.StartTask 0
```

Output A of the RCX P-Brick is controlled by the VBA Parent/Child/Method object of DigPnlA.PBrickCtrl.On (MotorA). To execute the embedded task of the relay switch contact closure detection the VBA Parent/Child/Method object of DigPnlA.PBrickCtrl.StartTask0 is used for the RCX P-Brick electromechanical switching function. These two lines of code rely on the simple event of the user clicking the rocker switch. The overall design of the GUI is simple but highly effective in Mindstorms robot control.

The system block diagram is an important tool that provides a lot of help in designing a Real-Time Control interface for Mindstorms robots. As shown in Figure 10-12, the system block diagram provides a roadmap that shows how the basic building blocks are interconnected. As seen in the block diagram, the closed loop is formed by the RCX output A wired to the input of the Basic Stamp. The transistor relay driver controlled by the Basic Stamp is wired to input 1 of the RCX P-Brick. Thus, all the connected circuit blocks form a closed loop. This closed loop is initiated by the dual function control of the rocker switch. The panel switch is another control that can substitute the rocker switch. The panel switch is an ActiveX Real-Time Control that has three positions. It uses the VBA format of iSwitchPanelX. Positions 2 and 3 are working contacts on the switch and are used to perform the dual robot control tasks that were discussed earlier. The VBA syntax used to set the position property is as follows:

```
"iSwitchPanelX.Position = switch position
```

Switch position equals 2 and 3. Build the new GUI with the panel switch, as shown in Figure 10-13. Set the position captions of the panel switch under the Properties window of the VBE. Use the captions shown in Figure 10-13. The code for the revised control panel is as follows:

```
Dim MotorA As String
Private Sub iSwitchLedX1_OnChange()
If DigPnl.iSwitchLedX1.Active = True Then
DigPnl.iSwitchLedX1.Caption = "ON"
DigPnl.PBrickCtrl.InitComm
Else
DigPnl.iSwitchLedX1.Caption = "OFF"
DigPnl.PBrickCtrl.CloseComm
```

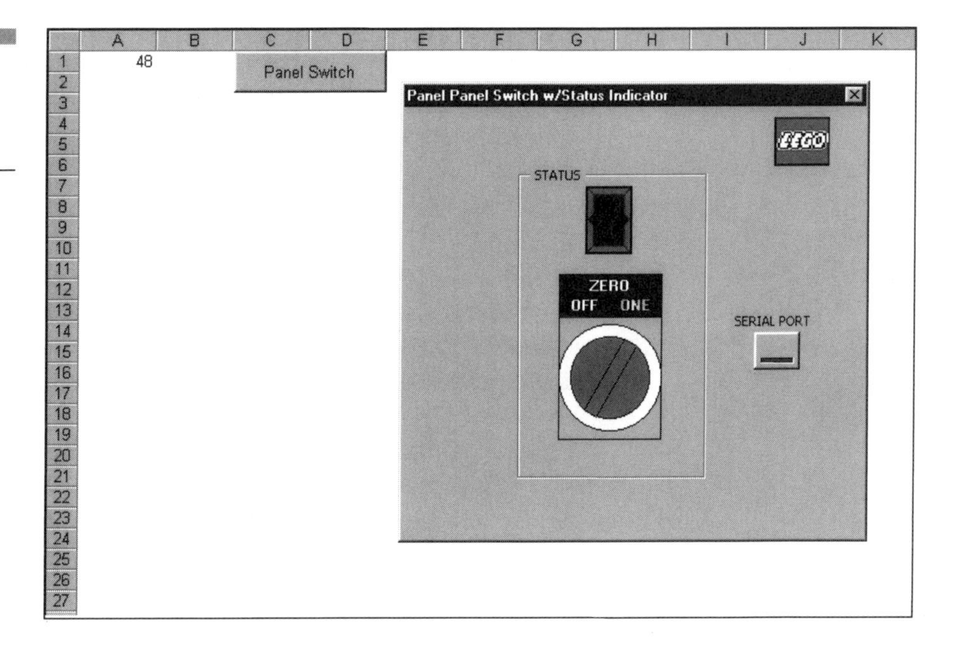

Figure 10-13 Modified GUI panel with rotary switch control

```
End If
End Sub

Private Sub iSwitchPanelX1_OnPositionChange()
If iSwitchPanelX1.Position = 2 Then
DigPnl.iSevenSegmentCharacterX1.Value = ssc1
MotorA = "0"
DigPnl.PBrickCtrl.On (MotorA)
DigPnl.PBrickCtrl.StartTask 0
Range("A1").Value = DigPnl.iSevenSegmentCharacterX1.Character
Else
DigPnl.iSevenSegmentCharacterX1.Value = ssc0
MotorA = "0"
DigPnl.PBrickCtrl.Off (MotorA)
DigPnl.PBrickCtrl.StopTask 0
Range("A1").Value = DigPnl.iSevenSegmentCharacterX1.Character
End If
End Sub
```

The code is almost identical to the second control panel design except that a different object is used. The ActiveX object used in this new panel design is iSwitchPanelX1.Position. The operating function for the control panel is the same as in the second design except the rotary switch is used instead of the rocker switch. The control panel is not cluttered with a lot of controls and the design requirements have been met with a simple solution.

Further Thoughts and Experimentation

VBA is a great programming language to use for quick prototypes used in POC studies. The control panels presented are not just the ordinary controls used with Windows applications. The Real-Time Controls are just a few of the many devices that can be used to build Mindstorms robot controllers. The idea behind designing and building real-time controls is to limit the amount of switches and buttons in the design layout phase. Embedding simple diagnostics tools using VBA is quite easy, as illustrated in the three control panel designs. In addition, the ability to link a virtual panel with equivalent hardware makes VBA a good front-end design tool.

A *Virtual Test Box* (VTB) that uses electronic circuit equations for calculating voltage and decibels is discussed and developed in the next chapter. A VTB can be built for robot control using math as the core control element. The next chapter illustrates how math is a powerful tool for Mindstorms robotic development.

CHAPTER 11

Virtual Test Box (VTB) Development: Math-Based Controllers for Mindstorms Robots

In Chapter 10, "Virtual Prototyping and Control Using ActiveX Controls," real-time ActiveX controls are used to demonstrate virtual prototyping techniques for building advanced control panels for controlling Mindstorms robots. Three types of real-time controls are explored through the construction of an improved master-slave control panel. The lever switch, the rocker switch with the *Light-Emitting Diode* (LED) indicator, and the panel switch are used along with simple switching control elements to operate the *Robot Command Explorer* (RCX) *programmable brick* (P-Brick) that is interfaced to the Basic Stamp. The RCX P-Brick is hardwired to the Basic Stamp with the help of the 7805 voltage regulator *Integrated Circuit* (IC) circuit. The panel switch is a single component with two switching control circuits. Thus, it is possible to create a dual function controller using this switch.

The technique of *multiplexing* has been introduced where dual or multiple functions can be performed using a single component such as a panel switch. The real-time control interface design involves the process of laying out a Mindstorms robot control panel with minimal controls. The selection of controls should be directly related to the target robot's task or behavior. This concept eliminates control complexity and clutter, enabling the target robot to be controlled effectively and easily.

This chapter is a continuation of the previous one and emphasizes the use of math as the primary engine for driving the control panel's physical link to the Mindstorms robot's behavior or task. By writing a small math-based module known as a *User-Defined Function* (UDF), a Mindstorms robot's task or behavior can be linked to a mathematical equation. The equation is animated through a calculator-based control panel called a *Virtual Test Box* (VTB). The VTB has the capability to operate a Mindstorms robot based on the numerical response of the embedded math equation. This chapter discusses the following topics:

- What is a UDF and how is it used with a VTB?
- Assigning variable names to multiple cells
- Building a Decibel Meter VTB
- Building a Frequency Cutoff Detection VTB

This chapter should really be called "Adventures in Mathematics with Mindstorms Robots" because the physical robot is controlled by an answer obtained from the embedded math equation being solved. It is really a *Proof Of Concept* (POC) study into developing a machine that is capable of making mathematics fun using Mindstorms robots. The robot can be any machine discussed in the previous chapters or a creation of one's own design. The main objective in this chapter is to learn how to do the following:

- To create low-cost virtual instruments using real-time ActiveX controls
- To create an analog processor
- To add multiplier elements to a UDF extension

The chapter starts with a discussion on UDFs and how they make setting up math equations in Excel easy.

What Is a UDF and How Is It Used with a VTB?

The VTB uses an *Application-Specific Math Equation* (ASME) that is embedded within its software structure. An ASME is a dedicated equation used to solve a specifically focused problem. In order to use the equation with the VTB, a subprocedure must be written. A subprocedure is a *Visual Basic for Applications* (VBA) module that has embedded an ASME that is known as a UDF. Using a UDF on an Excel spreadsheet enables developers to embed a commonly used equation anywhere within the thousands of cells inside the workbook. The key to creating a UDF is to assign a numeric value with a variable name. Once the assignment is made, that variable name can be used to build a function in a worksheet cell. The variable name can be used as input for building the UDF. The VTB links with the cell that has the embedded UDF. Whenever the VTB's input changes, the UDF cell's value changes as well. Figures 11-1a and 11-1b demonstrate how to give a variable name assignment to a single numeric value.

Assigning Variable Names to Multiple Cells

When dealing with math equations that have several variables, assigning names to multiple cells is quite easy. Select the cell range that contains the variable names to which the cells will be assigned. With the cell range highlighted, go to the main menu and click Insert, Name, and then Create. A dialog box appears on the screen asking you to Create Names in—Left Column. Click OK. Excel names the cells containing numeric values with the names highlighted. Figure 11-2 shows the assignment of multiple cells to several variable names.

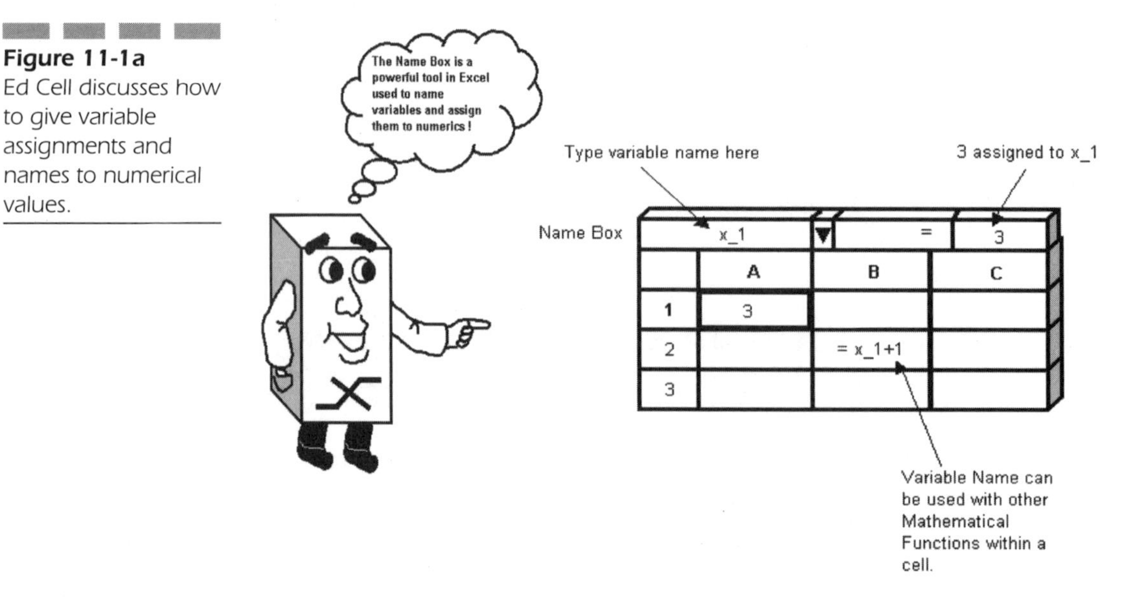

Figure 11-1a Ed Cell discusses how to give variable assignments and names to numerical values.

x_1 ▼	= 3	
Name Box	B	C
1	3		
2		4
3		

Figure 11-1b Using Name Box in Excel to make variable assignments to numerical values

	A	B	C	D	E	F	G
1	x_1	1					
2	x_2	2					
3	x_3	3					
4	x_4	4					

Create Names

Create names in
- [ ] Top row
- [x] Left column
- [ ] Bottom row
- [ ] Right column

OK Cancel

Figure 11-2 Creating names in the Excel Create Names dialog box used to assign a variable to numerical values in multiple cells

Multiple cells can also be assigned to a single variable name by first selecting the cells. Go to the Name Box, type the variable name, and hit the Enter key. The assignment of the variable name to the numeric value has been made. Figure 11-3 shows the assignment of multiple cells to a single variable name. This method is used when defining worksheet ranges as a *one-dimensional* (1D) array. Defining the cell range as a 1D array enables the array to pass through the argument list of the *User-Defined Name* (UDN). The math equation is set up using the array notation of variable (1), variable (2), and so on. To verify that the cell range has been assigned to the variable name, go to the main menu and click Insert, Name, and then Define. A dialog box appears on the screen.

Figure 11-4 shows the Define dialog box with controls for adding and deleting cells. Notice the Refers To section that shows the cell range. This location verifies the active cell range with the variable name displayed in the Combo Box. To clear the variable name shown in the Combo Box, click Delete. This resets the spreadsheet back to the standard *rows-columns* (R1C1) format.

The method of defining cell ranges with a variable name or creating a 1D array is very important when building a UDF. The UDF is solely dependent on this array because it provides input data that the subprocedure can use during the numerical processing phase. The UDF is the analog processor for the VTB because the ASME feeds data to the spreadsheet that has a direct link to the VBA instrumentation display movement. Another advantage to using UDNs is that the variable name and the equation can be written in textbook notation. This makes it easy to troubleshoot the equation during a debugging session of the VTB.

Figure 11-3 Defining names for multiple cell ranges

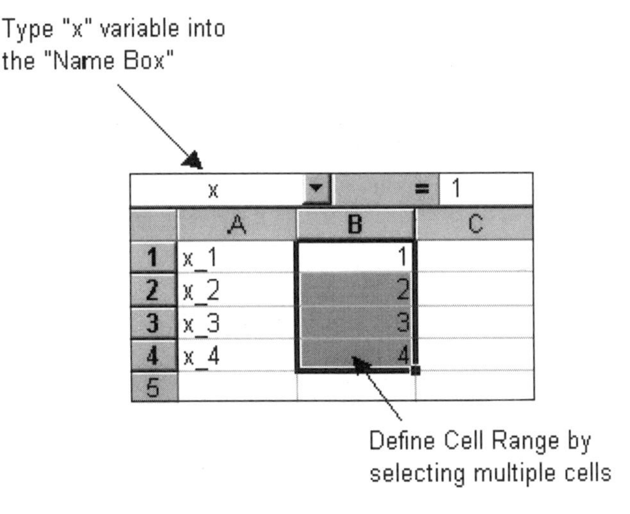

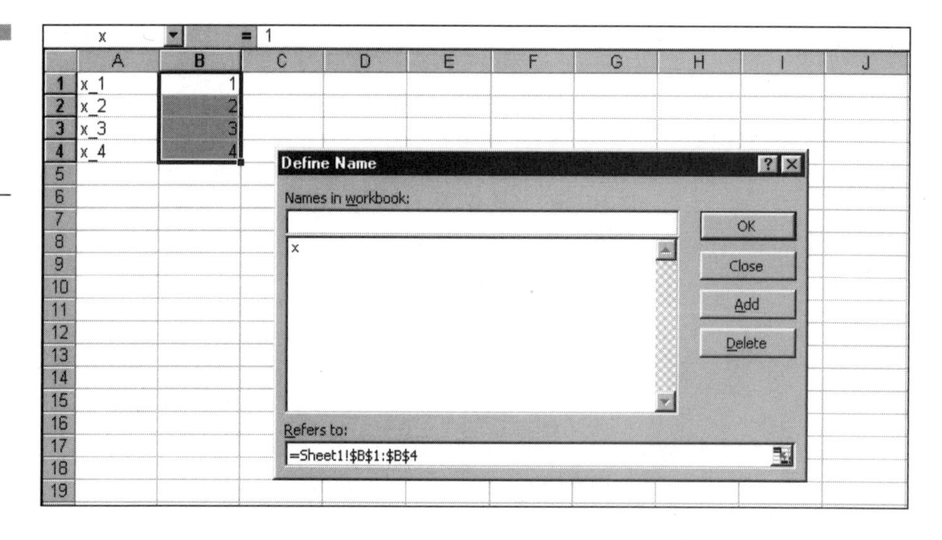

Figure 11-4 Defining the cell range using the Define Name dialog box

Building a Decibel Meter VTB

With a basic understanding of UDNs and UDFs, a VBA *graphical user interface* (GUI) panel can be built using real-time controls. The focus of this exercise is to build a GUI panel that has the capability to display the result of a math calculation using an analog gauge. The VTB consists of the VBA GUI panel with real-time controls placed on it that are linked to a cell-embedded UDF. The core equation in this project is used to calculate the decibel.

The decibel is generally a ratio of the output to the input of an audio amplifier. This ratio is generally known as the *gain* of the amplifier. The *decibel meter* is an instrument that measures the amplitude of the signal. The dBV represents the decibel-volt *root mean square* (rms), which is one of the most common references used to expressed the absolute measure of amplitude. The math equation used to calculate the dBV is as follows:

$$\text{'dBV} = 20 \times \log^{10} \times \text{Vrms}$$

Vrms refers to the voltage root mean square or a little higher than the average voltage of the signal. To calculate Vrms, use the following equation:

$$\text{Vrms} = \frac{Vppk}{2x\sqrt{2}}$$

V_{P-Pk} is the peak-to-peak voltage of the signal. An analog processor can be built with these two analysis equations. Figure 11-5 shows the building

Figure 11-5
Ed Cell explains the Analog Processor Block Diagram for calculating dBVrms.

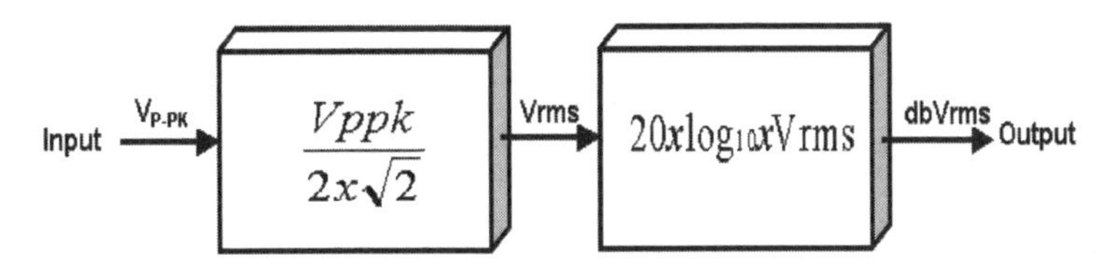

blocks for the dBVrms analog processor. A systems approach is used to build this analog processor. By defining the *Input/Output* (I/O) variables, the math engine can be easily built for test and application management. The overall process for building the Decibel Meter VTB consists of the following five steps:

1. Create the analog processor block diagram.
2. Design the spreadsheet.
3. Identify the design/analysis equations and define the UDFs.
4. Design the VTB.
5. Run and debug the software application.

Create the Analog Processor Block Diagram

This step mainly involves designing a high-level diagram that shows how each design/analysis equation is linked together to form the overall master equation that displays the calculated solution. As shown in Figure 11-5, a system block diagram is used to illustrate how the I/Os are connected with a processing block. The two equations are placed inside each block and the I/O data that pass from each math object are illustrated with labeled arrows.

Design the Spreadsheet

The spreadsheet should be designed with a systems appeal where cells are formatted with an input and output section. All columns should be labeled with reference designators, input values, and units. The rows use reference designators, the actual input data, and the embedded UDF. The key to laying out a spreadsheet for the analog I/O is to keep the layout as clean as possible. A lot of clutter can cause confusion when errors are debugged in the application. (See the section "Real-Time Control Interface Design Methods" in Chapter 10.) Build the spreadsheet for the decibel meter using Figure 11-6 as a layout guide.

Identify the Design/Analysis Equations and Define the UDFs

Following the analog processor block diagram, the equations shown in each of the processing blocks can be used to define the UDF. The method of building the UDF consists of writing a VBA subprocedure with a basic math equation embedded within the number-crunching block. Another name commonly used for this block is the *function module*. The arithmetic operations are kept simple by extracting basic math operations such as expo-

Figure 11-6 Layout for the decibel meter spreadsheet

	A	B	C	D	E	F	G	H
1						Decibel Meter		
2								
3	**REF**	**Value**	**Unit**					
4	V_{PPK}	2.72	volts	INPUT				
5	V_{RMS}	0.962445	volts					
6								
7								
8	**REF**	**Value**	**Unit**					
9	dB	-0.33248	decibels	OUTPUT				
10								
11								

nentiation, ×, /, +, and – from the main design/analysis equation. The function module or subprocedure is placed within an Excel cell that has the remainder of the equation for final calculation. This technique makes the math operation in the cell easier to manage and understand. To add a module to an Excel worksheet, go to the *Visual Basic Editor* (VBE), and click Insert and Module at the main menu. A blank text editor window within the VBE appears. The following code can be typed into the VB text editor:

```
Function Vrms(p)
Vrms = p(1) / 2.828
End Function
```

The Vrms equation shows a constant of 2.828 in the denominator. To maintain the design philosophy of keeping the function module math equations simple, 2.828 is the calculated value of $2\times\sqrt{2}$. Figure 11-7 shows the typed module in the VB text editor. To add the UDF onto the specified spreadsheet cell, go to the main menu and click Insert and Function. A dialog box appears on the screen. Next, scroll down until User Defined Function is found. Click it. The Vrms module is displayed in the adjacent window of the main dialog box. Figure 11-8 shows the dialog box with the Vrms and UDF modules. Paste the module into the two cells, as shown in Figure 11-9. Both cell locations (B5 and B9) change values automatically based on the decibel VTB control knob.

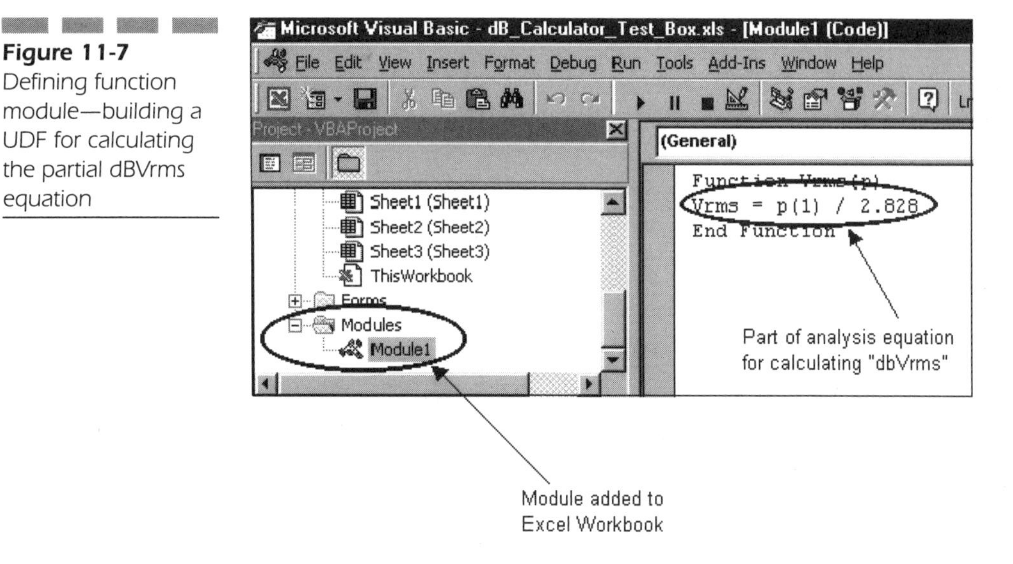

Figure 11-7 Defining function module—building a UDF for calculating the partial dBVrms equation

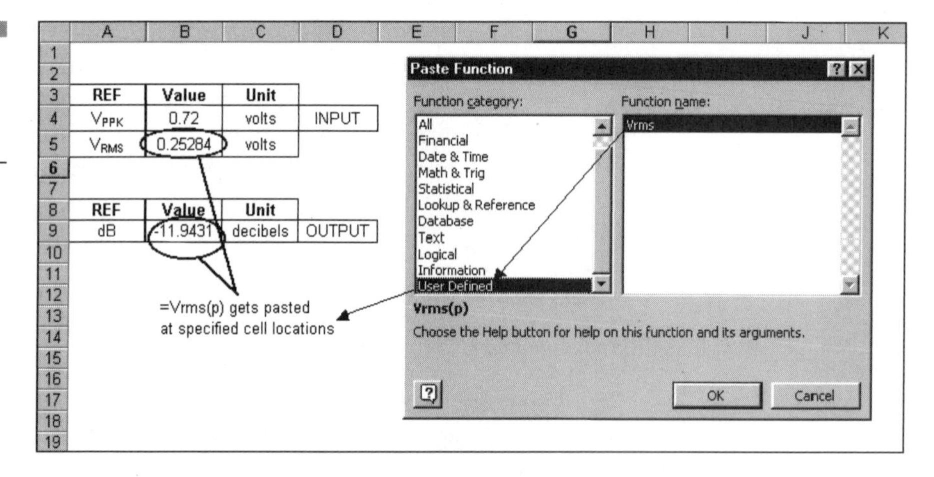

Figure 11-8 Placing UDF into the correct spreadsheet cells

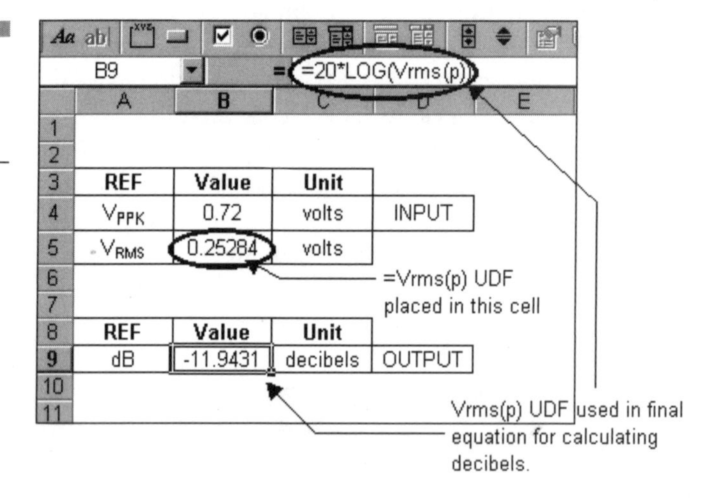

Figure 11-9 Using UDF in the final calculation for dBVrms analysis

Design the VTB

To build the VTB, the VBA panel needs to be designed with the appropriate controls. Build the panel as shown in Figure 11-10.

The following real-time controls are used to build the decibel meter:

- iKnobX is a *three-dimensional* (3D) knob that supports both mouse and keyboard control.
- iKnobX.Position specifies the current value represented by iKnobX. iKnobX.Position must be a value between PositionMin and PositionMax.

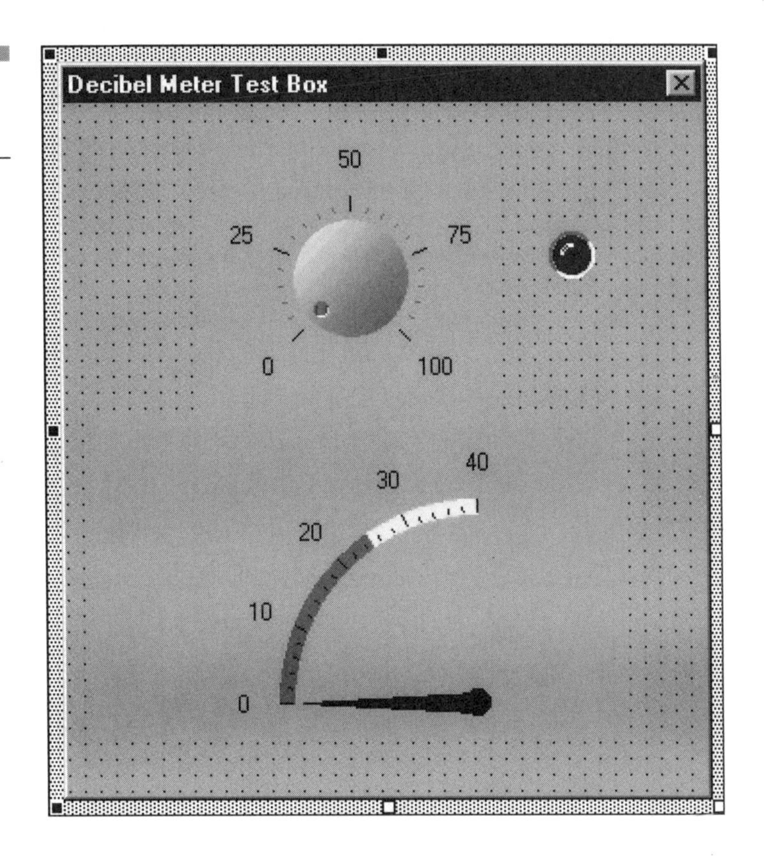

Figure 11-10 The final build of a Decibel Meter VTB

- iAngularGaugeX is a graphical linear gauge with an angular representation of the data. iAngularGaugeX.Position specifies the current value represented by iAngularGaugeX.
- iLedRoundX is an LED with a None, Raised, or Lowered bevel.
- iLedRoundX.Active indicates whether LED is active. When Active is True, the LED is illuminated with the active color. When Active is False, the LED is dimmed.

The main properties of the real-time controls used on the GUI panel are set as follows:

- iKnobX1.PositionMax=100
- iAngularGaugeX1.ArcRadius=90
- iAngularGaugeX1.ArcRangeDegrees=90
- iAngularGaugeX1.PointerStyle=3-iagpsTriangle

- iAngularGaugeX1.SectionColor1=&H00008080&
- iAngularGaugeX1.SectionColor2=&H0000FFFF&
- iAngularGaugeX1.SectionColor3=&H000000FF&
- iAngularGaugeX1.SectionColor4=&H00FF0000&
- iAngularGaugeX1.SectionColor5=&H00FFFF80&
- iAngularGaugeX1.SectionCount=4
- iAngularGaugeX1.SectionEnd1=25 / SectionEnd 2=50 / SectionEnd3=75 / SectionEnd 4 =100
- Transparent = True
- iLedRoundX1.ActiveColor=&H000000FF&
- iLedRoundX1.BevelStyle=2-ibsLowered
- iLedRoundX1.InActiveColor=&H00005500&

Add the two label ActiveX controls and set their properties as follows:

- Label1.Caption=dB
- Label2.Caption=Vppk

Although properties can be set using this information, it's always good to experiment with other object attributes and physical characteristics of the real-time controls used on the decibel meter VTB. Once the controls have been adjusted to the desired appearance, VBA code can be written to accommodate the embedded UDF and input data on the Excel spreadsheet. Using the VBE, type in the following code onto the text editor:

```
Private Sub iKnobX1_OnPositionChange()
Range("B4").Value = iKnobX1.Position
dB = Range("B9").Value
iAngularGaugeX1.Position = dB
If dB < 1 Then
iLedRoundX1.Active = True
Else
iLedRoundX1.Active = False
End If
End Sub
```

Run and Debug the Software Application

To check out the operation of the meter and its association with the spreadsheet, a Panel Display is required. Using the spreadsheet toolbox, place a CommandButton1 on the desirable location. Make sure the spreadsheet is in design mode to ensure proper placement and code development. After building the standard ActiveX control, add the following VBA code:

```
Private Sub CommandButton1_Click()
UserForm1.Show
End Sub
```

End the design-mode session by clicking the right-triangle icon on the Tools menu bar. Display the decibel meter VTB by clicking the Decibel Panel CommandButton. The VTB should be visible on the spreadsheet. Figure 11-11 shows the complete interactive analysis environment. By rotating the knob to the right (clockwise), all cells (B4, B5, and B9) should change their values. The analog gauge should be tracking the cell B9 data. If it is not, stop the applet and check the code for typing errors. To light up the LED, rotate the knob to the left (counterclockwise) until it approaches 0. Notice the negative values of cell B19. The LED is a visual warning indicator that alerts the operator that the applet is approaching an undefined solution. Rotate the knob slowly counterclockwise, making note of the B5 data. Continue rotating the knob until a #NUM is in the cell. Also, an error message is generated and placed on the dialog box. The error message "Type Mismatch" in this applet refers to a division by zero being undefined,

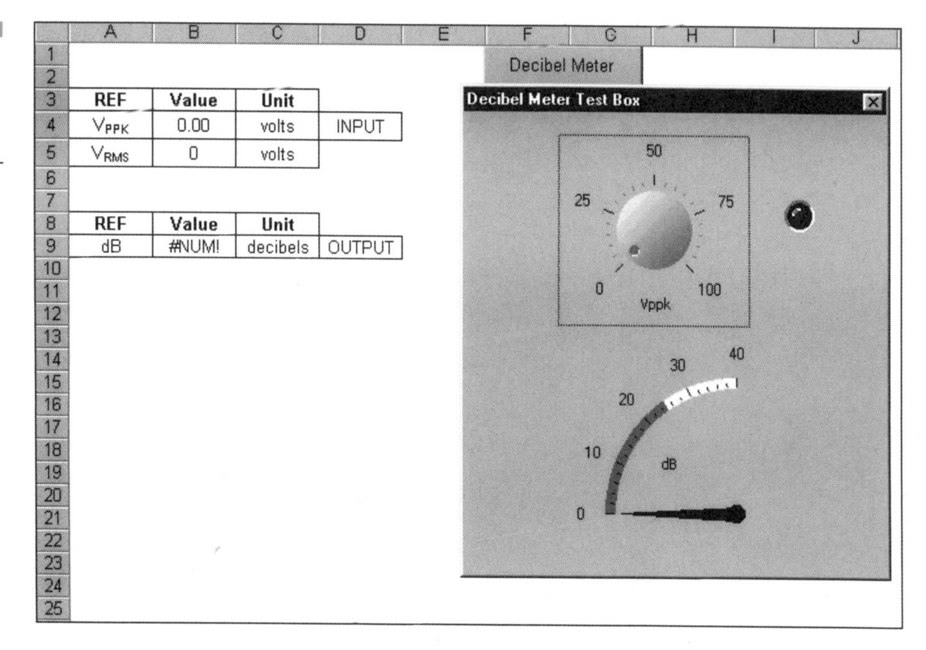

Figure 11-11 The interactive analysis environment on the spreadsheet

which constitutes this kind of fault detection. Figure 11-12 shows the "Type Mismatch" error.

Although the LED provides a visual indicator for an error, how can a robot be used to assist in this mathematical problem-detection scheme? A simple *Audio Robot* (A-Bot) robot can be built where the LED lights up when the VBA test box applet approaches zero. At this point, the A-Bot sounds an audible alarm. A-Bot is not a full-figure robot, but the RCX P-Brick provides a physical link to an electronic tone generator. A small LEGO tone generator can be used and placed on output B of the RCX P-Brick. If a tone generator is not available, you can create your own using the circuit described in the section "An Audible Tone Generator Lab Project Procedure" in Chapter 4, "Electronic Switching Circuits," or the Basic Stamp can be used as a sinewave tone generator.

A-Bot: A POC Development in Automated Front-End Designs

As stated earlier, A-Bot is not a robot in the sense that it has robotic features, but it is a POC development in automated front-end designs. A-Bot is a gateway between the decibel meter VTB and the sinewave tone generator. The RCX P-Brick is an automated front end for the Basic Stamp because it

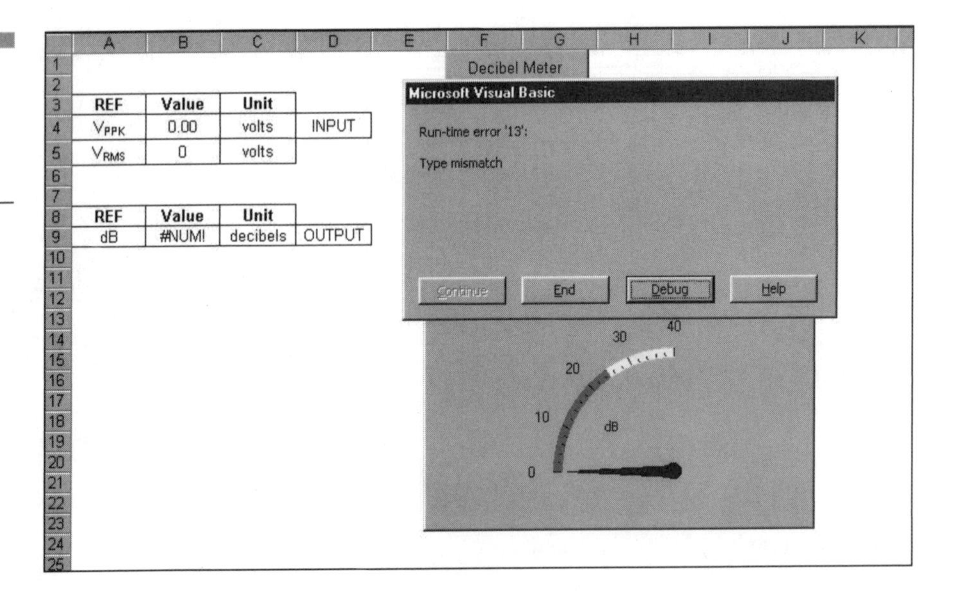

Figure 11-12 The dreaded "Type Mismatch" message caused by dividing by 0

has a link to the VBA GUI panel. The embedded firmware gives the RCX the capability to interpret and respond to the infrared commands sent by the VBA GUI. The automated process of interpreting and responding creates a seamless interface between the VBA GUI, RCX P-Brick, and Basic Stamp. The output of the RCX P-Brick is hardwired to the input of the Basic Stamp through a 5-volt voltage regulator circuit.

The sinewave tone generator only sounds off when the decibel meter's reading is less than 1. The Basic Stamp turns on the sinewave tone generator only after the knob on the meter is rotated and a relational condition of less than 1 is reached. Therefore, the RCX P-Brick provides a front end for the Basic Stamp, enabling it to be controlled indirectly with the VBA GUI. This technique extends the capability of both the RCX P-Brick and the Basic Stamp, enabling different software programming languages to have a common interfacing link to control a Mindstorms robot.

Building the A-Bot Automated Front-End Controller for the Sinewave Tone Generator

Building the A-Bot automated front-end controller for the sinewave tone generator is quite easy. The core building blocks are available from the master/smart slave controller that was discussed in Chapter 10. The only change to the hardware is that the transistor relay driver is swapped with a speaker and electrolytic capacitor. The DIGITAL _SW.BS2 software is modified to accommodate the speaker and electrolytic capacitor. Figure 11-13 shows the circuit diagram for the A-Bot automated front-end controller with the sinewave tone generator. The Basic Stamp uses the same I/O pins as in the master-smart controller project discussed in Chapter 10. As shown in Figure 11-13, output A of the RCX P-Brick is wired to the input circuit of the 7805 voltage regulator circuit.

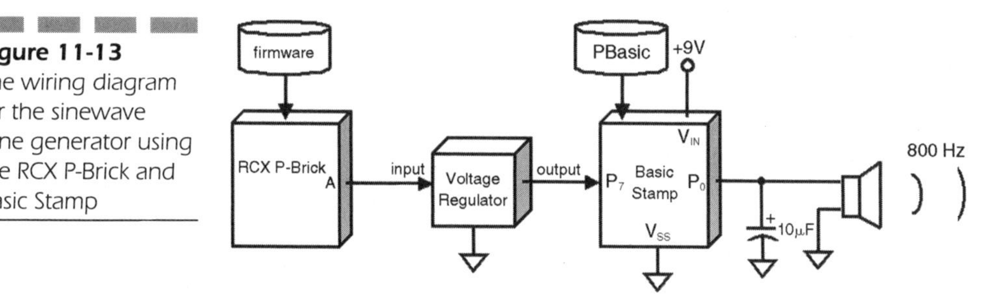

Figure 11-13 The wiring diagram for the sinewave tone generator using the RCX P-Brick and Basic Stamp

The 7805 voltage regulator circuit is fully drawn out in Chapter 10, Figure 10-9. The output circuit of the 7805 voltage regulator is hardwired to pin P7 of the Basic Stamp module. The output pin P0 of the Basic Stamp is wired to a speaker with a 10 µF electrolytic capacitor across its wires. The capacitor filters out any high-frequency noise associated with the speaker coil inductance and the 800 Hz signal generated by the Basic Stamp. The 800 Hz produced by the Basic Stamp is shown in Figure 11-14. After the hardware is completed, the software must be written for both the RCX P-Brick and Basic Stamp.

RCX P-Brick Software

As shown earlier, the decibel meter VTB software is the core element for A-Bot's automated front-end design. The firmware inside of the RCX P-Brick provides key functional elements like understanding direct commands from the Infrared Tower. The Spirit.ocx is the core software library for the RCX P-Brick relating to VBA *Object-Oriented Programming* (OOP) and has the capability to communicate with the Infrared Tower and its physical I/O layer. If the decibel meter GUI panel with output A is modified,

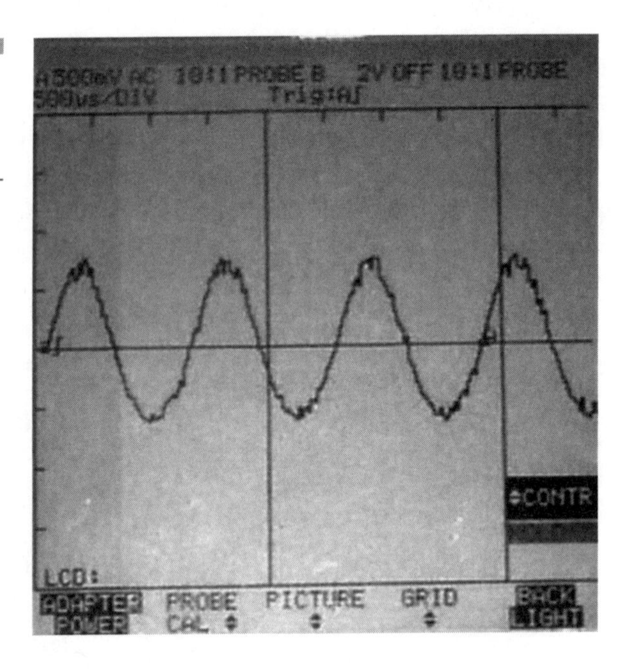

Figure 11-14 The 800 Hz signal produced by Basic Stamp 2

controlling the interface with the Basic Stamp becomes quite easy. The following is the modified code for the decibel meter VTB:

```
Dim MotorA As String
Private Sub iKnobX1_OnPositionChange()
Range("B4").Value = dBMeter.iKnobX1.Position
dB = Range("B9").Value
dBMeter.iAngularGaugeX1.Position = dB
If dB < 1 Then
dBMeter.iLedRoundX1.Active = True
MotorA = "0"
dBMeter.PBrickCtrl.Off (MotorA)
dBMeter.PBrickCtrl.On (MotorA)
dBMeter.PBrickCtrl.Off (MotorA)
Else
dBMeter.iLedRoundX1.Active = False
MotorA = "0"
dBMeter.PBrickCtrl.Off (MotorA)
End If
End Sub

Private Sub iSwitchLedX1_OnChange()
If dBMeter.iSwitchLedX1.Active = True Then
dBMeter.iSwitchLedX1.Caption = "ON"
dBMeter.PBrickCtrl.InitComm
Else
dBMeter.iSwitchLedX1.Caption = "OFF"
dBMeter.PBrickCtrl.CloseComm
End If
End Sub
```

The modification to the code consists of adding a serial port control using the iSwitchLedX1 real-time control. The VBA object PBrickCtrl.InitComm is used to initialize the serial port for outside communication. The opposite is true of PBrickCtrl.CloseComm, which, as the name states, closes the serial port from outside communication. To display the switch input request event, iSwitchLedX1.Caption is used to show the ON and OFF states of the virtual electromechanical component. The dBMeter.PBrickCtrl.Off (MotorA) Parent Object/Child Object/Method is used to turn off output A, which is referenced by (MotorA). To simulate an Active Low switch input to the Basic Stamp, create the trick of toggling between on and off using the following sequence of Method programming instructions:

```
dBMeter.PBrickCtrl.Off (MotorA)
dBMeter.PBrickCtrl.On (MotorA)
dBMeter.PBrickCtrl.Off (MotorA
```

If this sequence of switching the P_7 input pin of the Basic Stamp is not implemented, the decibel meter does not provide the correct signal for the Basic Stamp to be synchronized with the lighting of the LED with the correct relational condition. The change can be implemented by modifying the *Parallax Basic* (P-Basic) code as well as rewiring the switch to accommodate an Active High switching input. The following is an example where software is a better solution than hardware because of the additional steps required to make one change in hardware functionality. Sometimes software changes can make hardware modifications easy by flipping a bit.

Once the software is added to the GUI, a simple test can be carried out to test the virtual panel with the physical hardware. This is done by attaching a *Direct Current* (DC) Voltmeter or *Digital Multimeter* (DMM) to measure the voltage at the output of the 5-volt voltage regulator. Take the positive test lead of the DC voltmeter or DMM and attach it to pin 2 of the 7805 voltage regulator. Take the negative test lead of the measurement instrument and connect it to ground. Turn on the RCX P-Brick and run the decibel meter applet. Rotate the knob slowly to the right to initiate an alarm or red LED alert. The output A short pulse measuring approximately 500 mV will quickly be displayed on the DMM. This short pulse is accomplished with the three lines of code toggling between off, on, and off. The P_7 input pin of the Basic Stamp needs a momentary signal to get the proper output turn-on time for activating the tone generator to sound off. If a 500 mV output signal was not measured using a DMM or DC voltmeter, recheck the wiring and try testing the virtual panel with RCX P-Brick again. If the GUI software is running correctly and is properly interfaced to the RCX P-Brick, the Basic Stamp software can be written and tested.

The Sinewave Tone Generator Software for the Basic Stamp

A tone generator can be built using the Basic Stamp with minimal parts. As shown in Figure 11-13, a speaker, an electrolytic capacitor, and software are required to make a sinewave tone generator with the Basic Stamp. The key to generating sound from the Basic Stamp 2 is the FREQOUT instruction. The P-Basic programming style syntax is as follows:

```
FREQOUT, pin, duration, freq1, freq2
```

Pin is a variable or constant (0 through 15) that specifies the I/O pin to use. This command puts the specific pin into output mode during the generation of the tones and remains in that state after the instruction finishes. The duration parameter is also a variable or constant that specifies how long in milliseconds (1 to 65,535) the output tones remain on. The freq1 and freq2 are variables or constants that specify the tone's output frequency in hertz (0 to 32,767). The FREQOUT programmed for the Basic Stamp in the A-Bot application is as follows:

```
FREQOUT, 7, 1000, 800
```

where 7 is pin P_7, 1000 is the 1,000 milliseconds or one second, and 800 is 800 Hz. As shown in the Figure 11-14, a sinewave is produced using this P-Basic instruction. The 800 Hz has enough output gain without typing it to an audio amplifier. The control software used in this applet is a modification of the DIGITAL_SW_BS2 code. The P-Basic code for tone generation is shown in the following code listing:

```
Loop:
   BUTTON SWpin,ALow,DlyVal,SWrate,btnWk,SPKRpin,noPress
   DEBUG "*"
   FREQOUT SPKRpin,Durate,FREQ
noPress: GOTO Loop
```

The code has the same operating characteristics as the DIGITAL_SW.BS2 program except that a relay on the internal sound generator can be called to operate using the FREQOUT P-Basic instruction. The complete code is shown in Figure 11-15. To save time when writing code, open the DIGITAL_SW.BS2 file. Type in the new changes using Figure 11-15 as a programming guide. Then save the changes as TONE_GEN.BS2. Download

Figure 11-15 P-Basic program listing for the Basic Stamp tone generator

```
'==========================================================================================
' File: TONE_GEN.BS2
'
' Provides a 800Hz Sinewave Tone. Speaker is connected to P0.
' A 10uF capacitor is wired across the speaker leads to reduce high frequency noise.
' =========================================================================================

 SPKRpin    CON        0              ' Speaker connected to Pin 0

 SWpin      CON        7              ' Switch connected to Pin 7

 DlyVal     CON        255            ' Delay Value

 ALow       CON        0              ' Active Low switch state

 SWrate     CON        255            ' Switch cycles between autorepeats

 Durate     CON        1000           ' Output On in milliseconds

 FREQ       CON        800            ' Frequency in Hertz

 btnWK      var        byte           ' Workspace for BUTTON instruction

 btnWK=0                              ' Clear the workspace variable

'------------------------------------------------------------------------------------------

Loop:
   BUTTON SWpin,ALow,DlyVal,SWrate,btnWk,SPKRpin,noPress     ' Goto to noPress
   DEBUG "*"                                                 ' Print asterisks on the Debug window
   FREQOUT SPKRpin,Durate,FREQ                               ' Emit 800Hz tone through pin 0 for 1sec
noPress: GOTO Loop                                           ' Repeat endlessely.
```

the new code into the Basic Stamp. Next, add an Active Low switch to input pin P_7 of the Basic Stamp. Push and release the button, and an 800 Hz tone should come from the speaker. If a tone is not heard, turn off the 9-volt power supply and check your wiring. If the wiring is okay, check the software listing for any incorrect constant values. Download and run the code again. If an 800 Hz tone comes from the speaker, then the RCX P-Brick's output A can be wired to the P_7 input of the Basic Stamp. Make one final review of the wiring before applying power to the Basic Stamp. Turn on the RCX P-Brick and run the decibel meter applet. Rotate the knob to the right and slowly adjust it to the left approaching zero. Upon reaching a positive number that is less than 1, the LED on the GUI should turn red and the 800 Hz tone should sound off. The tone lasts for about 1 second and then shuts off. Each time the knob's reading is less than 1, the tone will be on; each time the knob's position is greater than 1, the tone will be off.

Reflections and Thoughts about the Decibel Meter Project

The unique aspect of this project is that minimal parts are used to build a tone generator. A-Bot is truly a wonderful project that introduces hobbyists and experimentalists to a world of math fun and robotics development. The project shows how a math-based controller can be used to manipulate the RCX P-Brick's firmware to control a programmable device for sound generation. The final project in this chapter looks at creating another math-based robot using a low-pass filter circuit with a design/analysis equation for calculating the cutoff frequency of the passive network. The same design method for building A-Bot is used to develop the low-pass controller applet with the exception of the GUI interface. The real-time controls give the panel design more options for Mindstorms robot manipulation. Also, the type of control for operating the robot has been expanded to the point of really developing a good design with the aid of the system block diagram. By defining the type and number of inputs, a prototype control panel using VBA and real-time controls provide the opportunity to create unique and fascinating Mindstorms robots. The key to using the math-based controller technique is to define design requirements for the robotic system up front before building the machine. By completing this task first, a set of design documents are created that enable Mindstorms robots to be developed with no-cost tools that can be customized in a matter of minutes.

Building a Frequency Cutoff Detection VTB

This project enables you to build a robot that is controlled by a frequency cutoff value. Filters are passive circuits made up of resistors, capacitors, and inductors in special topologies or geometries that can pass certain frequencies while rejecting other signals. The common filter circuit topologies used in electronics are low-pass, high-pass, band-pass, and notch networks. The low-pass filter is the core filter circuit that is used to build the frequency cutoff detection VTB.

Low-Pass Filters

Two common topologies are used to build a low-pass filter: *Resistor-Capacitor* (RC) and *Resistor-Inductor* (RL) networks. Figure 11-16 shows the two circuit topologies. The low-pass filter circuit works as a frequency-sensitive voltage divider network. The reactive components (L and C) create the frequency sensitivity for the network. When a high-frequency signal is applied to the input of the RC filter, the capacitor's impedance (the equivalent *Alternating Current* [AC] resistance) decreases. The reduction in impedance causes input current to leak through the capacitor to ground. The output voltage approaches the input voltage when this input current leakage occurs. Therefore, the high-frequency signal is *attenuated*. Attenuation is a measure of how much of the input voltage is passed to the output. At low frequencies, the capacitor's impedance is large, reducing the amount of current that flows through it. The output voltage drop is small

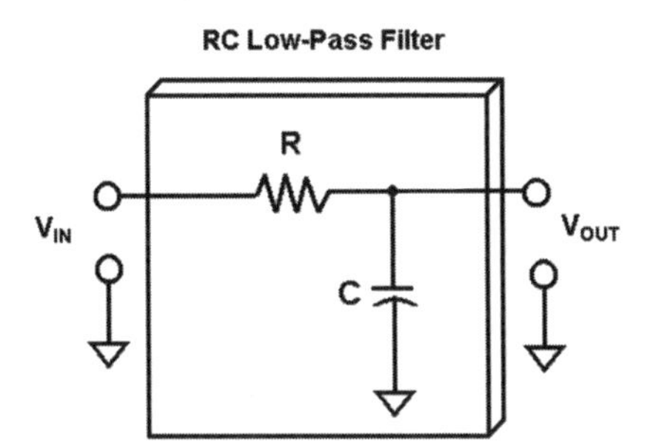

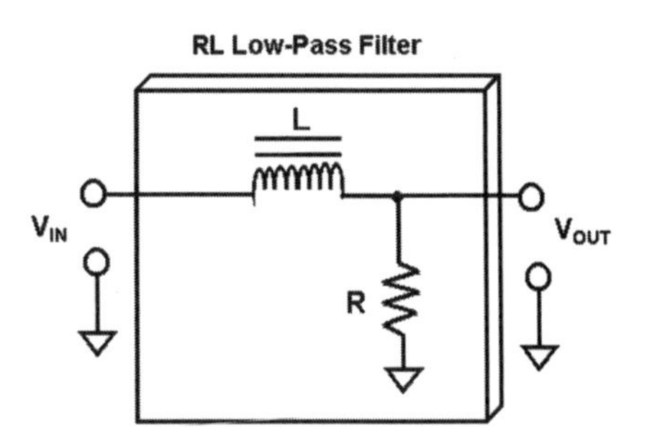

Figure 11-16 Common circuit topologies for low-pass filter

($V_{OUT} = I_T \times Z_C$). The low-frequency signals are greatly attenuated because the ratio of V_{OUT} / V_{IN} is less than 0.707.

In an RL network, the impedance of the inductor is large when a high-frequency signal is applied to the input circuit. A small amount of current can flow due to the large impedance in the circuit. Therefore, the output voltage is also small ($V_{OUT} = I_T \times R$). The RL low-pass circuit attenuates high-frequency signals and enables the input voltage to pass through the output at low frequencies. Figure 11-17 shows the low-pass filter attenua-

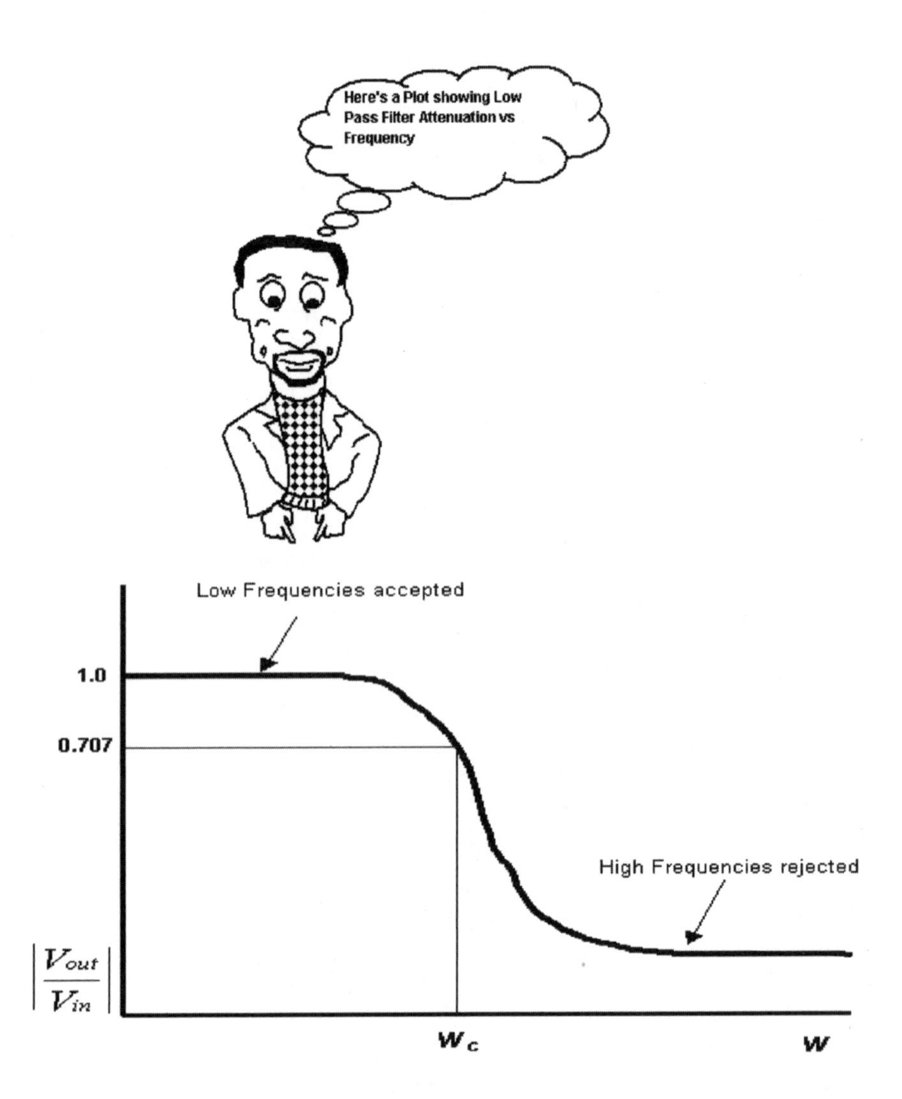

Figure 11-17 Mr. Don shows the attenuation versus the frequency plot.

tion versus the frequency plot. When the frequency of the input signal of an RC or RL circuit reaches its cutoff frequency (f_c), the output voltage drops to 0.5 volts. To calculate the cutoff frequency of an RC low-pass filter, the following equation is used:

$$fc = \frac{1}{2\pi RC}$$

where

f_c is the cutoff frequency in hertz.

R is the resistance in ohms.

C is the capacitance in microfarads.

The angular form for the cutoff frequency is as follows:

$$\omega c = \frac{1}{RC}$$

where w_c is the cutoff frequency in radians.

To find the attenuation of the low-pass circuit, the following analysis equation is used:

$$ABS\left(\frac{Vout}{Vin}\right) = \frac{1}{\sqrt{1 + (\omega cRC)2}}$$

The angular cutoff frequency and attenuation analysis equations are used to build the frequency cutoff detection VTB.

This project is musing with two frequencies: a calculated (cutoff frequency) frequency and a frequency selected with a slider real-time control. If the cutoff frequency is greater than or equal to the dialed input frequency, an LED turns on and an 800 Hz tone sounds off for 5 seconds. The five-step process used to build the decibel meter VTB is used to build the frequency cutoff detection VTB. The five steps are presented here as a review on the process of building a VTB:

1. Create the analog processor block diagram.
2. Design the spreadsheet.
3. Identify the design/analysis equations and define the UDFs.
4. Design the VTB.
5. Run and debug the software application.

Create the Analog Processor Block Diagram

The analog processor block for this project is quite simple. Figure 11-18 shows for a given value of R, the corresponding f_c can be calculated. The UDF can be built using the cutoff frequency. This value can be compared with a fixed frequency number. If the cutoff frequency is greater than the input frequency, an LED lights. The tone generator used in the previous project is used with this VTB to provide an audible indication that both signals are out of range.

Design the Spreadsheet

The spreadsheet layout is not much different than what was used in the decibel meter. The layout consists of a few dedicated cells for capturing a

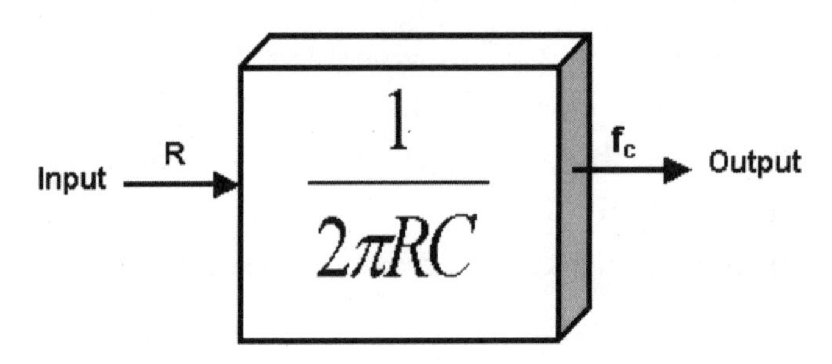

Figure 11-18 Ed Cell shows the cutoff frequency Analog Processor diagram.

resistor value (B4) that is input to the cutoff frequency analog processor block diagram. There is a dedicated cell (B5) for locating the cutoff frequency UDF. Cell B9 is used to capture the input frequency obtained from the slider real-time control labeled Freq-In. The other cells are used as headers for labeling calculation parameters and identifying the INPUT and OUTPUT sections for data entry and calculations. Figure 11-19 shows the layout of the spreadsheet. A CommandButton is added to the spreadsheet to display the GUI control panel. The software is very easy to understand and program. The following is CommandButton code that displays the decibel meter VTB:

```
Private Sub CommandButton1_Click()
UserForm1.Show
End Sub
```

Right-click the CommandButton and select View Code. Enter the three lines of code into the VBE. Change any properties that are necessary to achieve proper operation as well as experiment with new items.

Identify the Design/Analysis Equations and Define the UDFs

The design/analysis equation for the analog processor is quite simple. The cutoff frequency is found with the following equation:

$$fc = \frac{1}{2\pi RC}$$

The UDF is placed in cell B5 of the Excel spreadsheet. Input data R is used to calculate the cutoff frequency with a fixed capacitor value of 0.1 μF.

Figure 11-19 The spreadsheet layout for the frequency cutoff detection applet

	A	B	C	D	E	F	G	H
1						Frequency Generator		
2								
3	REF	Value	Unit					
4	R	2	ohms	INPUT				
5	f_c	605351	hertz	OUTPUT				
6								
7								
8	REF	Value	Unit					
9	f_{in}	50806	hertz					
10								
11								

To provide automation with data entry, a slider real-time control labeled *R* on the GUI panel is used. This ActiveX control is linked with cell B4. By moving the control up or down, the frequency changes based on the UDF located in cell B5.

Design the VTB

The frequency cutoff detection VTB is built using two slider controls, an LED, a switch with LED indicator, and three label controls. To use the slider controls in this applet, the following VBA programming style syntax is used:

```
'iSliderX.Position
```

iSliderX is a real analog output real-time control component. A mouse and/or keyboard can control the movement of the ActiveX object.

Position is a property used to get or change the value of the control. Position must be a value between PositionMin and PositionMax. By setting the frequency slider control to a known value and adjusting the resistor control labeled *R*, a calculated frequency assisted by the UDF is located at cell B5. If the calculated frequency is greater than the slider frequency, the LED turns on. Use Figure 11-20 to assist in laying out the controls.

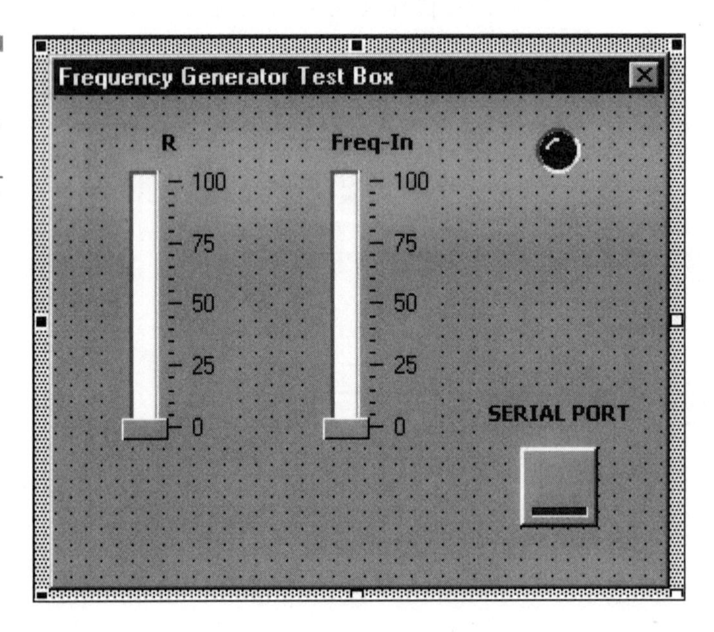

Figure 11-20 Final build of the Frequency Generator Test Box

With the controls placed on the GUI panel, software can now be added to the VTB. The code's structure is similar to the decibel meter virtual test bench. The code activates the GUI panel:

```
Private Sub iSliderX1_OnPositionChangeUser()
Range("B4").Value = UserForm1.iSliderX1.Position
InBound = Range("B5").Value
If InBound >= Range("B9").Value Then
UserForm1.iLedRoundX1.Active = True
Else
UserForm1.iLedRoundX1.Active = False
End If
End Sub

Private Sub iSliderX2_OnPositionChangeUser()
Range("B9").Value = UserForm1.iSliderX2.Position * 1000
End Sub
```

Open the VBE and type the code into the text editor. After the code has been entered, save the file and exit to the Excel spreadsheet. Click the Frequency Generator button and the Frequency Generator test box should appear on the spreadsheet. Next, set Frequency in Slider Control to a value of 50,806 hertz and adjust the *R* by sliding the control pointer upward. Continue moving the slider's pointer upward until the LED turns on. Read the value of cutoff frequency from the spreadsheet. The value should be greater than the input frequency. If not, check the code listing and make any necessary changes. Run the program again. The cutoff frequency should be greater than the input frequency.

Next, modify the code to add a software interface for controlling a Mindstorms robot:

```
Dim MotorA As String
Private Sub iSliderX1_OnPositionChangeUser()
Range("B4").Value = UserForm1.iSliderX1.Position
InBound = Range("B5").Value
If InBound >= Range("B9").Value Then
UserForm1.iLedRoundX1.Active = True
MotorA = "0"
UserForm1.PBrickCtrl.On (MotorA)
Else
UserForm1.iLedRoundX1.Active = False
MotorA = "0"
UserForm1.PBrickCtrl.Off (MotorA)
End If
End Sub

Private Sub iSliderX2_OnPositionChangeUser()
Range("B9").Value = UserForm1.iSliderX2.Position * 1000
End Sub

Private Sub iSwitchLedX1_OnChange()
```

```
If UserForm1.iSwitchLedX1.Active = True Then
UserForm1.iSwitchLedX1.Caption = "ON"
UserForm1.PBrickCtrl.InitComm
Else
UserForm1.iSwitchLedX1.Caption = "OFF"
UserForm1.PBrickCtrl.CloseComm
End If
End Sub
```

The UserForm1.PBrickCtrl.Off (MotorA) ParentObject/Child Object/Method is used to turn off output A referenced with (MotorA). UserForm1.PbrickCtrl.On (MotorA) is used to turn output A on. Although output A is used for switching control, the other two outputs can be used as well. The program is structured like the decibel meter test box. The modifications to the code consist of adding a serial port control using the iSwitchLedX1 real-time control. The VBA object PBrickCtrl.InitComm is used to initialize the serial port for outside communication. The opposite is true of PBrickCtrl.CloseComm. As the name states, it closes the serial port from outside communication. To display the switch input request event, iSwitchLedX1.Caption is used to show ON and OFF states of the virtual electromechanical component. A-Bot could be used to validate the additional code changes or a robot of one's choice or design. The fundamental operating principle behind the frequency cutoff detection VTB is its capability to turn on an over-the-limit warning based on a calculated frequency versus a control input signal.

Reflections and Thoughts about the Frequency Cutoff Detection Project

The frequency cutoff detection VTB is a technical musing project that focuses on a simple electronics equation used to analyze low-pass filters. The Analog Processor Block Diagram is a concept that is rooted in systems engineering. The I/O format used to build the spreadsheet and UDFs enables unique controls to be developed for manipulating a Mindstorms robot. An educational concept that can be developed using math-based controllers is the construction of an *animatronics* toy. By having students design and build robots that respond to certain answers or detect limits (such as the frequency cutoff project), their knowledge of solving and/or manipulating equations can be enhanced with this type of instruction.

An artistic aspect of building GUIs is another educational concept that can be explored. An image can be drawn and imported to a button or a UserForm. If imported onto a UserForm, the image can be placed where the controls become part of the image—for example, a cartoon character holding a lever switch in its hand. LEDs can be used as eyes to construct an image of robot with a physical link to real LEDs (see Chapter 10). The possibilities of math-based controllers using VTBs for Mindstorms robots are only limited by one's imagination.

The final chapter of this book discusses advanced techniques in GUI design and programming tips. This chapter explains how VBA prototypes can be used in POC studies for robotic inventions.

CHAPTER 12

VBA Prototypes: Developing Mindstorms Tools with Advanced Programming Techniques

Visual Basic for Applications (VBA) is a great software development tool for building Mindstorms robot control panels. As discussed in Chapter 2, "Developing GUIs: Software Control Basics," Chapter 10, "Virtual Prototyping and Control Using ActiveX Controls," and Chapter 11, "Virtual Test Box (VTB) Development: Math-Based Controllers for Mindstorms Robots," ActiveX technology makes designing and building *Graphical User Interfaces* (GUIs) easy and fun. The methods discussed in those three chapters can help you understand the software development activity that is important in robotics *Proof Of Concept* (POC) engineering studies.

This chapter dives deep into the advanced programming techniques that can be used to build VBA prototype control panels. Various programming models associated with using ActiveX controls in a VBA environment are explored. Also, VBA enables you to develop customized tools for testing Mindstorms robots. Tools development is a fascinating industry because of the software interfaces that enable electronics hardware to be tested using GUI-based virtual instruments. Sometimes an off-the-shelf tool may not have the desired interfaces to develop a particular task or behavior for the target Mindstorms robot. By building a customized VBA-based tool, hobbyists and experimentalists can have the testing feature. This is the roll-your-own approach to prototype development.

Your imagination can run free in this chapter because you have already encountered the material discussed in the previous chapters and have learned how it can be applied to VBA prototyping and Mindstorms robotic interfacing. This chapter takes more of a do-it-yourself approach in that basic concepts are presented through VBA prototyping. Basic block diagrams with code snippets are provided to enable you to develop software control and test boxes for Mindstorms robots. This chapter is really designed for hobbyists and experimentalists who want to develop their own robots with minimal instruction. The previous chapters in this book are an important resource for gaining information on developing the advanced POCs in programming and building the sophisticated software controls discussed in this chapter. The circuit blocks used in this chapter can be turned into actual interface circuits using the information presented in Chapter 3, "Electromechanical Controls Interfacing," and Chapter 4, "Electronic Switching Circuits." Therefore, the following VBA prototyping topics are discussed in this chapter:

- Mindstorms tools development using real-time controls
- Using arrays in Mindstorms robot applications
- Debug functions for VBA-assisted troubleshooting of Mindstorms robots

- Inserting pictures using the image ActiveX control onto the VBA prototype panel
- Technical tools musing projects

Mindstorms Tools Development Using Real-Time Controls

As explained in Chapter 11, real-time controls use the design interface process of laying out a Mindstorms robot control panel with minimal controls. The selection of controls should directly relate to the target robot's task or behavior. This concept eliminates control complexity and clutter, enabling the target robot to be controlled effectively and easily. Software tools development is the art of creating virtual instruments for the purpose of testing embedded code using the *Robot Command Explorer* (RCX) *programmable brick* (P-Brick) as the target unit to validate the software. Software validation is the testing process of checking if the *Device Under Test* (DUT) is okay using a test procedure document.

Traditional tools used in microcontroller software development include *In-Circuit Emulators* (ICEs), logic analyzers, and diagnostics readout boxes. An ICE is a box that contains hardware and software that acts like the real microcontroller under development. A logic analyzer is an electronic instrument that is capable of reading logic states, transitions, digital levels, and clock signals on a *Cathode-Ray Tube* (CRT) computer-based instrument. A diagnostics readout box is a software tool used to check out an *Electronic Control Unit's* (ECU's) *Input/Output* (I/O) using text-based and diagnostic messages. The LEGO Mindstorms *Software Development Kit* (SDK) has all of the commands for the RCX and its internal memory and I/O physical structure. For example, to create a tool that checks out an area of the RCX's I/O structure, a simple applet can be written using VBA. The tool can be built where a command is sent to the target RCX P-Brick turning on one or all of its outputs or reading a sensor connected to its input. Basically, a diagnostics readout box tool can be built to check the physical I/O of the target robot. A simple diagnostics tool was built in Chapter 2 to test V-Bot's hardware interface using a software-based control panel. The robot's internal driver output circuit and the voltage regulator were validated using the control panel. An improvement of the tool would be to provide a feedback control line to an available input, where the RCX P-Brick's output could be used as diagnostics sense or control line for one of its inputs. The VBA code could then read the detected input signal and display the value on the spreadsheet. Figure 12-1 shows the Functional Block Diagram for a Diagnostics Control Readout Box for the RCX P-Brick.

To make the design and development process of the diagnostics control readout box real, a VBA software tool applet will be built and validated for Robo-Fly. Robo-Fly is a hybrid design between a non-LEGO Mindstorms kit and the RCX P-Brick. Additional information about Robo-Fly is provided on page 373 in this chapter. When completed, the tool should be able to start and stop Robo-Fly's beating wings, and perform the basic diagnostics verifying that the robotic fly is flapping its wings. The diagnostics control readout box is a VBA applet where the transistor relay driver's *Normally Opened* (NO) contacts and the series 1-kilo-ohm resistor could provide a control signal into input 1 upon receiving a command from the control panel. The RCX P-Brick turns output A on, providing an input voltage around 7.4 volts *Direct Current* (DC) to the 7805 voltage regulator *Integrated Circuit* (IC). The 7805 voltage regulator IC reduces the input voltage of 7.4 volts DC to 5 volts DC. This control signal is used to switch on the transistor relay driver circuit, enabling it to close its NO contacts closed. The 1-kilo-ohm resistor is then electrically connected to the internal 10-kilo-ohm resistor, enabling approximately 0.45 volts to be sensed by the RCX P-Brick (see the section "DC Circuit Analysis" in Chapter 3). The following code could be reused in this application:

```
Dim SENSOR1 As Integer
Dim SENTOUCH As Integer
Dim SENVAL As Integer
Dim n As Integer
Dim b As Integer
Dim Status As Integer
```

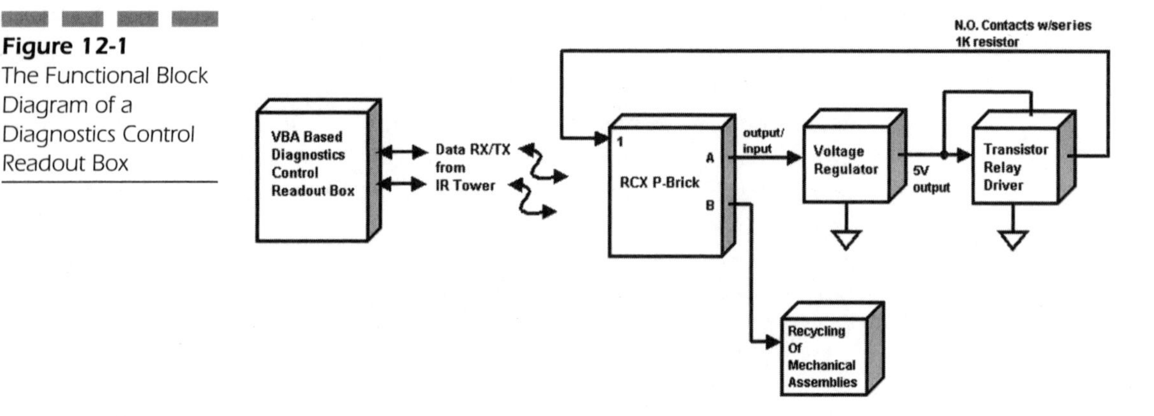

Figure 12-1 The Functional Block Diagram of a Diagnostics Control Readout Box

```
Dim Reading As String
Dim MyReading(30) As String
Private Sub Read_Click()
SENSOR1 = 0 'Read Input 1 of RCX
SENTOUCH = 1 'Reading Data from a Touch Sensor
SENVAL = 9 'Retrieve value from Touch Sensor at Input 1
PBrickCtrl.SetSensorType SENSOR1, SENTOUCH
Status = PBrickCtrl.Poll(SENVAL, SENSOR1)
If Status = 1 Then
TRMON.TextBox1.Value = "Close"
Else
TRMON.TextBox1.Value = "Open"
'TRMON.TextBox1 = PBrickCtrl.Poll(SENVAL, SENSOR1)
Reading = TRMON.TextBox1.Value
MyReading(n) = Reading
If n = 30 Then
MsgBox "Maximum Data Points Have Been Reach", vbExclamation,
"Warning"
Else
n = n + 1
End If
End If
End Sub
Private Sub CommOn_Click()
PBrickCtrl.InitComm
TRMON.COMMSTATUS.Caption = "ON"
End Sub
Private Sub CommOff_Click()
PBrickCtrl.CloseComm
TRMON.COMMSTATUS.Caption = "OFF"
End Sub
Private Sub LogCmd_Click()
n = 0
End Sub
Private Sub PrintCmd_Click()
For i = 0 To n
Cells(i + 3, 1).Value = i
Cells(i + 3, 2).Value = MyReading(i)
Next i
End Sub
Private Sub ClearCmd_Click()
For b = 0 To 30
Cells(b + 3, 1).Clear
Cells(b + 3, 2).Clear
Next b
End Sub
Private Sub ResetCmd_Click()
TRMON.TextBox1 = " "
End Sub
```

To start the tools development project, import the VBA applet that used the previous code into the new spreadsheet. The VBA transistor relay monitor from the section “Transistor Relay Monitor: VBA–C-Bot Application” in Chapter 4 becomes the .frm, which is imported with its associated code to be used in the tools development project. Import the file using Steps 8 through 15 outlined in the section “Developing a New GUI Using an Existing UserForm Procedure” in Chapter 10.

Once the .frm is imported into the new worksheet, the following modifications are made to the GUI UserForm. The two command buttons for COMM are replaced with the iSwitchLedX1.Active VBA object. The two command buttons for READ and RESET are replaced with iSwitchLeverX1 and X4, respectively. The Parent Object TRMON is changed to ROBOMON. The array code information for recording relay contact closure data will not be used at this time and can be removed from the code listing. The following lists the new code with the suggested design changes:

```
Dim SENSOR1 As Integer
Dim SENTOUCH As Integer
Dim SENVAL As Integer
Dim Status As Integer
Dim MotorA As Integer
Dim MotorB As Integer

Private Sub iSwitchLedX1_OnChange()
If ROBOMON.iSwitchLedX1.Active = True Then
ROBOMON.iSwitchLedX1.Caption = "ON"
ROBOMON.PBrickCtrl.InitComm
Else
ROBOMON.iSwitchLedX1.Caption = "OFF"
ROBOMON.PBrickCtrl.CloseComm
End If
End Sub

Private Sub iSwitchLeverX1_OnChange()
SENSOR1 = 0 'Read Input 1 of RCX
SENTOUCH = 1 'Reading Data from a Touch Sensor
SENVAL = 9 'Retrieve value from Touch Sensor at Input 1
PBrickCtrl.SetSensorType SENSOR1, SENTOUCH
Status = PBrickCtrl.Poll(SENVAL, SENSOR1)
If Status = 1 Then
ROBOMON.TextBox1.Value = "1"
Range("A5").Value = ROBOMON.TextBox1.Value
Else
ROBOMON.TextBox1.Value = "0"
Range("A5").Value = ROBOMON.TextBox1.Value
End If
End Sub

Private Sub iSwitchLeverX2_OnChange()
If ROBOMON.iSwitchLeverX2.Active = True Then
MotorA = "0"
MotorB ="1"
ROBOMON.PBrickCtrl.On (MotorA)
ROBOMON.PBrickCtrl.On (MotorB)
Else
ROBOMON.PBrickCtrl.Off (MotorA)
ROBOMON.PBrickCtrl.Off (MotorB)
End If
End Sub

Private Sub iSwitchLeverX4_OnChange()
ROBOMON.TextBox1 = " "
```

```
Range("A5").Value = ""
End Sub
```

One feature about the new applet code is that the listing is smaller compared to the TRMON listing shown earlier. The use of lever switches makes controlling the Robo-Fly easy because of the dual state of the electromechanical switch. Now bear in mind that this example uses the robotic insect that I built; any Mindstorms robot created by hobbyists and experimentalists will work as well. Depending on the number of electromechanical actuators used in the Mindstorms robot, the lever switch count and associated software should be the only items to modify. Figures 12-2 through 12-4 show different views of Robo-Fly. In Figure 12-4, the touch sensor is attached to input 1 of the RCX P-Brick.

Figure 12-2 Robo-Fly is a hybrid design consisting of an older data set with a Mindstorms RCX P-Brick and touch sensor.

Figure 12-3 Detail showing the pulley drive system for actuating Robo-Fly's wings

Testing the New Code with the VBA Prototype Diagnostics Controller

This sensor is used to test the diagnostic READ feature of the VBA prototype diagnostics controller panel. The code used can be changed to accommodate another input for the diagnostics of the Mindstorms robot. Turn on the RCX P-Brick to test the VBA prototype diagnostics controller panel.

Create a new spreadsheet in Excel and go to the *Visual Basic Editor* (VBE). Open a new UserForm and rename it ROBOMON. Place the real-

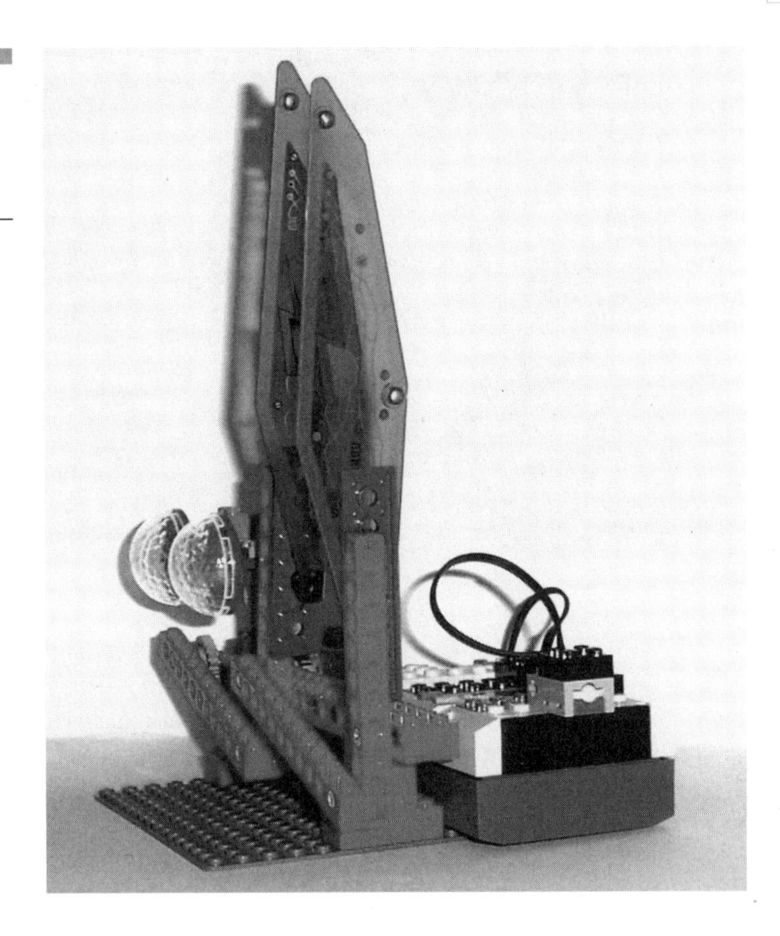

Figure 12-4
Detail showing the RCX P-Brick and touch sensor attached to input 1

time controls onto the UserForm using Figure 12-5 as a guide. Change all the appropriate properties of the software applet accordingly using Figure 12-5 as a guide. Type the new code listing using the VBE text editor. Check for typographical and syntax errors before running the software applet. Run the code using the Run Sub/UserForm command from the main menu within the VBE window. Click the Switch with LED Indicator button on the Robo-Fly test panel. The green indicator LED should be on. Then click the FLAP/STOP WINGS lever switch. Robo-Fly should start flapping its wings. Click the lever switch again. The wings should stop flapping. To check out the READ diagnostics lever switch, click the switch without pressing the touch sensor attached to input 1 of the RCX P-Brick. In the READ text box control, a binary value of 0 should be displayed. At cell location A5, the same binary digit is displayed. Click the RESET lever switch, and both cell A5 and the READ text box control are cleared of any binary digital value.

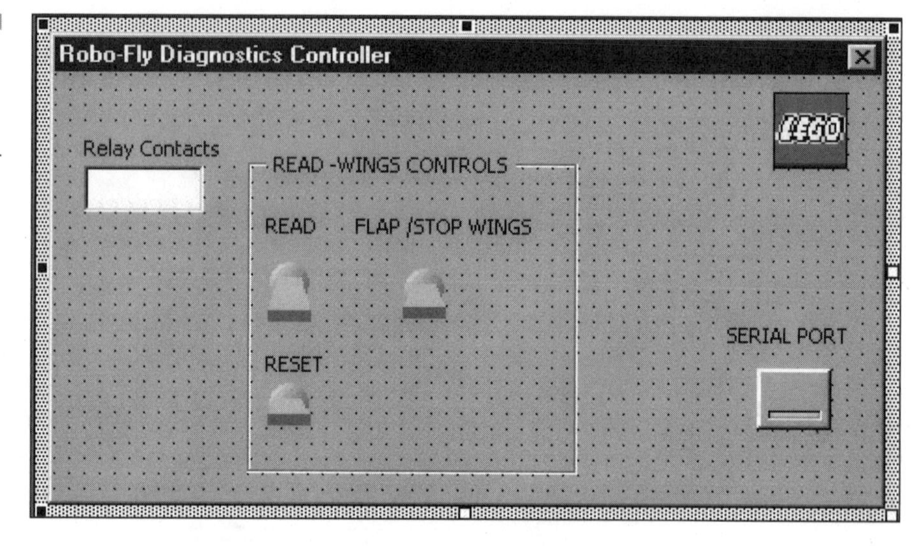

Figure 12-5 Control panel design for the Robo-Fly diagnostics controller

While pressing the touch sensor, click the READ diagnostics switch. Cell A5 and the READ text box ActiveX control should have a binary 1 displayed at both locations. Reset both data display locations by clicking the lever switch again. The data is erased.

An ActiveX CommandButton and label are added to the spreadsheet. The CommandButton is used to display the VBA test panel, providing convenience to the operator when running the Mindstorms robot applet. Figure 12-6 shows the placement of the CommandButton and label using a text box. The cell's borderlines were made thicker within the Format Cells—Borders control box. The CommandButton's Robo-Fly caption was created using the Properties window. The following shows the code used to display the panel:

```
Private Sub CommandButton1_Click()
ROBOMON.Show
End Sub
```

To add this code to the ActiveX control, a design-mode session must be running. While the session is running, click, drag, and resize the control to make it match what is shown in Figure 12-6. Next, right-click the control and select View Code. The VBE text editor appears on the screen. Type in the code, save the spreadsheet, and return to the active worksheet. Exit out of design mode. To show the diagnostics controller on the spreadsheet, click the Robo-Fly button. The spreadsheet now displays the diagnostics controller test panel on the screen, as shown in Figure 12-7.

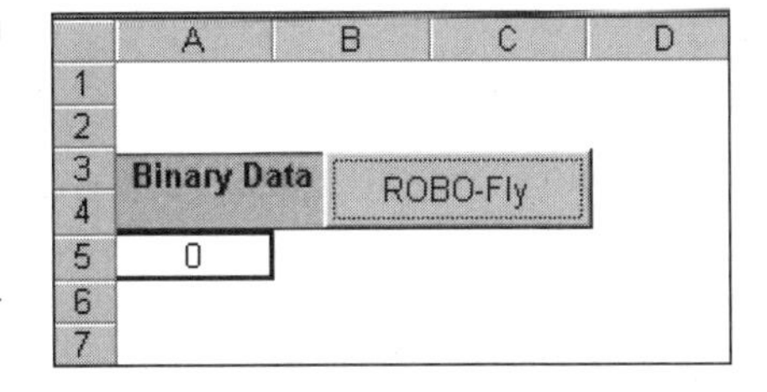

Figure 12-6
The placement of the CommandButton and label using a text box.

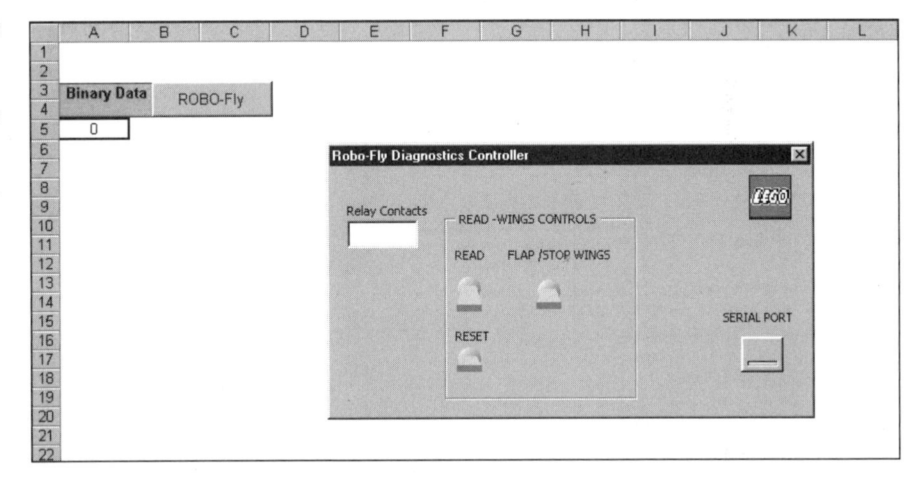

Figure 12-7
The diagnostics controller test panel is displayed on the spreadsheet's screen.

If Robo-Fly and the VBA prototype controller panel run correctly in manual mode, the voltage regulator transistor relay driver circuit can be added to perform the automatic diagnostics feature of the project. Wire the RCX P-Brick diagnostics feedback circuit (as shown in Figure 12-8) using a Radio Shack Electronics Learning Lab breadboard or equivalent. When connecting the electric wire to output B of the RCX P-Brick, make sure the wires are pointing up to achieve proper circuit interface operation for the voltage regulator circuit. The +5-volt DC source of the relay can be wired to the output of the 7805 voltage regulator. The output circuit provides a stable +5.01 volts DC that has sufficient drive capability for the transistor relay driver. The modified LEGO electric wire with the series 1-kilo-ohm resistor attaches to input 1 of the RCX P-Brick. As stated previously, when the RCX P-Brick receives the infrared command for Robo-Fly to flap its wings, the voltage regulator circuit turns on, providing a control signal to the transistor relay driver. The relay contacts close, enabling a closed circuit to exist with the series 1-kilo-ohm resistor and the internal 10-kilo-ohm interface circuit. The 0.45 volts received at input 1 are interpreted as a binary 1 by

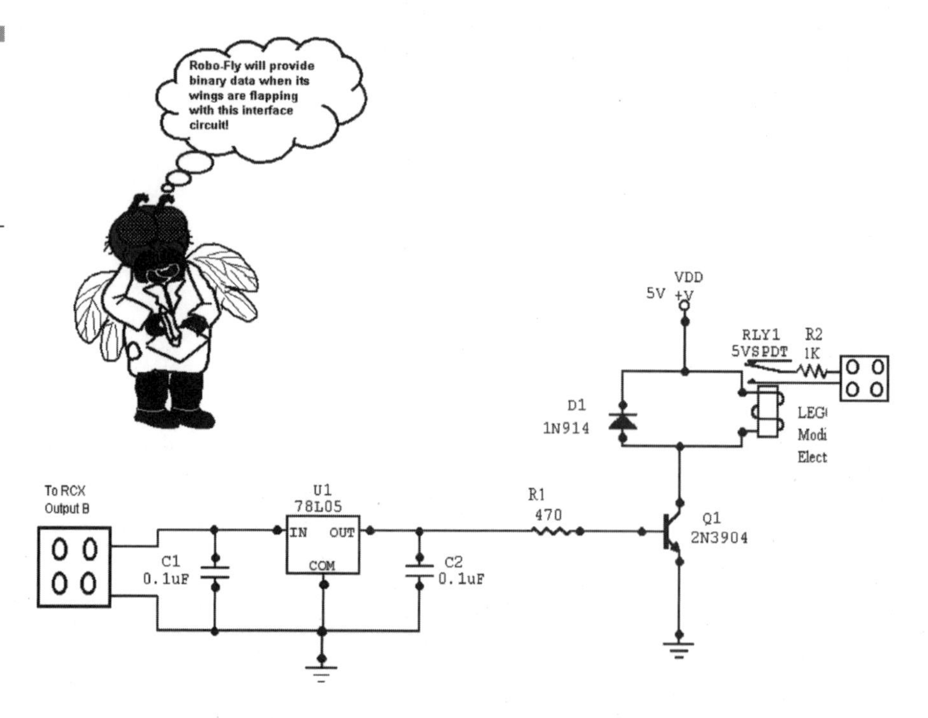

Figure 12-8 Wiring diagram of the RCX P-Brick diagnostics feedback circuit introduced by Philbo Von Fly

the RCX P-Brick's microcontroller. This latched signal is read by the VBA instruction line PBrickCtrl.Poll(SENVAL, SENSOR1). This value is then read by the integer variable Status. The VBA instruction line for reading the binary data is as follows:

```
" Status = PBrickCtrl.Poll(SENVAL, SENSOR1)"
```

The VBA applet establishes a condition where the Status equals 1; it should display the binary data on the text box and the spreadsheet cell A5. If it does not equal 1, it should display a binary 0 in the text box and on the spreadsheet. The following lines of code show the logic for reading the binary data and controlling outputs A and B:

```
Private Sub iSwitchLeverX1_OnChange()
SENSOR1 = 0 'Read Input 1 of RCX
SENTOUCH = 1 'Reading Data from a Touch Sensor
SENVAL = 9 'Retrieve value from Touch Sensor at Input 1
PBrickCtrl.SetSensorType SENSOR1, SENTOUCH
Status = PBrickCtrl.Poll(SENVAL, SENSOR1)
If Status = 1 Then
ROBOMON.TextBox1.Value = "1"
Range("A5").Value = ROBOMON.TextBox1.Value
Else
```

```
ROBOMON.TextBox1.Value = "0"
Range("A5").Value = ROBOMON.TextBox1.Value
End If
End Sub
```

The following lines of code control outputs A and B using the real-time lever switch:

```
Private Sub iSwitchLeverX2_OnChange()
If ROBOMON.iSwitchLeverX2.Active = True Then
MotorA = "0"
MotorB ="1"
ROBOMON.PBrickCtrl.On (MotorA)
ROBOMON.PBrickCtrl.On (MotorB)
Else
ROBOMON.PBrickCtrl.Off (MotorA)
ROBOMON.PBrickCtrl.Off (MotorB)
End If
End Sub
```

The unique feature of the VBA tool is its capability to read when Robo-Fly is flapping its wings based on the binary control signal received at input B. The number of controls used does not clutter the test panel, providing a good interface for quick diagnostics and control of Robo-Fly.

A Few Words about Robo-Fly

As mentioned earlier in the discussion of this project, Robo-Fly is a hybrid design consisting of a wiper model built from an old set of LEGO instructions, the RCX P-Brick, and a touch sensor. By mixing two designs, Robo-Fly becomes a hybrid. Robo-Fly's main mechanical drive for flapping its wings functions through a pulley. Figure 12-3 shows Robo-Fly's pulley-driven mechanical drive assembly. The lesson to learn here is that unique Mindstorms robots can be created by using older construction sets and building an electronic brain on the smart machine using the RCX P-Brick. By adding the old with the new, a hybrid design is born.

Using Arrays in Mindstorms Robot Applications

The ability to record data during a diagnostics session and then play back the information provides great flexibility in a Mindstorms robot application. Robo-Fly produces binary data when it flaps its wings by driving a transistor relay driver's NO contacts. The control signal is fed to input 1 of the RCX P-Brick. When this data is received, a binary 1 appears on the spreadsheet when the READ lever switch is clicked. The data is viewed on the text box and at cell A5.

After the number of times Robo-Fly flaps its wings is recorded, the data can be stored and retrieved using an *array*. An array can be thought of as a mathematical filing cabinet for storing data. The files in the cabinet hold documents or records that can be viewed by going to the appropriate drawer. An array is an electronic data storage system. A main variable can hold as much data as the developer allows it to maintain. The array manages the data through the index of the variable. Individual bits of data are stored to a subset of the main variable. The index is a counter that logs the data to a subscript of the variable. For example, the main variable for Robo-Fly could be *reading*. The subscript of reading is reading(1), reading(2), and so on. The index is a letter that is a variable assignment to the numerical counter. The counter runs free and should have a limit set for the amount of counts it can provide for recording the data. The FOR statement in VBA is used to read the variable assignments by using a counter to keep track of data received associated with the main variable name. To increment the index or change the variable assignment's subscript, the NEXT VBA instruction is used to increase the count value sequentially. A complete listing of the code that can be used to record and retrieve Robo-Fly's flapping-wing data is as follows:

```
Dim SENSOR1 As Integer
Dim SENTOUCH As Integer
Dim SENVAL As Integer
Dim n As Integer
Dim b As Integer
Dim Status As Integer
Dim Reading As String
Dim MyReading(30) As String

Private Sub LogCmd_Click()
n = 0
End Sub
Private Sub iSwitchLeverX1_OnChange()
SENSOR1 = 0 'Read Input 1 of RCX
SENTOUCH = 1 'Reading Data from a Touch Sensor
SENVAL = 9 'Retrieve value from Touch Sensor at Input 1
PBrickCtrl.SetSensorType SENSOR1, SENTOUCH
Status = PBrickCtrl.Poll(SENVAL, SENSOR1)
If Status = 1 Then
ROBOMON.TextBox1.Value = "1"
Reading = ROBOMON.TextBox1.Value
MyReading(n) = Reading
Else
ROBOMON.TextBox1.Value = "0"
```

```
Reading = ROBOMON.TextBox1.Value
MyReading(n) = Reading
End If
End Sub
Private Sub PrintCmd_Click()
For i = 0 To n
Cells(i + 3, 1).Value = i
Cells(i + 3, 2).Value = MyReading(i)
Next i
End Sub

Private Sub ClearCmd_Click()
For b = 0 To 30
Cells(b + 3, 1).Clear
Cells(b + 3, 2).Clear
Next b
End Sub
```

Observations and Notes about the VBA Data Record-and-Retrieve Applet

To develop an array assignment, the Dim MyReading(30) As String must be declared. If the data is not declared, a "Compile Error: Sub or Function not defined" message is displayed on the screen. Also, any subprocedure function that uses the array name will not work because its dependence on the variable has not been declared. Note that the maximum allotted data count is 30. By changing the subscript, the number of data points recorded can be bigger or smaller depending on the application of the Mindstorms robot. The three lines of VBA code for initializing the counter are noted by the Private Sub LogCmd_Click() event. Prior to logging the flapping-wing data, the event should be initiated in order to ensure that the correct sequence of binary values is recorded. Also note that the ActiveX controls have names that are related to the functions they will perform when the event is initiated. To change the name of the ActiveX control, go to the Properties window and type a new name for the control object.

Debug Function for VBA-Assisted Troubleshooting of Mindstorms Robots

When writing code to record data, a quick visual check of the captured data is sometimes necessary. Debug.Print is a VBA instruction that can tremendously assist in validating the recording feature of the VBA diagnostics controller or any other software prototype tester. This handy VBA tool prints the variable assignment names in the Immediate window within the VBE. To obtain access to the Immediate window, go to View at the main menu and click Immediate Window. At the bottom of the VBE, another text editor box labeled Immediate will be visible. By inserting Debug.Print with the target variable names, this VBA instruction prints the values to the Immediate window. Developing VBA code that can be used to monitor and print can be accomplished in two ways.

Method One for Using Debug.Print

One method of using Debug.Print is to embed this VBA instruction within the target code for validation or troubleshooting. The following is an example of embedding the code within the actual Robo-Fly VBA applet:

```
Private Sub iSwitchLeverX1_OnChange()
SENSOR1 = 0 'Read Input 1 of RCX
SENTOUCH = 1 'Reading Data from a Touch Sensor
SENVAL = 9 'Retrieve value from Touch Sensor at Input 1
PBrickCtrl.SetSensorType SENSOR1, SENTOUCH
Status = PBrickCtrl.Poll(SENVAL, SENSOR1)
If Status = 1 Then
ROBOMON.TextBox1.Value = "1"
Reading = ROBOMON.TextBox1.Value
MyReading(n) = Reading
Debug.Print MyReading(n)
Else
ROBOMON.TextBox1.Value = "0"
Reading = ROBOMON.TextBox1.Value
MyReading(n) = Reading
Debug.Print MyReading(n)
End If
End Sub
```

The two lines of code highlighted in boldface type print the values of the array MyReading(n) upon initiating a READ diagnostics event with the lever switch on the panel. To view the data after executing the control, open the Immediate window within the View menu. The data should consist of a running list of values collected from the ROBOMON.TextBox1.Value for the binary digits 1 and 0.

Method Two for Using Debug.Print

Another method of using Debug.Print with the VBA diagnostics controller for Robo-Fly is to embed the instruction within a real-time ActiveX control.

When the event for the control has been initiated, the Debug.Print instruction instantaneously logs the data onto the Immediate window. The following three lines of code carry out this Debug.Print feature:

```
Private Sub iSwitchLeverX5_OnChange()
Debug.Print MyReading (n)
End Sub
```

Debug.Print provides a seamless tool for validating a record-and-retrieve operation for a Mindstorms robot. Also, this tool ties in with the prototyping methods that VBA can offer when developing Mindstorms robotic control.

Musing with Message Boxes (MsgBox)

Another visual tool within VBA that conveys important items regarding the diagnostics of a Mindstorms robot is the *Message Box* (MsgBox) function. The MsgBox is an excellent debugging tool because the VBA function can be inserted at any time to pause the active running code and display the result of the diagnostics. The MsgBox function returns a value and displays a dialog box that you can respond to. The MsgBox function can even be used if no response is required. The main purpose of this tool is to display a status message of the Mindstorms robot diagnostics being executed. The VBA programming style and syntax for using the MsgBox function are as follows:

```
MsgBox prompt ,buttons, title, helpfile, context
```

The following terms require further explanation:

- *prompt* The message displayed in the active popup box. This VBA programming parameter is required to achieve the proper operation of the MsgBox function.
- *buttons* An icon that appears in the active popup box. The following is a partial list of buttons that can be used with the MsgBox to display diagnostics information for a Mindstorms robot:
  - *VbOKOnly* Display the OK button only on the MsgBox function.
  - *VbOKCancel* Display the OK and Cancel buttons on the MsgBox function.
  - *VbAbortRetryIgnore* Display the Abort, Retry, and Ignore buttons.
  - *VbYesNoCancel* Display the Yes, No, and Cancel buttons.
  - *VbYesNo* Display the Yes and No buttons.

- *VbRetryCancel* Display the Retry and Cancel buttons.
- *VbCritical* Display the Critical Message icon.
- *VbQuestion* Display the Warning Query icon.
- *VbExclamation* Display the Warning Message icon.
- *VbInformation* Display the Information Message icon.

- *title* The text that appears in the MsgBox's title bar. The default title is Microsoft Excel. This VBA programming parameter is optional.
- *helpfile* The name of the help file associated with the MsgBox. This VBA programming parameter is optional.
- *context* The ID of the help file. This VBA programming parameter is a specific help topic and is optional.

The MsgBox function can be used by itself to display a message or with a special "error" icon. The following sections provide some examples of using the MsgBox function in VBA.

Example 1

```
Private Sub CommandButton1_Click()
MsgBox "Click Ok to continue"
End Sub
```

In this example, clicking the CommandButton calls the MsgBox function. By clicking the OK button under the Click OK to Continue string, the MsgBox vanishes from the screen.

Example 2

```
Private Sub CommandButton1_Click()
MsgBox "This is an example of the Exclamation button",
vbExclamation
End Sub
```

When the code is executed using the CommandButton, the text string message is displayed along with the Exclamation icon. The warning tone sounds like a doorbell. Figure 12-9 shows an example of the MsgBox with the Exclamation button icon.

Example 3

```
Private Sub CommandButton1_Click()
MsgBox "This is an example of the Critical Message button",
vbCritical, "Critical"
End Sub
```

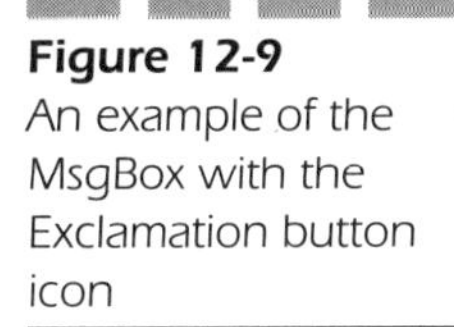

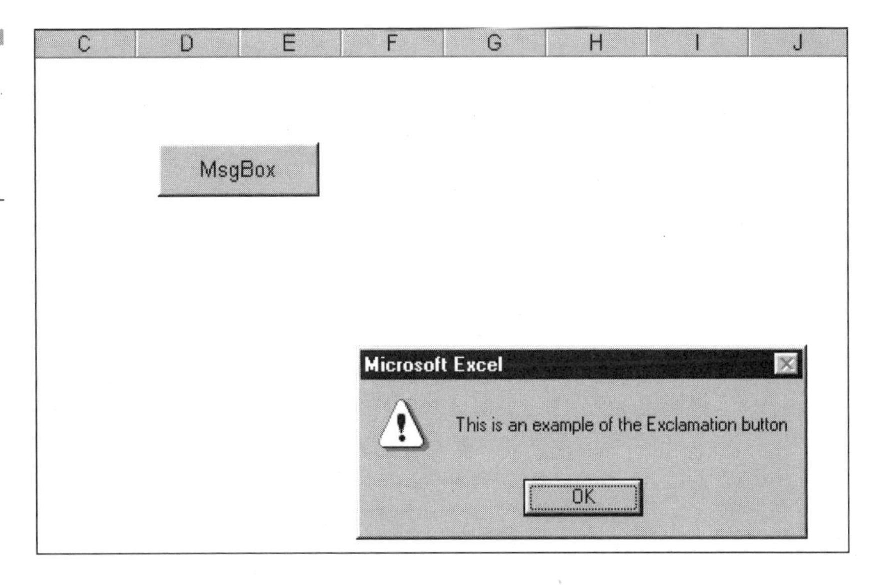

Figure 12-9
An example of the MsgBox with the Exclamation button icon

When this code is executed, not only will the string text and Critical button icon be displayed on the MsgBox, but the title of Critical will also be visible instead of the default Microsoft Excel. The following shows the modified code for the Robo-Fly test panel using the MsgBox to display the binary status of its flapping wings:

```
Private Sub iSwitchLeverX1_OnChange()
SENSOR1 = 0 'Read Input 1 of RCX
SENTOUCH = 1 'Reading Data from a Touch Sensor
SENVAL = 9 'Retrieve value from Touch Sensor at Input 1
PBrickCtrl.SetSensorType SENSOR1, SENTOUCH
Status = PBrickCtrl.Poll(SENVAL, SENSOR1)
If Status = 1 Then
ROBOMON.TextBox1.Value = "1"
Reading = ROBOMON.TextBox1.Value
MyReading(n) = Reading
MsgBox "Binary value is 1", vbExclamation
Else
ROBOMON.TextBox1.Value = "0"
Reading = ROBOMON.TextBox1.Value
MyReading(n) = Reading
MsgBox "Binary value is 0", vbExclamation
End If
End Sub
```

Hopefully, the previous examples spark inspiration and creativity inside those individuals building VBA prototype diagnostics test panels for Mindstorms robotic controls applications. The messages and icons provided with the MsgBox function provide the opportunity to develop sophisticated message-displaying tools for Mindstorms robots.

Inserting Pictures Using the Image ActiveX Control onto the VBA Prototype Panel

Adding images to a VBA prototype panel brings the control object to the physical realm of the target Mindstorms robot. An image added to a diagnostics panel enables the developer to make the VBA prototype an interactive tool for manipulating and monitoring Mindstorms robots. A visual image of the component adds a touch of realism to the prototype tool and enables the selection controls to be chosen with a central design. To assist in this process, a visual image of sorts, a freeware program called *Michael Lachmann computer-aided design* (MLCAD), enables *three-dimensional* (3D) models of LEGO creations to be created on a desktop PC or notebook computer. The software can be downloaded from the MLCAD homepage at http://mlcad.ldraw.org or Don Dueck's web site at www.kaejae-worx.com/~don/lego/setup. Don has all of the necessary software for setting up the MLCAD at his web site, including the latest version of MLCAD. Complete instructions for setting up MLCAD and the associated software are available on his web site. Figure 12-10 shows a screenshot of Don Dueck's "Creating Your Own LEGO Model" web page.

After MLCAD has been installed and all of the supporting software has been loaded onto the desktop PC or notebook, a 3D model can be created. Using the program is pretty intuitive and, like most software applications, requires little experimentation on the user's part. A tutorial written by Sebastian Stein is available online at www.hpfsc.de/mlcd_tut/Default.html. Before entering the web site, Sebastian has provided documentation for a variety of foreign languages. Figure 12-10 shows the various foreign languages the document is written in.

With the image built using MLCAD and rendered in *Persistance of Vision, Version 3 Software* (POV3) (see Don Dueck's web site for more information), the 3D LEGO model can be used on a VBA panel. Figure 12-11 shows a concept design for C-Bot's conveyor system. The raw file created in POV3 shows the conveyor and less text information. Also, the model must be saved as a bitmap file onto the desktop PC or notebook's hard drive. Saving it as a bitmap file enables a painless importing session to occur for the image ActiveX control.

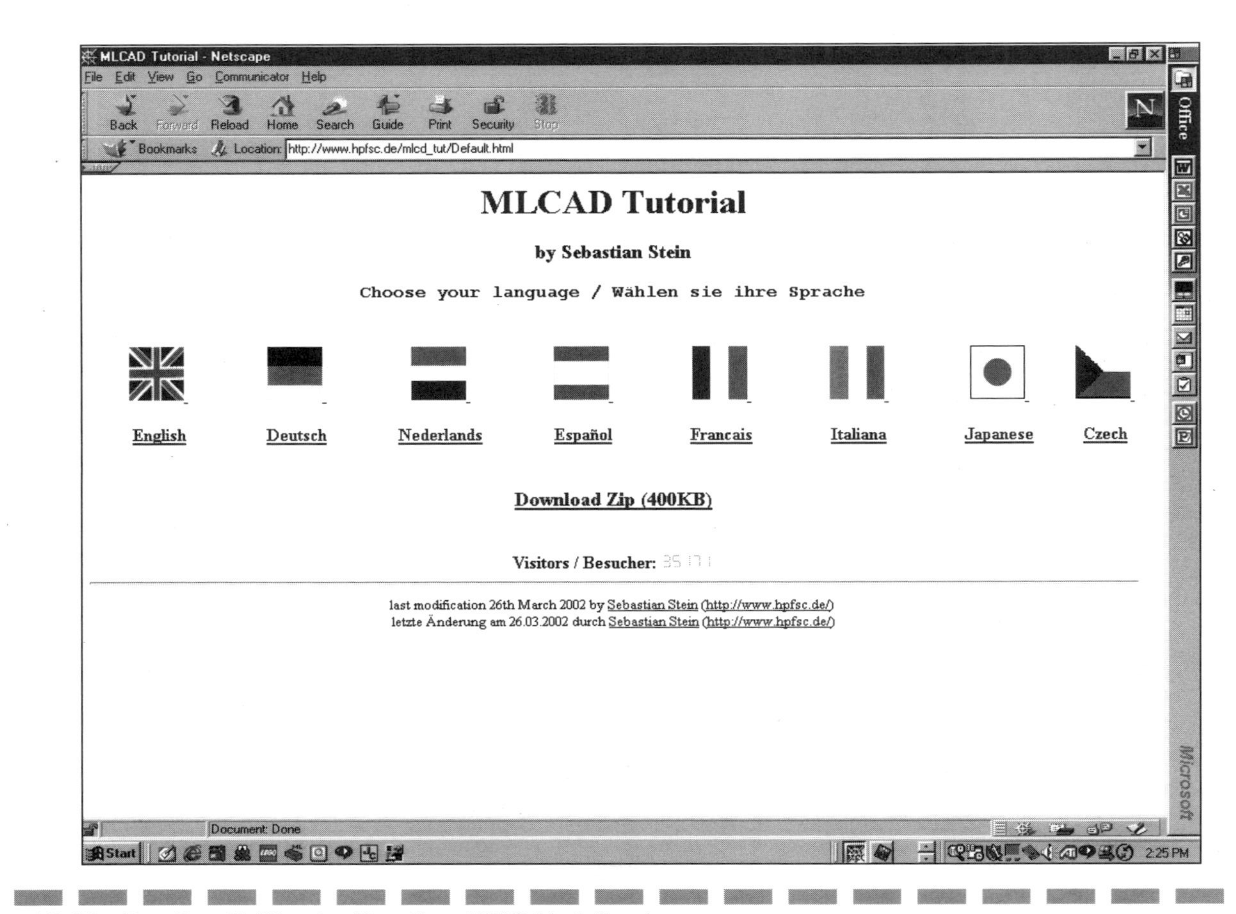

Figure 12-10 Don Dueck's "Creating Your Own LEGO Model" web page

To use the image control either in design mode on the active spreadsheet or with a UserForm, select the Image Control icon from the toolbox. Drag and drop the control onto the spreadsheet or UserForm. Right-click the object if you are using the control on a spreadsheet or click it once if you are placing the control on the UserForm to obtain the Properties window. Scroll down until you see the Picture property and click the None designator. A small command box with three dots appears to enable you to browse into the folder where the image is stored. Once the image has been located, double-click the picture to embed it inside of the image control. Resize the image by dragging one of the appropriate handlebars surrounding the image control. Figure 12-12 shows the conveyor placed on a UserForm using the image control. The image control properties such as BorderColor, BorderStyle, PictureAlignment, and Scrollbars can be explored via simple experimentation. Using images on a Mindstorms Robot test panel makes the virtual prototype real because of the 3D model that is placed on the software GUI. If the controls are correctly laid out, the 3D image enhances the interaction experience for hobbyists and experimentalists.

Technical Tools Musing Projects

In the final section of this book, your imagination can really run wild. The information presented in this book enables hobbyists and experimentalists to explore building robots with a systems approach. POC techniques for

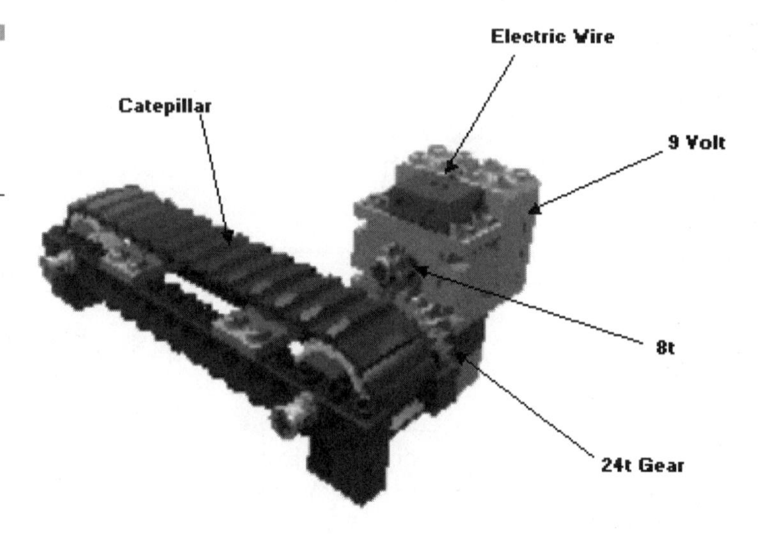

Figure 12-11 A concept design for C-Bot's conveyor system using MLCAD software

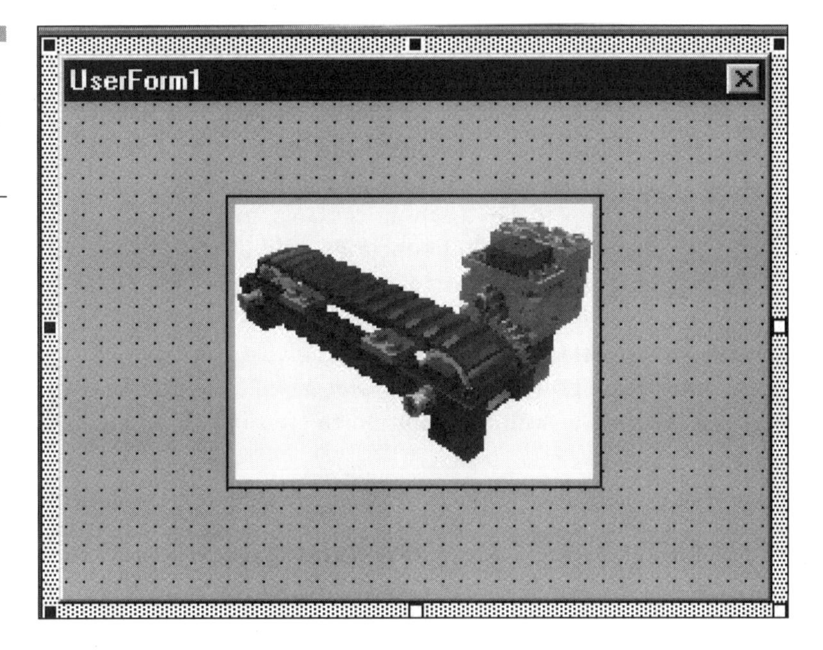

Figure 12-12 The conveyor placed on a UserForm using the image ActiveX control

building electronics, mechanics, and software were examined through hands-on lab projects and development approaches for building Mindstorms robots. The robots used to illustrate the basic ideas used in POC engineering studies enable hobbyists and experimentalists to explore techniques and tools for building Mindstorms robots. These ideas can now be unleashed by using the methods discussed in this book for further exploration. In the following section, I will present some of the robotic concepts I'm currently exploring with LEGO Mindstorms. The examples outlined will illustrate the only limitation to using this construction kit is the imagination of the builder.

Basic Stamp/RCX Controllers

As discussed in Chapter 9, "Remote-Control Techniques," a front-end design enables the Basic Stamp to be programmed through VBA language. Using image and real-time controls, VBA prototype panels for small Lego robots with Basic Stamp actuators can be built quite easily. The VBA panel could be designed where the three inputs of the Basic Stamp are programmed through the assistance of three outputs of the RCX P-Brick. This VBA panel can be used to develop a robotic machine interface to provide the advanced control and diagnostics of a Mindstorms robot.

A Robotics Training System

The Basic Stamp/RCX controller along with electronics and test equipment can be developed into a tabletop robotics training system. The test procedures used in Chapters 1 through 4 provide validation techniques that can be adopted for teaching controls, mechanical interfaces, and software for high schools, adult education, and universities. Using alternative software languages such as *Interactive C Version 4.0* (IC4) and *Not Quite C* (NQC) gives students the opportunity to explore new and exciting programming technologies without spending vast amounts of money. The *Texas Instrument* (TI) graphics *calculator-based laboratory* (CBL) introduced in Chapter 4 is a welcoming addition to creating the robotics training system of the future.

Robotic Continuity Tester

Passive sensing provides a wealth of unusual robotic creations using the Mindstorms RCX and Scout P-Bricks. Using the touch sensor as a surrogate for sensing close circuits, a continuity tester of sorts can be built. By adding test leads to the modified LEGO electric wire, a robotic continuity tester can check wires, cables, and simple wiring harnesses. A VBA panel can be built where a pass/fail can be displayed. An LED can also be added to the panel to visually indicate a good wire versus a bad one. A small program written in NQC or RCX code would monitor the target wire under test for an open or closed condition in the electrical component. The program would be located inside of the RCX P-Brick memory location (Programs 1-5) used to check the electrical wire under test.

INDEX

Symbols

A

B

C

F

G

H

I

K–L

M

R

S

W

SOFTWARE AND INFORMATION LICENSE

The software and information on this diskette (collectively referred to as the "Product") are the property of The McGraw-Hill Companies, Inc. ("McGraw-Hill") and are protected by both United States copyright law and international copyright treaty provision. You must treat this Product just like a book, except that you may copy it into a computer to be used and you may make archival copies of the Products for the sole purpose of backing up our software and protecting your investment from loss.

By saying "just like a book," McGraw-Hill means, for example, that the Product may be used by any number of people and may be freely moved from one computer location to another, so long as there is no possibility of the Product (or any part of the Product) being used at one location or on one computer while it is being used at another. Just as a book cannot be read by two different people in two different places at the same time, neither can the Product be used by two different people in two different places at the same time (unless, of course, McGraw-Hill's rights are being violated).

McGraw-Hill reserves the right to alter or modify the contents of the Product at any time.

This agreement is effective until terminated. The Agreement will terminate automatically without notice if you fail to comply with any provisions of this Agreement. In the event of termination by reason of your breach, you will destroy or erase all copies of the Product installed on any computer system or made for backup purposes and shall expunge the Product from your data storage facilities.

LIMITED WARRANTY

McGraw-Hill warrants the physical diskette(s) enclosed herein to be free of defects in materials and workmanship for a period of sixty days from the purchase date. If McGraw-Hill receives written notification within the warranty period of defects in materials or workmanship, and such notification is determined by McGraw-Hill to be correct, McGraw-Hill will replace the defective diskette(s). Send request to:

Customer Service
McGraw-Hill
Gahanna Industrial Park
860 Taylor Station Road
Blacklick, OH 43004-9615

The entire and exclusive liability and remedy for breach of this Limited Warranty shall be limited to replacement of defective diskette(s) and shall not include or extend any claim for or right to cover any other damages, including but not limited to, loss of profit, data, or use of the software, or special, incidental, or consequential damages or other similar claims, even if McGraw-Hill has been specifically advised as to the possibility of such damages. In no event will McGraw-Hill's liability for any damages to you or any other person ever exceed the lower of suggested list price or actual price paid for the license to use the Product, regardless of any form of the claim.

THE McGRAW-HILL COMPANIES, INC. SPECIFICALLY DISCLAIMS ALL OTHER WARRANTIES, EXPRESS OR IMPLIED, INCLUDING BUT NOT LIMITED TO, ANY IMPLIED WARRANTY OF MERCHANTABILITY OR FITNESS FOR A PARTICULAR PURPOSE. Specifically, McGraw-Hill makes no representation or warranty that the Product is fit for any particular purpose and any implied warranty of merchantability is limited to the sixty day duration of the Limited Warranty covering the physical diskette(s) only (and not the software or information) and is otherwise expressly and specifically disclaimed.

This Limited Warranty gives you specific legal rights; you may have others which may vary from state to state. Some states do not allow the exclusion of incidental or consequential damages, or the limitation on how long an implied warranty lasts, so some of the above may not apply to you.

This Agreement constitutes the entire agreement between the parties relating to use of the Product. The terms of any purchase order shall have no effect on the terms of this Agreement. Failure of McGraw-Hill to insist at any time on strict compliance with this Agreement shall not constitute a waiver of any rights under this Agreement. This Agreement shall be construed and governed in accordance with the laws of New York. If any provision of this Agreement is held to be contrary to law, that provision will be enforced to the maximum extent permissible and the remaining provisions will remain in force and effect.